Diagnosis

Interpreting the Shadows

Diagnosis
Interpreting the Shadows

By

Pat Croskerry, Karen S. Cosby,
Mark L. Graber, Hardeep Singh

CRC Press
Taylor & Francis Group
Boca Raton London New York

CRC Press is an imprint of the
Taylor & Francis Group, an **informa** business

Cover illustration. [Author unknown]. Les allegories de Platon: la caverne de Platon [The allegories of Plato: Plato's cave]. In: Charton ME, editor. Le Magasin Pittoresque; volume 23. Paris: [publisher unknown]; 1855. p. 217. Illustrator: Chevignard. Located at: Special Collections Research Center, University of Chicago Library.

CRC Press
Taylor & Francis Group
6000 Broken Sound Parkway NW, Suite 300
Boca Raton, FL 33487-2742

© 2017 by Taylor & Francis Group, LLC
CRC Press is an imprint of Taylor & Francis Group, an Informa business

No claim to original U.S. Government works

Printed on acid-free paper

International Standard Book Number-13: 978-1-4094-3233-3 (Paperback)

Library of Congress Cataloging-in-Publication Data

Names: Croskerry, Pat., author. | Cosby, Karen S., author. | Graber, Mark L., author. | Singh, Hardeep (Physician), author.
Title: Diagnosis : interpreting the shadows / Pat Croskerry, Karen S. Cosby, Mark L. Graber, Hardeep Singh.
Description: Boca Raton : Taylor & Francis, 2017.
Identifiers: LCCN 2016054076| ISBN 9781409432333 (paperback : alk. paper) | ISBN 9781138743397 (hardback : alk. paper) | ISBN 9781472403292 (ebook)
Subjects: LCSH: Diagnosis, Differential. | Clinical medicine--Decision making.
Classification: LCC RC71.5 .C76 2017 | DDC 616.07/5--dc23
LC record available at https://lccn.loc.gov/2016054076

Visit the Taylor & Francis Web site at
http://www.taylorandfrancis.com

and the CRC Press Web site at
http://www.crcpress.com

Contents

Section I Models of Diagnosis

Section II Informal and Alternative Approaches to Diagnosis

Section III The Elements of Reasoning

Foreword

The exponential growth of medical knowledge in the last half of the twentieth century led to dramatic improvements in the ability of physicians to diagnose and treat disease. Today, patients coming to a physician with a new symptom typically assume the physician will be able to figure out what is wrong with them, and "fix it." While this confidence is a great compliment to our profession and the research that underlies these advances, it has often proven to be misplaced. We caregivers make many mistakes in applying that knowledge. Our diagnoses are not always correct, and errors are made in applying treatments, even in such seemingly simple tasks as identifying the correct patient or the correct medication.

The patient safety movement began with evidence that a huge number of patients were being harmed by errors in their care. This led to the appreciation of insights from human factors engineering, that most errors can be prevented by designing processes and systems to reduce reliance on human thought processes that are known to be unreliable. It was a logical step to try to apply these industrial concepts to healthcare, a prospect enthusiastically embraced by doctors, nurses, pharmacists, and others whose personal experiences often provided strong motivation for change.

An immense voluntary improvement effort arose—governments have given far too little support—resulting in significant improvements in specific areas, such as reductions in medication errors and hospital-acquired infections. However, these successes have made little dent in overall patient safety. Hundreds of thousands of patients are still injured each year because of our mistakes. Many are errors in diagnosis, the failure to recognize a serious medical condition or take appropriate measures to identify it so it can be treated in a timely fashion.

Diagnostic errors are, in fact, a huge problem. Some have estimated that 12 million Americans experience a diagnostic error each year, of which 40,000–80,000 are fatal. Diagnostic errors are the leading reason for malpractice suits. It is urgent, therefore, to ask why, if we know so much, do we have so much trouble applying that knowledge appropriately.

From the beginning of the patient safety movement, there were those, especially the authors of this book, who realized that the issues were more complicated and the causes of errors much deeper than failures of systems. More precisely, most errors result from the dysfunctions of human thought processes. These are the failures that our systems need to compensate for. And we haven't done very well.

The reasons are that cognitive failures are many and complex, and preventing or intercepting them requires multiple strategies—changes in multiple types of "systems," if you will. We have learned a great deal about cognitive functioning, and it is well explicated in what you are about to read: how we solve problems, the power of cognitive and affective biases, the difficulties we have in understanding probabilities, the individual and external factors that affect our thinking, and the importance of rationality, to highlight only some factors.

Armed with this information, the conscientious physician will ask, "What can I do?" What "systems" can I put in place for reducing biases, for minimizing the effect of individual variations in personality, gender, aging, and beliefs, for enhancing rational thinking? The authors do not disappoint. They offer advice both for better training of physicians and methods to help physicians in practice.

Substantial changes are needed in how we educate our doctors. By all accounts, modern medical schools do an excellent job in equipping their graduates with the knowledge they need—or how to get it—for diagnosis and treatment, the "what" of medicine. They have done a far less impressive job in teaching doctors the "how" of applying that information, both in terms of the critical thinking needed for accurate diagnosis, and in communicating with colleagues and patients. Learning these skills has typically not been the focus of formal education, but has largely been left to the old-style apprentice system, in which it is assumed that the student will learn from the examples of their mentors. Sadly, the mentors too often suffer from the same limitations of biased, intuitive, and irrational thinking.

To address this problem, the authors call for medical schools to emphasize the crucial role of analytic and rational thinking by teaching it! They call for adding behavioral science as a new basic medical science, with particular emphasis on cognitive psychology, clinical reasoning, and above all, rationality. They point out that rationality has the most powerful impact on diagnostic decision making. Medical students must learn early the extent to which we are all susceptible to the cognitive and affective biases that can lead to cognitive failure.

For the established physician, attention turns to methods for mitigating cognitive biases.

Combating cognitive biases is a never-ending task, even for those who appreciate its necessity and are committed to it. The authors have a plethora of suggestions, 19 in all, from methods to increase awareness to establishment of decision rules and forcing functions.

Is information technology (IT) a critical part of the solution to improving diagnostic accuracy? Certainly. Whatever detractors may say about the electronic health record (EHR), few would disagree that its capacity to provide the information needed, at the time it is needed, in a form that can be used, is one of IT's great strengths. As the EHR becomes the norm, now finally nearing 90% of hospital and office records, this resource is more widely available. Decision aids are increasing in sophistication and usability, including methods to identify and compensate for biases that have the potential to significantly improve diagnostic accuracy.

In short, the authors provide a great deal of information about the "what" of cognitive dysfunction, and practical guidance concerning the "how" of compensating for it. But they leave the reader with the most difficult challenge: changing the culture of medicine so that the focus on how we think is given comparable status to the emphasis on the "what." Fortunately, they provide ample ammunition for that battle.

Lucian L. Leape

Preface

Diagnosis is the core of medicine. To be a good diagnostician is to have diagnostic acumen, the essential characteristic of the effective clinician. Diagnostic success leads to more therapeutic success because, with an accurate diagnosis, an appropriate treatment can be offered. Patients have a better chance of a cure and improved outcomes. Improved diagnosis means that a variety of derivative benefits will accrue to clinicians, such as the satisfaction of having helped the patient, the esteem of colleagues, self-gratification, and perhaps professional success. It has been this way since its primitive beginnings in the hands of shamans. Over the years, medical diagnosticians have become increasingly sophisticated in their approach, and advances in knowledge and technology have allowed remarkable diagnostic differentiation of disease in all areas of medicine with equally sophisticated treatments. The foci of a cerebrovascular accident or a myocardial ischemic event can be diagnosed and delineated within millimeters, and timely interventions can target the offending lesion.

However, before these impressive technologies can be successfully applied, some differentiation of the disease needs to be made. Therein lies the challenge. Often, the initial presentation of disease may be incomprehensible, subtle, obscure, abstruse, and occasionally completely misleading. Critical signals need to be separated from considerable noise. For example, pneumonia (a respiratory diagnosis) and migraine (a neurological diagnosis) may both present as abdominal pain, which suggests a gastrointestinal diagnosis. Patients do not present themselves to doctors with neat labels saying what they have, instead they have signs and symptoms that may or may not be helpful in establishing an accurate diagnosis. Diagnosis is not like prospecting for gold where an unambiguous nugget presents itself. The initial manifestation of disease may be a fairly thick end of a large wedge that eventually narrows down to the business end, and that narrowing down process may be a hazardous journey with signposts along the way that only sometimes point in the correct direction. Often, physicians are dealing with indistinct, shadowy possibilities of where the correct path might lie.

The illustration on the cover of this book characterizes the uncertainty of the diagnostic process. Published in 1855, it depicts Plato's Allegory of the Cave, first described in *The Republic* in 380 BCE. Three prisoners have been chained at the head and neck since birth so that they can only look at a wall on which shadows cast by their cave fire reflect the shapes of objects behind them which they can never directly view. Their reality is only shadows. The issue for Plato lay in establishing the nature of reality. Can we ever be said to know true reality if the world that we see is only a shadow of the form that the world represents?

So it is with diagnosis. Diseases may be entirely invisible. At the outset, we rely on signs and symptoms that are one or more steps removed from the actual nugget. Often what the clinician sees is only a shadow of a covert process, and if they blindly trust what they see, they may be effectively blind. As Sir Zachary Cope noted, "Spot diagnosis may be magnificent, but it is not sound diagnosis. It is impressive but unsafe." Trusting shadows may be a dangerous business.

Diagnosticians have to deal with many shadows. They must learn not to trust shadows that are ill-formed or unfaithful representations of reality, or shadows that their perceptual processes misrepresent to them, or second-hand shadows that have been passed on to them from others. Part of being a well-calibrated diagnostician is treating shadows with the respect they deserve and, along the way, acquiring some intellectual humility—being aware of one's own failings in perceptions and beliefs and the vulnerability of the human brain toward bias. Bias may distort the shadowy constructs of reality.

We might well misjudge our abilities and our successes if it weren't for the undeniable fact that too often we fail, and too often we err in our efforts at diagnosis. What is even more difficult is to determine how we arrived at the wrong conclusion in the first place, and how we can perform better given similar circumstances again. Although we may analyze and critique these events, frankly we are often blind to any accurate understanding of why our clinical reasoning and our system of care failed. We try to learn from these failures, but without real insight it is likely that we merely appease our apprehension without necessarily improving our thinking, our science, or our systems of care.

If this book had been written a century ago, most of its content would probably have focused on how physicians think, reason, and decide. In some medical curricula, as early as the tenth and eleventh centuries (for example, at the medical school at Salerno in Italy), philosophy, logic, and critical thinking were considered essential elements of training. In contrast, in modern curricula, these topics receive scant treatment. Similarly, in assessment of the diagnostic process, there is a bias toward medical orthodoxy. This takes the form of examining the failure of medical systems in medical terms and often in the language of medicine. How physicians think is an area that has attracted less than reasonable attention. There is a perceptible deficit in addressing some of the cognitive processes that underlay the Salerno tradition. Readers will notice a distinct emphasis here on cognitive aspects of the diagnostic process which have typically not received sufficient attention in modern treatments. This will entail incorporating the language of cognitive science. If we are to understand and embrace this area in our discussion of diagnosis, these unfamiliar terms and concepts will need to be introduced and incorporated into our lexicon. Likewise, we need to acknowledge that there are invisible traps in diagnostic processes—latent defects that can create situations and moments of vulnerability that may be unrecognizable at the time, but ultimately manifest with risk and failure due to system and organizational deficiencies. We cannot afford to remain blind to these faults, lest they ensnare us and derail the diagnostic process.

The Allegory of the Cave depicts the different responses of those allowed to escape the cave and view the sun. Some cower in fear, pained by the light and perhaps confused. Others recover to celebrate the daylight. Similarly, clinicians may react with distrust and suspicion to ideas that challenge traditional models of thinking, training, and methods of diagnosis. But we hope that the ideas shared in this book will enlighten the reader and energize a renewed interest in and commitment to excellence in diagnosis.

In recent years, several new books have appeared which have made laudable contributions to our understanding of the process of clinical reasoning that underlies diagnosis. Our intention in writing this book is to add to that burgeoning literature while addressing some of the shadowy areas of the diagnostic process. Of these, cognitive processes loom large, but we will also try to flesh out some of the more straightforward areas with which we do pretend some familiarity. Several topics, not usually part of the medical discussion on diagnosis, will be addressed for the first time. We hope this book will spawn further interest and research into some of these areas.

Reference

Cope S. The principles of diagnosis in acute abdominal disease. In *Cope's Early Diagnosis of the Acute Abdomen*, Revised by W. Silen. 15th ed. New York: Oxford University Press; p.5.

Authors

Karen Cosby is a senior emergency medicine physician at Cook County Hospital and associate professor at Rush Medical School in Chicago, Illinois. Her career interests include medical education, clinical decision making, and patient safety. She has chaired the fellowship subcommittee of the Society to Improve Diagnosis in Medicine and the patient safety interest group for the Society for Academic Emergency Medicine, where she led efforts to produce a curriculum for patient safety. She was a co-investigator of an Agency for Healthcare Research and Quality grant on diagnosis error. She has edited three textbooks (including *Patient Safety in Emergency Medicine*). Her contributions to patient safety include a framework for classifying factors that contribute to error in emergency medicine, and a 15-year review of patient care management problems identified in mortality and morbidity reviews. Her main career focus is investigating and understanding medical failure and improving education to meet the needs of an increasingly complex healthcare system.

Pat Croskerry is professor of emergency medicine and in medical education at Dalhousie University in Halifax, Nova Scotia, Canada. He has been a contributor to the annual conference of the Society for Improvement in Diagnosis since its inception. He has published over 80 journal articles and 30 book chapters in the areas of patient safety, clinical decision making, and medical education reform. In 2006, he was appointed to the Board of the Canadian Patient Safety Institute, and in the same year received the Ruedy award from the Association of Faculties of Medicine of Canada for innovation in medical education. He is senior editor of *Patient Safety in Emergency Medicine* (2009). He was appointed director of the new critical thinking program at Dalhousie Medical School, and a fellow of the Royal College of Physicians of Edinburgh in 2012. In 2014, he was appointed to the US Institute of Medicine Committee on Diagnostic Error in Medicine.

Mark L. Graber is professor emeritus, Stony Brook University in Stony Brook, New York; a senior fellow at RTI University in Research Triangle Park, North Carolina; and founder and president of the Society to Improve Diagnosis in Medicine. With Ilene Corina of Long Island, Dr. Graber was the architect of "Patient Safety Awareness Week," now recognized internationally during the second week of

March. He has authored over 100 peer review publications and many foundational papers on diagnostic error. Dr. Graber originated the Diagnostic Error in Medicine conference series in 2008 and the journal *DIAGNOSIS* in 2014, and was a member of the Institute of Medicine committee responsible for "Improving Diagnosis in Health Care." In 2014, he was the recipient of the National Quality Forum's John M. Eisenberg Award for Individual Achievement.

Editor and Consultant to Authors

 **Hardeep Singh** is a general internist and patient safety researcher based at the Michael E. DeBakey Veterans Affairs (VA) Medical Center and Baylor College of Medicine, Houston, Texas. He leads a portfolio of federally funded multidisciplinary research in understanding and reducing diagnostic errors in healthcare. In 2012, he received the AcademyHealth Alice S. Hersh New Investigator Award for high-impact research and, in 2014, he received the prestigious Presidential Early Career Award for Scientists and Engineers from President Obama for pioneering work in the field. In addition to being highlighted by major national newspapers, his work has made substantial public and health system impact by influencing national patient safety initiatives and policy reports, including those by the National Academy of Medicine (formerly the Institute of Medicine), the US Department of Health and Human Services, National Quality Forum, American Medical Association, Agency for Healthcare Research and Quality, and the World Health Organization.

Section I

Models of Diagnosis

1

What Is Diagnosis?

Karen Cosby

CONTENTS

Defining Diagnosis

Interested readers who pick up this book do not need the concept of diagnosis explained. However, it is likely that each of us has a different perspective and understanding of what a diagnosis is, and what we mean in our discourse about diagnosis. Most would agree that a diagnosis is an explanation of a pathological condition with determination of the underlying cause(es) and pathophysiology. A complete and accurate diagnosis explains physical manifestations, predicts the natural course and likely outcome, anticipates potential complications, and leads to suggested treatment options. An accurate diagnosis is essential to a successful medical encounter. A missed or delayed diagnosis may result in diminished opportunity to intervene to change the natural course of the disease. A misdiagnosis may lead to ineffective or even dangerous actions that may complicate the illness.

Depending on the setting and our own particular role in medicine, we may have widely varying ideas about what constitutes a diagnosis, and the degree of precision, accuracy, and certainty carried by a diagnostic label; examples of types of diagnoses are shown in Figure 1.1. An office-based

Initial diagnosis	Preliminary diagnosis
Preoperative diagnosis	Postoperative diagnosis
Working diagnosis	Differential diagnosis
Symptom-based diagnosis	"Rule-out" diagnosis
Prenatal diagnosis	Screening diagnosis
"Suspected" diagnosis	Diagnosis of exclusion
Admitting diagnosis	Discharge diagnosis
Nursing diagnosis	Billable diagnosis
Self-diagnosis	"Diagnosis-by-Internet"
Clinical diagnosis	Tissue diagnosis
Laboratory diagnosis	Computer-aided diagnosis
CT diagnosis	Radiology diagnosis
Pathology diagnosis	Histology diagnosis
Evolving diagnosis	Phone diagnosis
Alternate diagnosis	Remote diagnosis (telemedicine)
Retrospective diagnosis	Leading diagnosis
Principal diagnosis	Final diagnosis
Confirmed diagnosis	Postmortem diagnosis

FIGURE 1.1
Examples of types of diagnoses, reflecting different phases of the diagnostic process, as well as differing degrees of accuracy, certainty, and finality.

primary care doctor may be comfortable inferring the diagnosis of urinary tract or kidney infection based on a clinical syndrome of flank pain, fever, and dysuria. If that same patient visits an Emergency Department, the diagnosis may be confirmed by urinalysis. If there is a suspicion of complicated disease, the clinician may order a computed tomography (CT) scan to rule out perinephric abscess or infected renal stone with obstruction. If admitted, a hospitalist might rest his final diagnosis on the results of a urine culture to identify the exact organism and sensitivities to antibiotics. An infectious disease researcher might require an immunofluorescence study of urinary sediment to identify antibody-coated bacteria to distinguish between upper and lower urinary tract disease. Each of these clinicians is correct, but all with varying degrees of certainty and precision. Each has a risk of being wrong if information is incomplete or results are misinterpreted.

The Challenge of Making a Diagnosis

Most medical encounters revolve around screening for diseases or conditions, solving problems that patients present with, or taking actions on a presumed or established diagnosis. It is vitally important to recognize the natural limitations in our ability to detect and diagnose disease. There are a number of reasons why the diagnostic process is often challenging, inexact, or even wrong.

Biological Systems Are Complex

Biological systems have limited expressions of disease. The classical description of inflammation defined by the Roman scholar Celsus in the first century AD still applies today. The human body has four main ways to respond to injury or illness: *calor* (heat, warmth), *dolor* (pain), *rubor* (redness), and *tumor* (swelling) [1,2]. These general responses are largely nonspecific.

The characterization of most diagnoses begins with recognition of syndromes – signs and symptoms that form recognizable patterns. With time and further study, we may come to better understand the underlying pathophysiology. At the bedside, however, diagnosis still tends to rely on pattern recognition. Unfortunately, many patterns overlap. A classic example is a patient with a painful, red, swollen leg. The diagnosis may be a fracture or soft tissue injury, underlying abscess, hematoma, gout, arthritis, venous stasis, deep venous thrombosis, cellulitis, lymphedema, or rash. The symptoms may be caused by edema from heart failure, nephrotic syndrome, nutritional deficiency, hypothyroidism, or even a side effect of medications. Some cases are a combination of problems, such as cellulitis from a minor abrasion in a leg with chronic edema from heart failure. Experienced physicians may be able to distinguish between these diagnoses by examination alone, but some will require testing to rule in or rule out some possibilities. Eventually, most diagnoses are confirmed with time, and are often declared by how a patient responds or fails to respond to various interventions.

Biological organisms are made of multiple systems that exist in equilibrium. Perturbation in one system may manifest symptoms or problems in another. One diseased state can upset that equilibrium, especially in patients with chronic disease. A minor respiratory ailment may cause decompensation of heart failure in patients with underlying heart disease. The initial diagnosis of a simple respiratory ailment may lead to a serious deterioration and more important diagnosis in another system.

Manifestations of disease may surface distant from the underlying problem. The painful red eye of iritis may be due to any number of underlying multisystem diseases. Patterns of illness may be incomplete or atypical. Some presentations are so atypical they defy explanation, such as painless aortic dissections, pyelonephritis without pyuria, acute coronary syndromes without chest pain, and asymptomatic pulmonary emboli, to name just a few examples. Illness is dynamic. Anyone witnessing the fulminant course of meningococcal meningitis or necrotizing fasciitis can attest to how quickly initially benign presentations can progress to devastating illness.

Emerging Infections

Diseases caused by infectious agents inevitably change as organisms evolve. Just as new strains of influenza emerge each year, new infectious agents may occur sporadically with little warning or time for preparation, such as Severe Acute Respiratory Syndrome (SARS) and Middle East Respiratory Syndrome (MERS). With global travel, clinicians may encounter organisms they have only read about, such as occurred with the first case of Ebola in the United States [3]. Fear of vaccination complications has led to the recurrence of once common diseases that were well controlled by standard preventive care, diseases that many clinicians have never seen and have no experience with, including polio and measles. Threats of bioterrorism raise the possibility of manufactured biological weapons and the remote possibilities of conditions that would likely challenge any clinician or healthcare system. As science improves, new diagnoses are defined and characterized. The interminable progress of science and the remarkable variety of infectious diseases make it difficult to reliably know, let alone recognize, all possible disease states.

Calibration Is Difficult

Clinicians often do not get feedback about their diagnoses [4] and may not recognize when they are wrong [5]. One of the difficulties in conducting diagnostic evaluations is that the actual diagnosis is not known upfront, even when both patient and provider may feel quite certain about it after limited evaluation. A wrong diagnosis may only be apparent after a condition worsens or complications arise. The mismatch between certainty and accuracy betrays both provider and patient: the provider experiences the sting of remorse and failure; the patient experiences (perhaps) avoidable consequences of the delay and, at the very least, inconvenience.

Common Misperceptions about Diagnosis: What Does My Diagnosis Mean?

The cognitive underpinnings to clinical reasoning and system influences on the diagnostic process are discussed in detail throughout this book. However, it is useful to pause to ask how our method of diagnosis impacts what we mean by a diagnostic label. From one point of view, not all diagnoses are equal. In fact, many diagnoses are neither certain nor final.

The Value of "No Diagnosis"

First-line physicians do not always make a formal diagnosis for all complaints. A patient who presents with chest pain will typically have a clinical evaluation (history and physical examination) and basic tests (electrocardiogram, chest x-ray, troponin, d-dimer). Absent any clear explanation, they may be offered a diagnosis of exclusion, such as nonspecific chest pain, implying a condition likely of no consequence. No true diagnosis may be reached. Both provider and patient may be satisfied to accept the absence of significant disease as an end point itself without reaching a specific diagnosis or explanation for their symptoms. In fact, the willingness to avoid a premature label for a clinical syndrome is desirable, since an inaccurate label may mislead both the patient and other clinicians at future encounters. In cases where there is no clearly established diagnosis, a label of "not yet diagnosed" (NYD) may serve as a marker of need for future testing should the condition persist or worsen.

Presumed Diagnosis

In some cases, diagnoses are only *presumed* based on local epidemiology, a patient's risk profile (comorbidities, family history, social history), and how well their symptoms match typical patterns. Formal testing is often not necessary. When testing is done, it is often ordered to "be safe," to rule out something more serious. Many diagnoses are never confirmed. And for many conditions, testing is unnecessary, since most healthy people recover from limited illness in spite of lacking a formal, specific diagnosis. Most benign viral infections are presumed. A classical presentation of a common condition is often presumed but not tested.

Diagnostic (and Treatment) Thresholds

At times, a clinical presentation is sufficient to trigger treatment even without testing. The benefits of treatment may simply outweigh the potential side effects. If the treatment is straightforward and cost-effective, it may be simply more pragmatic to treat. A diagnostic and treatment threshold has been met, although few would say that a firm diagnosis has been proven or tested [6,7]. One example is a decision to treat a febrile patient with an exudative pharyngitis with antibiotics without formal testing.

The Diagnostic Trajectory

Diagnoses may be made with a single visit for short-lived or self-limited conditions. For others, however, a diagnostic process may evolve over time with repeated visits and reassessments. It is useful to refine how patients and doctors think about the diagnostic process. Instead of prematurely placing diagnostic labels that infer certainty and finality, it is more realistic to think of the process as a *diagnostic trajectory*.

On one end of the trajectory is a single visit at one period of time. The quality of diagnosis from a limited evaluation is determined by the quality of information obtained and how typical the illness is. But with limited evaluation, there will always be some degree of uncertainty. For minor conditions, the consequence of this uncertainty is usually harmless. For persistent or severe symptoms, a more aggressive approach to diagnosis will need to be pursued. As the diagnostic trajectory is followed, additional testing, imaging, and consultation may be necessary. The diagnostic trajectory begins with little information and high uncertainty, but progresses to become a more refined, accurate, and complete diagnosis. This process works better if patients are aware of their role in following the path that leads to a more sophisticated assessment. It also only works when providers recognize when they need to advance their differential diagnosis to consider alternative possibilities and if necessary, seek specialty consultation and additional tests. If the diagnosis does not become progressively more certain and specific over time, or the patient fails to improve or even worsens unexpectedly, the initial diagnosis should be reconsidered, because it might be wrong. Discussions about diagnosis often do not involve communication about where providers are in this diagnostic trajectory. A diagnosis may be presumed based on symptoms alone, confirmed by preliminary screening tests or imaging, or fully established having considered a broad differential and extensive, specific testing.

It is important to recognize that this process should proceed at a pace dictated by the severity of illness. In patients with acute illness, the trajectory may require hospitalization and rapid mobilization of intensive emergency resources. For others, the process may evolve over weeks and months. It is essential that patients understand the process; those who don't may skip from doctor to doctor, each time finding lack of satisfaction as each provider duplicates the efforts of previous physicians.

The Patient's Perspective on Diagnosis

Patients seek a diagnosis in hopes of finding an explanation for their symptoms and treatment options to help them feel better and live longer. For them, a diagnosis may mean relief, reassurance, and hope; for some, a bad diagnosis may bring despair. Some diagnoses can have profound implications on personal decisions for career and work (Should I change jobs?), family (Should we move closer to relatives?), finances (Should we take that dream vacation now or save more for life insurance?), personal goals (Can I train for a marathon?), and even reproduction (Can we, or should we, have a baby?). Patients seek and need clarity and certainty for their diagnoses. Unfortunately, diagnosis is often not as clear or certain as some expect. Indeed, patients often think that a diagnosis is a fact, a simple black and white determination, when it is probably better described as a conclusion that is based on assumptions and imperfect evidence, and one that can be challenged, questioned, reassessed, and revised as needed. Patients may not be informed about how certain their diagnosis is, or aware of the limitations of diagnostic labels. A change in their diagnosis may be unsettling to them, and may even cause them to lose trust in their provider or suspect that they may have been misled. Worse, if patients do not understand how precarious a diagnosis can be, they may fail to seek reevaluation when needed, or may make choices they later regret.

Diagnostic Error

Our current understanding of disease and our methods of diagnosis are imperfect. There is a growing awareness of the frequency and nature of diagnostic error. The Institute of Medicine report, *Improving Diagnosis in Health Care*, defined diagnostic error as "a failure to establish an accurate and timely explanation of the patient's health problem(s), or communicate that explanation to the patient" [8]. Although simply enough stated, criteria for "accurate" and "timely" are not easily defined or standardized. A diagnosis of "chest pain" is accurate, though imprecise; that label might do well enough for a chest wall strain, but will hardly help a patient with a heart attack. What about "timely"? If we are to intervene successfully, a heart attack should be recognized within minutes, but we could reasonably accept a few weeks to figure out a diagnosis of porphyria. Determinations of diagnostic errors and delays are all contextual and based on a final diagnosis that often can't be known with certainty at the outset.

The topic of diagnostic error is difficult to understand, study, and solve. Once a diagnosis is known, there is a tendency to judge every evaluation that preceded the moment of certainty to be wrong or unnecessarily delayed. It is difficult, perhaps impossible, to reconstruct history to understand how a diagnosis wasn't made earlier. It is helpful to recognize that as diseases and conditions evolve, patterns become more complete and more specific, abnormal physical findings become more apparent, and the growing body of evidence adds to a more complete and accurate picture.

Although the numbers may be disputed, physicians may be largely unaware of how common diagnosis error is, or of limitations in their own diagnostic accuracy [5]. The *Society to Improve Diagnosis in Medicine* (SIDM); *Best Doctors, Inc.*; and *In Need of a Diagnosis* are examples of organizations committed to improving diagnoses [9–11]. In dealing with diagnostic error, it would be wise for clinicians to remain humble about their diagnostic conclusions, recognizing that diagnoses are too often inaccurate. Patients should be warned that a diagnosis is only good as long as it advances them toward health or at least provides a reasonable explanation of their condition; unexpected worsening or failure to improve should prompt reassessment and further inquiry.

The Evolving Science of Diagnosis

Biomedical science and technology are rapidly evolving. Conditions that once were described only as a clinical syndrome are now characterized and understood on a genetic, cellular, and molecular basis. Genomic analysis can now characterize even exceptionally rare conditions. Tumors can be reclassified even as they change in response to treatment. With improvements in diagnostic tests, conditions can be detected earlier, some even before they cause symptoms.

Entire industries exist to develop new treatments, and as more effective treatments arise, standards for screening and diagnosis may need to be adjusted. Changing recommendations for controlling blood pressure and blood glucose are just two examples of new definitions for the diagnosis of hypertension and diabetes driven by changing standards for treatment.

Improved methods of detection and better treatments make previous diagnostic labels outdated. What might have been unknown, undetectable, or untreatable a decade ago might well be labeled a diagnostic error today if missed or delayed.

Conclusion

While our technology is impressive, it is important to acknowledge that limitations and challenges in achieving timely, accurate, and complete diagnoses persist and deserve our attention. Some involve our most basic skills in thinking and medical reasoning, the purview of providers. Some of the challenges require that we recruit help from outside traditional domains of medicine, including those who can help design robust and reliable systems of care. And increasingly, our efforts at making diagnosis accurate and timely may benefit from the voices of philosophers and ethicists (How much diagnosis do we want?) and even considerations about public policy (How much diagnosis can we afford?). This book is our attempt to grapple with these concepts.

SUMMARY POINTS

- Diagnosis is often imprecise and imperfect.
- Biological systems are complex and have limited expressions (signs and symptoms) of disease.
- Providers often make presumptive diagnoses with limited evidence.
- Diagnosis may require time and repeat assessments; it is important for both providers and patients to recognize where they are in the diagnostic trajectory.
- Diagnostic failure and diagnostic errors are common.
- Although medical science is advanced, the field of diagnosis is still evolving.

References

1. Scott A, Khan KM, Cook JL, Duronio V. What is "inflammation"? Are we ready to move beyond Celsus? *Br J Sports Med*. 2004 Jun;38(3):248–9.
2. Kumar V, Abbas AK, Fausto N, Aster JC. *Robins & Cotran: Pathological Basis of Disease*. 9th ed. Philadelphia: Elsevier Sanders; 2015.
3. Upadhyay DK, Sittig DF, Singh H. Ebola U.S. Patient Zero: Lessons on misdiagnosis and effective use of electronic health records. *Diagnosis (Berl)*. 2014;1(4):283.

4. Croskerry P. The feedback sanction. *Acad Emerg Med*. 2000 Nov;7(11):1232–8.
5. Meyer AN, Payne VL, Meeks DW, Rao R, Singh H. Physicians' diagnostic accuracy, confidence, and resource requests: A vignette study. *JAMA Intern Med*. 2013 Nov 25;173(21):1952–8.
6. Kassirer JP, Kopelman RI. *Learning Clinical Reasoning*. Baltimore: Williams & Wilkins; 1991:5–6.
7. Pauker SG, Kassirer JP. The threshold approach to clinical decision making. *N Engl J Med*. 1980 May;302(20):1109–17.
8. National Academies of Sciences, Engineering, and Medicine. *Improving Diagnosis in Health Care*. Washington, DC: The National Academies Press; 2015.
9. Society to Improve Diagnosis in Medicine, at: www.improvediagnosis.org.
10. Best Doctors, at: www.bestdoctors.com.
11. In Need of Diagnosis, at: web.archive.org/web/20151031133116/http://www.inod.org.

2

Medical Decision Making

Karen Cosby

CONTENTS

Introduction

The primary work of diagnosticians involves investigating the source of illness, collecting evidence, drawing conclusions, and communicating results. The process of diagnosing typically involves a series of steps requiring a variety of specific skill sets and expertise. Diagnosis is not a singular or necessarily linear process. From patient to patient, the process can differ, depending on circumstances, how simple the problem is, how direct a cause is assumed, and whether or not an intervention or treatment is identifiable and without controversy or contraindications. Although diagnostic pathways are sometimes reduced to algorithms, in reality, medical decision making is often less ordered, and frankly, even a bit messy. There are endless variations of illness, and the task of considering, weighing, and testing the possibilities relies on a foundational knowledge of bioscience and epidemiology, reasoning skills, and something we vaguely attribute to clinical reasoning and judgment, the skills gained from exposure and practice. This chapter explores the largely unseen, even unspoken, mental process of clinical reasoning and the making of a diagnosis.

Models of Clinical Reasoning: Analytical versus Intuitive

Diagnosis has been described as a sequential process that leads from input data (patient manifestation of disease) to output (diagnosis) [1]. There are two intrinsically different models for how clinicians process clinical data to determine a diagnosis: an analytical model (the rational quantitative approach) [2,3], and a nonanalytical model (the intuitive approach) [4–6].

The Analytical Model

The *rational quantitative analytical approach* is based on the *hypothetico-deductive* model that resembles the work of scientists who gather facts, then develop and test hypotheses [2,3]. The process is slow, deliberate, and methodical. Kassirer and Kopelman describe five phases of an analytical approach to diagnosis in their classical textbook *Learning Clinical Reasoning* [3], including the gathering of facts, generating hypotheses for a differential diagnosis, testing, refining, and then finally verifying the final diagnosis (Figure 2.1).

Because of its foundation in the scientific method, the rational analytic method has enjoyed respect and has been viewed as a method of serious thinkers. But the analytical approach is tedious and slow, and on a practical level, too inefficient for the rapid decisions demanded of most clinical settings. An analytical approach is traditionally viewed as a beginner's approach, when experience is limited and reasoning must by necessity be thorough and cautious. Experienced clinicians may use detailed analysis when working with new or unfamiliar problems. Analytical methods may also be useful for complex problems that require multiple specialty consultations or in situations where time and circumstances allow, such as for

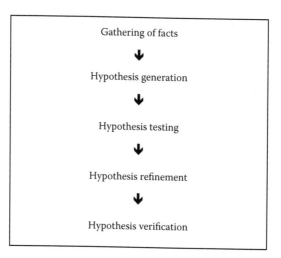

FIGURE 2.1
Five phases of diagnosis. (Based on Kassirer, J.P. and Kopelman, R.I., *Learning Clinical Reasoning*, Lippincott Williams & Wilkins, Philadelphia, 1991 [3].)

problems referred to tertiary care centers specializing in particularly difficult or rare cases.

Classically described as an exhaustive method, the hypothetico-deductive approach need not necessarily be an all-or-nothing technique. Physicians have adopted abbreviated forms, generating short lists of diagnostic considerations based on prevalence and characteristic features of illness, then ruling out a few alternatives to be sure they don't miss atypical variants. This is a practical but reasoned adaptation to the more formal method that helps catch outliers and mimics of common conditions.

The Intuitive Model

The *intuitive model* contrasts sharply with the analytical mode [5]. With expertise, and with practice, clinicians begin to recognize common patterns. On encountering a case, they assess whether or not they recognize the problem. With time and exposure to a variety of cases, clinicians assimilate similarities and discriminating features that allow them to use pattern recognition much of the time. The intuitive mode is almost instantaneous. Because it is quick and energy efficient, most prefer to use the intuitive mode rather than labor in an analytical mode. While clinicians may not like to admit it, some have been known to suppress a painful groan when encountering cases that force them into analytical mode; for example, when needing to evaluate the weak and dizzy elderly patient.

While clinicians might like to consider themselves rational and believe they use analytical thought for diagnosis, there is good evidence that most diagnoses are made not through a reflective analytical approach, but rather, simply by rapid pattern recognition [7]. The ability to recognize a diagnosis with limited information uses cognitive skills similar to those of recognizing different human faces. The rapid, almost instantaneous recognition of a familiar face happens with little conscious thought. The judgment is not based on listing a set of features and drawing the conclusion; rather, there is an appreciation for the essence of the person (their general appearance but also their mannerisms, their gait, their smile), characteristics that one may not even be aware of noticing. Experts perceive something bigger than the sum of the parts, features that somehow bring recollection of similar cases in their past and how they relate to the present one before them [5]. The ability to rapidly recognize a clinical syndrome is sometimes referred to as a *hallway diagnosis*, a term aptly describing conditions that clinicians recognize with a casual glance absent any objective information. The dramatic appearance of renal colic, a woman in labor, and flash pulmonary edema are a few such examples common to acute care medicine.

The acquisition of knowledge in the biosciences (biology, anatomy, pathology, and pathophysiology) forms an essential foundation for clinical reasoning in medicine; but clinical expertise also requires a kind of diagnostic sixth sense that arises not so much from medical "book knowledge," but rather, from the set of clinical experiences encountered over time [5]. Experts have a large repository of disease "faces" and thus are able to expand their recognition of even subtle variations in disease presentation.

In reality, clinicians rely on *both* analytical and intuitive reasoning [6,8,9]. When cases are simple and straightforward, pattern recognition is efficient and accurate; when problems are unfamiliar, complex, or difficult, most physicians downshift to a more methodical, slower, reflective pattern—a practice that likely improves diagnostic accuracy [7]. The challenge for the clinician is to know when to use which strategy, and to recognize signals that should trigger a switch into the alternate mode. This back and forth strategy is also a feature of the *dual process model* for clinical reasoning described in depth in Chapter 3: the analytical mode is analogous to System 2; the intuitive mode is analogous to System 1 [10–12]. These concepts are now widely recognized and popularized in recent best-selling books, including Daniel Kahneman's *Thinking, Fast and Slow* [13], Malcolm Gladwell's *Blink: The Power of Thinking without Thinking* [14], and Michael R. Legault's *Think!: Why Crucial Decisions Can't Be Made in the Blink of an Eye* [15].

Physicians typically don't think much about which type of reasoning they use; the choice is natural and largely unconscious. Having concluded that their patient has acute pulmonary edema, they might be hard pressed to explain how they know what they know. If questioned, they might describe the picture they are so familiar with: the patient has dyspnea, frothy sputum, and diaphoresis, is anxious and sitting upright, and has severe uncontrolled hypertension; but if urged to explain more, they might be—well, annoyed with the request. Intuition is easy and energy efficient, and allows them to proceed with a course of action with minimal delay. In fact, if asked to slow down or reconsider the diagnosis, some would argue that the diagnosis is obvious and doesn't merit much thinking at all. However, should something unexpected happen, or a new fact emerge that confounds the diagnosis, clinicians may need to rethink their diagnosis and resort back to analytical mode.

Logical Reasoning in Diagnosis

The formulation of a diagnosis can be viewed as the development of an argument. Evidence is collected and analyzed, and a conclusion (or diagnosis) is reached. If diagnoses are rational and objective, rules of logic might inform us of how valid our conclusions are, how well we reason, and where we might err in the process [16–20]. There are two principal types of logical reasoning: deductive and inductive.

Deductive Logic

Deductive reasoning begins with a general premise, and leads to the specific. If the premise is true, then the conclusion is certain. Deductive proofs in math are a good example:

a. If $x = 3$

b. And if $y = 4$

c. Then $3x + y = 13$.

If the premises (a and b) are true, c is necessarily true. Deductive proofs like this are reliable and accurate, leaving no uncertainty.

Similarly, a *deductive syllogism* (a natural language equivalent of the mathematical example) looks like this:

Premise 1: A treatment ought to be covered by insurance companies if it has been proved to save lives.

Premise 2: Treating hepatitis with antivirals saves lives.

Conclusion: Insurance companies ought to cover the cost of treating hepatitis.

The conclusion is valid (it follows from the premises); however, one can question the premise and thereby refute the conclusion: should insurance companies cover the cost of treatment for all conditions, independent of the burden of the cost and availability of treatments?

While deductive logic is certain and accurate, it is not generative; that is, it does not improve understanding or generate new theories or knowledge. Thus, deductive reasoning does not adequately describe or fit the clinical work of diagnosis, and it cannot solve diagnostic questions. In diagnostic reasoning, we don't start our reasoning from premises of known general scientific knowledge. We start with a problem, an ill-defined set of signs and symptoms, and reason backward to postulate a cause. Thus, diagnosis needs a different logic; our initial steps in diagnosis are better described with inductive logic.

Inductive Logic

Inductive reasoning begins with specifics, and concludes with the general. The conclusions of an inductive process are *probably* true considering the available evidence, but there is no way to guarantee truth, since one can never be sure of having seen all possible evidence, including facts that might invalidate the hypothesis. Inductive reasoning is inexact and always leaves some amount of uncertainty. Unlike deductive reasoning, inductive reasoning is generative and leads to a new hypothesis that may or may not be true.

One example of inductive reasoning is demonstrated in criminal court cases, in which the jury is expected to conclude a verdict in the face of incomplete evidence. The case begins with facts discovered through investigation, and the jury is charged with declaring the guilt or innocence of the accused. The jury makes their verdict without ever knowing or being able to verify the accuracy of their judgment. Did they see all the evidence? Did someone lie or deceive? Was the defense attorney a weak advocate, or might the attorney have offended the jury and biased them against the defendant? Jurors are expected to perform to some standard of reasonable certainty, but there is no pretense of absolute certainty.

Inductive reasoning is also used by naturalists, who observe phenomena and attempt to find explanations. Similarly, reasoning from symptoms to determine cause (diagnosis), as we do in medicine, relies on inductive reasoning. Based on background knowledge in biology, pathology, and pathophysiology, we collect evidence (signs and symptoms of disease) and reason backward to possible causes. We can postulate and propose causes, but we can never be perfectly certain.

For example, a clinician is faced with a patient with chest pain. The triage nurse has ordered a troponin, and the result is reported as positive. We use induction to postulate a diagnosis:

Premise: My patient with chest pain has an elevated troponin.

Premise: Patients with myocardial infarctions (MIs) have elevated troponins.

Conclusion: My patient probably has an MI.

The conclusion is *likely* true, but it is not absolutely certain. Why? Not all elevations in troponin are due to MIs. Not all patients with MIs have chest pain. Finding an elevated troponin in a patient with chest pain makes it *likely* that he has an MI, but clinical reasoning then proceeds from a hypothetical explanation to the next phase of clinical reasoning using testing and *probability* to add another element of diagnostic certainty. A patient with chest pain and elevated troponin might have a pulmonary embolus. Or, perhaps, the specimen is hemolyzed, and the troponin is a false positive result. The clinician needs to ask, "What is the probability of MI in this patient and should I add alternative explanations to the differential diagnosis?" Inductive reasoning allows the clinician to simultaneously consider multiple competing hypotheses by weighing their relative probabilities.

Neither deduction nor induction quite adequately describes how clinicians actually make diagnoses. A third form of logic, less well known, has been described that is a better fit for clinical reasoning: abductive reasoning.

Abductive Logic

Abductive reasoning begins with specifics (in our case of medical diagnosis, symptoms and signs of diseases) and hypothesizes (invents) a plausible explanation [19,21,22]. Abductive reasoning accommodates the imperfections typical of clinical practice: it accepts that evidence and facts are incomplete, and requires the thinker to develop acceptable and "most likely" explanations for the observed findings [23]. The clinician may admit that a hypothesis fails to account for *all* the evidence, but looks for the best fit that can reasonably be supported.

Abductive reasoning provides a plausible explanation for the facts available in the moment. An important part of abductive reasoning is the recognition that an explanation is tentative and useful only as long as a better explanation doesn't present itself. The ability to provisionally accept the hypothesis allows one to take action in the face of imperfect knowledge. Conclusions from abductive reasoning allow one to move forward in the diagnostic trajectory and treatment strategies without becoming mired in doubt and indecision. Abductive reasoning may be useful for some situations when there is a drive to act in the face of partial information. An example of abductive reasoning is given in Box 2.1.

Abductive reasoning allows tentative diagnoses to be considered and acted on, but remains flexible and even skeptical, looking for evidence for or against the leading theory. Abductive reasoning is dynamic and actionable even in the face of uncertainty. The clinician need not be convinced of the theory, only that it is plausible. The strength of an abductive argument isn't necessarily in being accurate, only in being plausible. The concept of abductive reasoning adds value to the discussion of medical reasoning by recognizing the dynamic nature of clinical decisions, all the while acknowledging uncertainty while still moving forward in testing and treating.

These three models of logical reasoning inform the diagnostic process. A more complete model of diagnosis can be described with a model that

BOX 2.1   EXAMPLE OF CLINICAL REASONING USING ABDUCTIVE LOGIC

A patient with cancer had chemotherapy 2 weeks ago. The patient has an episode of loss of consciousness and is found on the floor of her home. Her clinician is most concerned about the possibility of neutropenic sepsis, knowing that the timeframe is right for the nadir of her expected leukopenia. She might just be dehydrated. She might have had a syncopal event from a dysrhythmia. Perhaps the unwitnessed event was a seizure, and the patient has unrecognized metastatic disease in the brain. Perhaps she just slipped and hit her head, and has a traumatic brain injury. But if she is neutropenic, she is at risk for bacteremia and neutropenic sepsis. If she is septic, delay in antibiotics can worsen her outcome. The physician reasons that the most plausible explanation is sepsis, and orders blood cultures and antibiotics. Her doctor may not know the white blood cell count or whether she has a fever, but there is a reasonable concern that she might be septic. The need to act and the relative risks argue for action. The clinician doesn't ignore other possibilities, but acts rapidly on the most plausible ones. Should a period of observation prove that she doesn't have a fever, her laboratory tests exclude neutropenia, and her blood cultures are found to be sterile, the hypothesis will be revised.

combines all three methods of logical reasoning [19]. A clinician gathers evidence and proposes a likely diagnosis using abductive logic. (My patient has chest pain, maybe he has an MI.) He then tests his theory using deductive logic. (If my patient has an MI, he will likely have wall motion abnormalities on an echocardiogram; let's request one.) Then, he uses the result of his test to confirm or refute his idea using induction. The process follows an iterative path until a diagnosis is developed to some reasonable degree of certainty dictated by the need or desirability of intervention. Understanding each type of logic can help clinicians assess the strength of their arguments and the limitations of their conclusions, features that are summarized in Table 2.1.

TABLE 2.1

Characteristics of Formal Models of Logic

Features	Deductive	Inductive	Abductive
Accuracy	Certain	Probable	Plausible
Rules	Formal, fixed	Generative (generates new understanding)	Generative (creates new and even novel understanding)
Characteristics	Conclusions true given premises	Weighs possibilities based on probability	Considered tentative, even skeptical, willing (even expected) to be revised as new information becomes available

Collecting Evidence: The Narrative

A complete framework for diagnostic reasoning is still not fully explained by these models of clinical reasoning and logic. Something essential is still missing. The missing piece can be described as the difference between the theoretical framework of "work as imagined" and "work as realized." Aristotle describes *phronesis* as a practical rationality that describes the "knowing and doing" that occur in the clinical encounter [24]. The foundational underpinning of diagnosis involves an interaction between doctor and patient, and an interpretation of the patient's subjective experience of illness. The primary source of information comes from the patient, and the primary method of collecting evidence begins with an interview and a narrative [25,26].

Schleifer and Vannatta argue that the most powerful diagnostic tool a physician has is the patient interview, the "History of Present Illness (HPI)" [27]. Like Sherlock Holmes, clinicians must gather clues from the victim and the suspects, assessing the story as well as the storytellers [28]. Experienced clinicians know that collecting good evidence from the patient may require a nuanced approach perfected over time and with experience, and is affected by a number of factors that influence how people of differing age, gender, personalities, backgrounds, and cultures interact. The skill required for interviewing patients has more in common with the work of anthropologists, ethnographers, historians, and sociologists; we all share the need to construct meaning from a narrative account.

The physician must sort and sift through the symptoms patients describe, filtering what seem to be those relevant to a potential diagnosis from those that are likely to be trivial, inconsequential, or unrelated to the illness at hand. There are challenges to determining the essential facts from a patient's history. Not all relevant information is always available. Not all available information is relevant. Some details patients provide are misleading distractions that are unrelated to the actual diagnosis. Mamede demonstrated that when salient distracting clinical features were introduced early in a case presentation, the time required to formulate a diagnosis increased, and diagnostic accuracy fell [29]. The ability to discern the relevant signal and ignore or discount the irrelevant is partly gained with expertise, but may also be a matter of good judgment, or dare we say, simply luck. One such example is seen in an illustrative case in Box 2.2.

The answers patients provide to questions may depend on how the question is asked, their willingness to share personal information, and their ability to recall accurately and communicate effectively. Patient information is not always accurate, complete, or consistent. Sometimes, patients do not understand why specific information is important, and give vague responses when we seek details that are meaningful to us, such as "When did you first become ill?" and "Which symptom came first?" Patients do not always give a factual account; it's not that they intend to deceive as much as they may hope

BOX 2.2   THE VALUE OF THE NARRATIVE

A 30-year-old man presented to an Emergency Department with abdominal distension and hypotension. A brief history revealed that he had just been discharged from another hospital a week after major surgery on his pancreas. He was alert but diaphoretic with a tender, distended abdomen. His initial blood count revealed a significant anemia. He must be bleeding, but where? I quickly checked for other clinical findings of blood loss, but found none. The distended abdomen was likely harboring the source of his hemorrhage. Perhaps a ligature came loose; strange, I thought. A week is a bit delayed for that explanation. But I had nothing better to account for his condition. His condition was critical and time was of the essence. I called the surgical team and ordered blood. The surgeon rushed to the ED and whisked him away.

In the rush, a woman appeared. His mother? She pulled on my arm, anxious to tell me something. I explained that he was bleeding and needed emergency surgery. She kept interrupting me, and I, her. "No, no!" I said. "You must listen to me, this is important. He's very sick. He's going to the Operating Room. We have to hurry!"

The following day I went to check on the patient, expecting to find him recovering from surgery. To my surprise, as I approached the ICU, a team of surgeons came running down the hall pushing a stretcher with the patient! "Where are you going, what is wrong?" I asked. "We're going to the OR! He has a ruptured spleen!"

Later, I pondered: what went wrong? I had expressed my concern that he was bleeding; how was that missed? The surgical team concluded that he must have hemorrhagic pancreatitis; surgery in complicated pancreatitis is thought to worsen outcome. Their goal, all the while, was to medically stabilize the patient and avoid surgery at all costs.

The problem? The message the mother had tried so hard to give me … what was that, I began to recall. As the fog lifted from my memory I slowly realized … She was trying to tell me that he had fallen down the stairs. At the time I thought, well, of course. He is hypotensive and orthostatic. It's a wonder he could stand at all. But in my sense of urgency, I failed to connect the dots. He didn't fall because he was bleeding; he was bleeding because he fell. The devil's in the details! And the narrative.

that you will give them the judgment (and reassurance) they seek. Families who bring a relative to the hospital may have cajoled them into a visit, only to have them deny the very problem that prompted their visit. Awareness of the dynamic at the bedside and recognition of others who have information to add to the encounter is important. Sometimes, it's not what people say, but rather, how they tell their story and how they behave, that is most meaning-ful. Interestingly, sometimes the clue that provides the key to understand-ing a patient's illness comes not from the intentional directed interview, but rather, from the casual discourse that happens during their visit. The skill to gain the confidence of the patient and the ability to sense subtle cues depend not on academic knowledge but on relational skills.

In some cases, patients may be too ill to reasonably provide needed infor-mation. Someone struggling to breathe is unlikely to engage in extended con-versation; a patient in pain is unlikely to be cooperative until after their pain is better controlled. In the most extreme example, an unconscious patient will be unable to speak at all. In those instances, other methods of investigation will have to suffice, such as examination of the patient's wallet and pockets, a search for medical alert tags, interviews with police and paramedics, and clues taken from the scene of an accident. The difficulty in finding and deter-mining facts in extreme cases is illustrated by the misidentification of two young blond women involved in a horrible accident [30]. While one family mourned and buried the victim declared dead on the scene, the other family kept vigil at the bedside of the second comatose woman, who was covered in bandages and suffering a serious head injury [30]. The two families of the victims did not recognize the true identity of the surviving victim until weeks later, when bandages were removed from the head and face of the unconscious survivor. As incredible as this story may seem, many healthcare workers will attest to the difficulty in identifying and tracking people, their blood samples, and even their organs—as evidenced by wrong site surgery and even mix-ups in organ donations. What is fact versus assumption can be even more difficult in particularly emotion-laden settings when people's lives and futures are in the balance, especially when urgency for action gives little opportunity for redundancy in process.

Illness as experienced and described often falls outside the frameworks of disease presented in textbooks. Experienced clinicians learn to recognize common variants that defy usual textbook descriptions. The classification of angina, myocardial ischemia, and MI is artificially categorized under acute chest pain syndromes. Yet, clinicians know that patients who suffer from cardiac ischemia will sometimes deny chest pain altogether, often describ-ing instead a sense of uneasiness, a foreboding, an uncomfortable feeling, or oppressive discomfort. Others may experience an unusual ache in their left jaw, fatigue, dyspnea, indigestion, dizziness, or palpitations. Nonetheless, these complaints are often reduced to a classification fitting an electronic template and International Statistical Classification of Diseases (ICD)-10 code for "chest pain."

The skill of gathering evidence is important for all subsequent clinical reasoning, since the facts assumed from the history lay the groundwork for all the diagnostic questions and tests to follow. Some information from an interview provides hard facts that are usually reliable, assuming the patient has an intact memory and is alert: the patient's name, their date of birth, and (usually) a current list of medications. Those are hard facts that can usually be accepted at face value. Other facts from the history sometimes deserve some critique. Patients may relate that they had "a cardiac workup" that was "normal." Without pursuing further questions or validating the account with actual records, any number of possibilities exist. Perhaps the patient had an electrocardiogram and was sent home from another hospital. Do they mean they had a stress test? A cardiac catheterization? Was the test result conclusive? If imaging was done, was the quality of the study adequate? Was the choice of technique and type of image appropriate for the question posed? The patient history is best done with an approach meant to question, rephrase, and verify. The growth of the electronic health record has improved our ability to validate facts, although it can also be a source for error. In one example experienced by the author, a Do Not Resuscitate (DNR) order that was featured prominently in the banner of the electronic medical record of a healthy young person caught the attention of medical staff; investigation revealed that it had been placed there by mistake.

There is more to the narrative account of a patient's illness than what can be captured by simply checking off boxes to "yes" and "no" questions or completing a preformatted template. Disease is often best understood in the narrative. The patient's story provides context and meaning that go beyond the mere collection of facts. Stories are contextual, and often personal. To reduce the story to an ICD-10 code loses the richness, the human drama, and the complexity of intricate details of illness and its manifestations.

The "story," the "narrative," is important in understanding and learning about diagnosis. In part, students learn about diagnosis through storytelling, piecing together common threads between clinical encounters to learn variations of presentations. Clinicians share their stories and often compare their experiences with their peers. The clinical account described earlier (Box 2.2) is one such example. Our narratives create the memories that likely make us better diagnosticians. Most clinicians can retell stories of patients that impacted them years after the fact. Whether they describe human tragedy, humor, or simply strange and bizarre events, these accounts are well embedded in our memories, and with them, the lessons they taught us. The power of a patient's narrative, "the story," is illustrated well in Oliver Sacks' *The Man Who Mistook His Wife for a Hat* and in Lisa Sander's *Every Patient Tells a Story*, collections of fascinating stories of medical mysteries explained through narrative accounts [31,32].

Collecting Data and Testing: The Physical Exam

The narrative leads to inquiry, and the physical examination is the first phase of testing for objective evidence of disease. The value of the physical exam has been questioned by some, who argue that newer and better-validated methods exist for detecting disease. But the clinician has at his hands the ability to test and revise his hypothesis using simple time-honored methods to detect evidence. Failing to conduct a thoughtful examination, one that intentionally looks for findings to support or refute diagnostic hypotheses, is similar to leaving the scene of a crime without collecting evidence. Akin to putting blinders on, it is as if one hoped to reason through a case without the benefit of evidence.

Aging but experienced mentors sometimes have to convince the cynical student that old-fashioned methods of physical examination still have merit and are worth the investment of time and effort. Exhortations about the importance of the physical examination may be heard resounding from the hallways and corridors of academic medical centers, such as:

- You can't see what you don't look for. You will never diagnose papilledema without a funduscopic exam.

- You aren't inclined to look for what you don't expect to see. If your clinical experience is limited you may never learn to recognize some of the occult manifestations of disease. Necrotizing fasciitis may be detected by palpating distant from the initial skin findings; the surprise soft crunch of soft tissue air may be felt before the erythema spreads. The presence of dilated vessels in the upper torso may provide the first clue of a superior vena cava syndrome, the finding of an underlying lung cancer. One of the thrilling parts of learning medicine is the unexpected clue, the recognition of its relevance to the matter at hand, and the delight in using and sharing the finding to "crack the case." I still recall the delight of an intern who detected the low rumbling diastolic murmur of mitral stenosis that helped establish the diagnosis of endocarditis, a finding and diagnosis missed by all the other members of the more senior examining team. The finding likely saved the patient's life.

- Success depends on the expertise and interest you bring to the exam. As described by Stanley and Campos, "sometimes even a minor detail—an inadvertent gesture; a change in voice or speech; the mood of a patient; a factor such as foreign travel … can lead the observer in a different direction for diagnosing" [19]. Those who minimize the value of the exam will likely never benefit from its value. Physical exam skills come from personal experience with patients and cannot be gained from sitting in the library reading a textbook. They are the gift patients give us to unlock their mystery.

- The absence of a finding is not necessarily proof of lack of disease. Many have witnessed firsthand the typical symptoms of zoster only to have the rash appear after the onset of pain, remote from the first exam. That fact can be acknowledged without undermining the argument to look for a rash and the willingness to revisit the exam on another day.

- The presence of a pathognomonic finding may be very specific for disease, and the presence of one cardinal finding may save unnecessary tests. A constellation of findings often provides very strong, if not conclusive, evidence of disease.

When novices declare that the physical examination was "unremarkable," the real question is: "How hard did you look?" and "Did you know what you were looking for?" Success at detecting signs of disease is dependent on the rigor with which one looks and the skill one has in detecting and recognizing physical signs of illness. Sherlock Holmes expresses this well in *"A Case of Identity"*:

> Watson says to Holmes that what the detective saw "was quite invisible to me". Holmes replies: "not invisible, but unnoticed, Watson. You did not know where to look, and so you missed all that was important."
>
> **(Sebeok and Umiker-Sebeok, 1983, p 21, citing "A Case of Identity,"
> as cited by Schleifer and Vannatta) [27].**

Use and Interpretation of Tests

Once the evidence has been collected, the diagnostic process continues to the next phase: using data to test hypotheses. When examining a differential diagnosis, there are formal mathematical methods to test the hypotheses under consideration. Formal diagnostic testing is a largely analytical process that is governed by rules of probability [33,34]. What is the likelihood of this patient having a condition based on the disease prevalence and the patient's unique set of risk factors? How good is this test for reliably ruling in, or ruling out, disease?

Tests are developed based on their ability to discriminate between patients with disease and those without disease [34–36]. Not all tests are equally helpful; the diagnostic performance of tests is illustrated in Figure 2.2. The effective use of tests relies on the establishment of a range of normal values that have sufficient ability to distinguish between normal and disease states.

The use of diagnostic tests depends on their operating characteristics—how well they predict the presence or absence of the condition in question [33]. A summary of these basic characteristics of medical tests is described in Figure 2.3 and includes:

> Sensitivity: the ability of a test to detect the presence of disease
>
> Specificity: the ability of a test to accurately predict the absence of disease

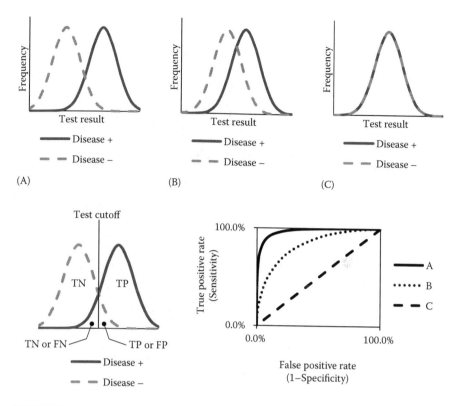

FIGURE 2.2
Diagnostic performance of tests and their receiver operator curves for tests with high (A), moderate (B), and poor (C) performance. TP, true positive; TN, true negative; FP, false positive; FN, false negative.

Sensitivity and specificity are useful concepts in determining whether a test is likely to be of value. A negative result of a very sensitive test is useful in ruling out disease; a positive result of a test with high specificity is good to rule in disease.

The decision to test, and what methods to apply, is a distinct phase of work in diagnosis. Once a test result is obtained, the interpretation and application of the result require additional reasoning. Given a result, how does the probability of disease change? This question is better addressed by predictive value, defined as

Positive predictive value (PPV): the probability of disease given a positive test result

Negative predictive value (NPV): the probability of disease being absent given a negative test result

The application of PPV and NPV is influenced by the prevalence of the condition being questioned in the population examined. This has profound

	Disease +	Disease −	
Test +	True positive TP a	False positive FP b	a + b
Test −	c FN False negative	d TN True negative	c + d
	a + c	b + d	

Prevalence of disease	= (a + c) / (a + b + c + d)
Sensitivity	= a / (a + c)
Specificity	= d / (b + d)
Positive predictive value (PPV)	= a / (a + b)
Negative predictive value (NPV)	= d / (c + d)
Positive likelihood ratio (+LR)	= Sensitivity / (1 − Specificity) =TP/FP
Negative likelihood ratio (−LR)	= (1 − Sensitivity) / Specificity = FN/TN

FIGURE 2.3
2 × 2 table with definitions for diagnostic test interpretation. TP, true positive; TN, true negative FP, false positive; FN, false negative.

implications for the potential of testing to change posttest probability of disease. As seen in Figure 2.4, a test with 95% sensitivity and 95% specificity gives a PPV of 99.4% in a population with disease prevalence of 90%, but the PPV falls to only 16.1% with a disease prevalence of 1% [33]. Common advice to young clinicians warns of this phenomenon: "When you hear hoof beats (in North America, outside of zoos), don't go looking for zebras." Likewise, the same test gives an NPV of 95% in a test group with a disease prevalence of 50% but falls to 16.1% with a prevalence of 99%. If the disease is very prevalent, testing may offer little help in changing the odds of disease. The use of testing can be treacherous and misleading if tests are misapplied to the wrong population.

Another method for assessing the likelihood of disease uses likelihood ratios (LRs) and Bayesian analysis [37,38]. Likelihood ratios are also referred to as *conditional probabilities* and reflect how the odds of disease change relative to the odds expected for the general population once a given test result is known. A positive likelihood ratio (+LR) is the odds of a positive test result in a patient with disease divided by the probability of a positive result in a patient absent disease; that is, true positive/false positive, or [sensitivity/

Prevalence	99%	90%	70%	50%	30%	10%	1%
Positive predictive value	99.9%	99.4%	97.8%	95.0%	89.1%	67.9%	16.1%
Negative predictive value	16.1%	67.9%	89.1%	95.0%	97.8%	99.4%	99.9%

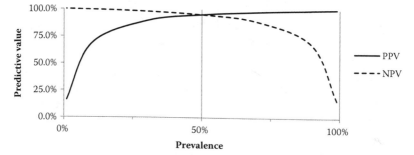

FIGURE 2.4

Effect of prevalence on predictive values based on a test with 95% sensitivity, 95% specificity. (Adapted from Sackett, D.L., et al., *Clinical Epidemiology: A Basic Science for Clinical Medicine*, Little Brown and Company, Boston, 1985 [33].) PPV, positive predictive value; NPV, negative predictive value.

(1−specificity)]. A negative likelihood ratio (−LR) is the odds of a negative test in a patient with disease relative to a patient without disease; that is, false negative/true negative, or [(1−sensitivity)/specificity]. An LR of 1 means that the test does not change the odds of disease; it has no discriminating value. An LR greater than 1 increases the odds of disease; the higher the value, the greater the odds of disease. Values between 0 and 1 argue against the diagnosis, or at least, lower the posttest probability. Using Bayesian analysis, LRs are used to modify *pretest probability* (also known as *a priori probability*) to yield *posttest probability*. An easy example is the evaluation of possible appendicitis. The presence of right lower quadrant (RLQ) pain has a +LR of 8 and a −LR of 0.28. The presence of vomiting in appendicitis has a +LR of 0.92 and a −LR of 1.12 [39]. Thus, the presence of RLQ pain should increase one's suspicion of appendicitis, and the absence of RLQ pain should decrease suspicion. The presence of vomiting has minimal impact on posttest probability.

Bayesian analysis is even more powerful when combinations of tests are used. In each case, the posttest probability of one test becomes the pretest probability for the next. A carefully constructed series of tests should lead to an increasing degree of certainty [33]. There may be multiple available tests for any given condition, and the likelihood of disease may be based on the presence or absence of typical physical findings and the value of selected tests targeting the diagnosis under consideration: thus, each test helps rule in, rule out, and reorder the likelihood of conditions in the differential diagnosis.

Diagnostic tests don't typically give simple yes or no answers to a question about a specific diagnosis. Before choosing a test, it is best to consider how the result will change the posttest probability and influence subsequent actions. A common example cited involves the use of stress testing for detection of coronary artery disease. Consider these examples [18,40]:

1. A young 35-year-old woman with nonanginal chest pain has a likelihood of underlying coronary artery disease of 0.8%. Stress testing has a sensitivity of 90% and a specificity of 85%. Using a stress test in this patient will yield a positive predictive value of only 6%, and the odds are 16:1 against a positive result being correct. Applying a test in this patient will more often than not result in unnecessary further testing. This patient should not undergo stress testing unless there is some identifiable factor(s) that modifies her unique risk differing from the general population of young women.

2. A 65-year-old man with typical angina has a likelihood of underlying coronary artery disease of 94.3%. Using stress testing in this patient is of no value; the odds are 2:1 against a negative test being correct. In this example, nothing short of a gold standard test (cardiac catheterization) is sufficient to rule out disease.

The use of Bayes' theorem and the analytical approach to testing has grown partly through the influence of evidence-based medicine (EBM). EBM is defined as "the integration of best research evidence with clinical expertise and patient values" [41]. The EBM movement has attempted to provide clinicians with the most current and best objective medical knowledge in a clinically relevant way. Students of EBM are encouraged to abandon routine medical texts (because they are often outdated) and rely more on evidence-based data bases such as *Uptodate®* that summarize current data, assess the strength of evidence, and provide guidance for practical decision making. EBM includes systematic reviews and meta-analyses and may be used in the development of clinical decision rules. The evidence-based approach to diagnosis uses the rational quantitative tools described here to evaluate diagnostic hypotheses. The technique is rigorous and methodical. While it can be slow, some questions recur in practice, and clinicians may find it particularly useful to build a repertoire of common questions and scenarios they face in their daily work.

The calculation of mathematical probabilities and the use of Bayesian analysis (or perhaps more appropriately, Bayesian inference) lends an aura of certainty to diagnostic reasoning, but we must be cautious in how much trust we place in the results. Posttest probability is predicated on pretest probability, which in turn, is determined by clinical judgment, or expert opinion—or put another way, an educated guess. And while one might think that these methods should reduce uncertainty and be met with enthusiasm, in fact, most clinicians don't have a working understanding of many of these statistical concepts, and even fewer use them at the bedside [42,43]. Studies of actual clinician performance demonstrated that most do poorly in their estimates of probability of disease [44], and there is significant interobserver variation in estimation of risk [45]. The accuracy of clinical judgment varies so much between individuals that one can question whether there is anything of value added in all this analysis, something that is explored more in Chapter 9 (Individual Variability in Clinical Decision Making and Diagnosis). That does not necessarily mean that statistical methods don't influence and improve care. The mathematical analysis of risk and probability factors into clinical decision rules and eventually affects standard of practice. In any given moment at the bedside, clinicians may not form questions, search data sets, or calculate odds, but they are likely familiar with evolving practice guidelines that are heavily influenced by these methods. Clinicians who actively seek feedback can hope to outperform the norm, but they need a supportive environment and tools to optimize their calibration and decision making [46,47].

There are other factors besides likelihood of disease that dictate how diagnostic testing is used based on the risk of testing, the risk of treating, and the potential gain of diagnosing and treating a condition. If a disease under consideration is very likely to cause death or serious harm, and an intervention that can mitigate that harm is available and without significant additional risk, clinicians may choose to test even if the likelihood of disease is low. They may even choose to treat without testing. On the other hand, if a condition is fairly harmless and not likely to benefit from any specific treatment, some may choose not to test at all, since it is unlikely to change treatment recommendations. The art of medicine is pragmatic. The idea of reaching a point where enough information is available to reasonably decide to test, or to treat, is described as a *threshold*. Independent of formal testing, a threshold may be reached where it is reasonable to determine a course without further delay [48]. In such cases, the goal isn't to achieve diagnostic certainty as much as to be safe.

Decisions about testing and treating are influenced by factors other than principles of epidemiology. Despite the evidence for or against a test, patients may favor an approach based not on numbers or risk, but on personal preference. A patient may argue for breast cancer screening, despite a risk of false positive results and unnecessary biopsies, to minimize all chances of missing or delaying a diagnosis of cancer, even when the risk of testing exceeds the potential benefit. Physicians are known to modify their practice to accommodate the worries of patients, even when practice guidelines argue differently. Many pressures affect real-life practice: patient preference, fear of litigation, and fear of missing a diagnosis [36].

Although formal methods of testing rely on scientific methods and mathematics, they are not fully objective and suffer from their own limitations. Even the math cannot remove uncertainty from the diagnostic process.

Informal Testing Strategies and Reasoning Guidelines

Scientific models of reasoning and formal rules of logic only partly explain how medicine is practiced. There are a variety of informal guidelines that describe how doctors act. Some are based on reason; others, perhaps, have their basis in superstition. They form some of the informal culture of work and wisdom shared between clinicians. While not validated, they do reflect the reality of "flesh and blood" decision making; that is, how decisions happen in the moment under real-life circumstances.

A few examples follow:

- When considering diagnostic possibilities, try the "loveliest explanation" (the one that is the best explanation) [49].
- Seek the simplest explanation that is a reasonable fit. When possible, follow the rule of Occam's razor (the rule of parsimony); that is, find a single fit for all the evidence.
- "Rule Out the Worst-Case Scenario." Always consider the worst case possible. If you can't be right, at least be safe.
- Beware of red flags, conditions that suggest the possibility of a missed diagnosis. Examples include repeat visits to the hospital and unexpected findings.
- Use rules of thumb. For example, anyone over 40 years of age with chest pain needs an electrocardiogram (EKG); always assume a woman of childbearing age is pregnant; and don't trust, verify.

Folklore and Aphorisms in Medicine

Medicine is filled with historical references to sage advice [25,50]. Some of them are firmly entrenched in the cultural heritage of medicine, preserving voices from the past. For those who promulgate medicine as a mostly scientific endeavor, anecdotal stories and catchy phrases seem silly and superstitious. Some might even chastise anyone who shares them as having a lazy mind or believing in a lesser science. But there is a reason why these timeless, enduring phrases are passed on to new generations. They are valued because they humor us, shock us, and perplex us. Many contradict themselves. But in doing so, they acknowledge how irrational, illogical, and unpredictable clinical work can be. They remind us that paradox is evident in actual clinical work. They emphasize the interpretive nature and inherent uncertainty in our judgments; that is, all rules are ultimately dependent on context and are situational. A rule that applies well in one case will fail in another. The maxims serve to warn clinicians of traps and

"When you hear hoof beats, think horses, not zebras!"

This maxim warns clinicians to pursue common things first, but in doing so reminds them that rare things do exist. Paradoxically, in considering this rule, the clinician must at least acknowledge that zebras exist, even if only to ignore the possibility until the situation suggests a reason to consider them again [51].

Occam's razor: The rule of parsimony. Look for a condition in which all the signs and symptoms can be explained by one disease, but

Hickam's dictum: A patient can have as many diseases as he pleases.

The lesson: try to find a single good explanation, but alternatively, consider as many as you need. Be flexible.

"Always do everything for every patient" versus

"Just don't do something, stand there."

The lesson: consider the options. Sometimes you need to push forward, but be willing to recognize when less is more.

"Listen, the patient is trying to tell you the diagnosis" versus

"Never trust the history." (e.g., "Could you be pregnant?", "Do you have a drug dependency?", "Does your spouse abuse you?")

The lesson: Listen, then verify.

FIGURE 2.5
Examples of medical maxims and aphorisms.

the less-than-perfect predictability of human disease and its manifestations. As diagnostic workups go, there is a process that is borne out over time, not uncommonly with twists and turns and unexpected events. The contradictory nature of maxims helps ground clinicians in a healthy skepticism, reminding them to remain on guard, inquisitive, and open to new explanations [51]. A few examples are described in Figure 2.5.

Critical Thinking

One might expect that having completed 20 or more years of formal education, doctors would know how to think, critically appraise evidence, write, reason, and communicate. However, the critical examination of reasoning skills may be neglected in an overburdened curriculum focusing on content over thought process. Those who engage in diagnostic work engender the trust of patients who rely on their reasoning skills and judgment; certainly, patients should expect that their physician be grounded in sound reasoning. Their lives depend on it. Critical thinking is defined as "the intellectually disciplined process of actively and skillfully conceptualizing, applying, analyzing, synthesizing, and/or evaluating information gathered from, or generated by, observation, experience, reflection, or communication" [52]. How interesting that these skills are similar to the skills required for diagnosis. Further, "critical thinking is that mode of thinking—about any subject, content, or problem—in which the thinker improves the quality of his or her thinking by skillfully analyzing, assessing, and reconstructing it. Critical thinking is self-directed, self-disciplined, self-monitored, and self-corrective thinking" [53]. The analysis of the diagnostic process deserves examination and requires a critical thinker (see Chapter 8, The Rational Diagnostician). The path to becoming an expert diagnostician requires continual renewal and improvement [54].

Optimizing Individual Performance

With experience, clinicians should gain expertise, as they have opportunities to learn and improve their skills in diagnosis. Each case adds to their schemata of illness, and with time and variety, a rich collection of illness scripts rounds out their representation of different diseases [54,55]. But not everyone learns from their experience. Depending on their practice setting, physicians often do not know the final outcome of their judgments. Dissatisfied patients often turn elsewhere for care or seek alternate opinions when the first opinion fails to answer their questions or improve their symptoms. Medical care is often fragmented, such that patients move through a line of providers and consultants without any of them ever seeing the full pathway or accuracy of their diagnoses. Without feedback, clinicians likely assume that all went well. Over time, even determined and cautious clinicians may allow sloppy thinking, easy short-cuts, or flaws to creep into their practice; without feedback, they may not even recognize the harm they cause or how their skills have degraded [44,45,56].

Methods to improve feedback have been suggested that would help inform practitioners, but few healthcare systems provide routine or reliable methods for feeding information and results to those involved. Some clinicians actively seek feedback. In an academic setting, they are likely the ones who offer up their cases for review at morbidity and mortality (M&M) conferences, eager to learn from each case. Others seem naively unaware of their limitations; clinicians have even been known to boast that they have escaped critique at M&M for a good part of their career, as if they had been untarnished. Dare we ask: if they have not seen anything unexpected come from their care, perhaps they haven't looked? Self-awareness is important to learning and improving. Reflective practice may improve performance, particularly in difficult cases [7,8].

Is it not a personal and professional ethos for clinicians to seek to know their accuracy and calibrate themselves? By nature of being human, they encounter limitations and flaws even with the reasoning they try so hard to perfect. When they do, are they not ethically bound to at least learn from them? We have an ethical imperative to drive the science of diagnosis to higher standards, and we are responsible for our own self-improvement [57].

Reckoning with Uncertainty and Imperfection

Despite our desire to achieve certainty and to standardize and simplify diagnosis, there remains an irreducible uncertainty in medical diagnosis. Diagnosis is an iterative process. At any stage in the process, if you clicked a stopwatch, you would likely be at a point that would be considered not yet quite right, "an error." Who stops the watch? Who judges the end point? For some conditions, the disease declares the end, when a critical illness threatens the immediate life or limb of the patient; in others, a slow, insidious course may not be recognized until late, as in an advanced cancer. In either case, a retrospective review will likely find missed opportunities, either in screening or early detection, or in recognizing the first symptoms that herald the onset of the disease. This review may come in the form of a critical judgment from those downstream in the course of illness, or in the bad news of an unexpected poor outcome.

Clinicians have been judged to be overly confident or lacking awareness of their flawed judgments [58–60]. Clinicians face a dilemma: they can never be certain, but they must be decisive. They cannot sit and ruminate over all possibilities until every hypothesis has been considered, examined, and tested. They must decide. Moreover, they must persuade the patient of their opinion enough to engage them in following their recommendations. And they must convince their peers and consultants. At some point, they turn from a position of inquiry, in which they remain flexible and thoughtful, to a position of advocacy to sway others to provide the care they feel is appropriate. A primary care doctor who examines a patient with right lower quadrant abdominal pain weighs his hypothesis, but may eventually have to implore a reluctant patient to go the hospital, and must convince a consultant to see the patient. The patient does not need a doctor who equivocates; the surgeon who operates either cuts the skin or stays home. He must take a stance. In making the argument, both doctors choose to accept evidence in favor of the diagnosis and disregard evidence that is contradictory.

Conclusion

Diagnosis is a complicated, complex process with different phases and a number of approaches. The fact that medical practice is grounded in scientific knowledge and uses analytical testing with mathematics and statistics gives a semblance of certainty, but these methods do not guarantee accuracy. In actual practice, clinicians rely on analytical methods, intuition, formal logic, and informal rules, all of which help describe and explain how diagnoses are made. Understanding these processes can help elucidate where diagnosis is reliable, and unfortunately, how it remains tenuous and imperfect.

SUMMARY POINTS

- Clinicians use a variety of thought processes in diagnostic work.
- Most diagnoses are made by rapid pattern recognition; complex or unusual cases require more deliberate, analytical approaches.
- While logical reasoning is necessary, medical investigations proceed backward from symptoms to cause and rely on abductive logic, a method that has inherent uncertainty.
- The narrative account of a patient's illness provides context and meaning, and physical findings provide clues: both are essential but often undervalued phases of clinical work.
- Medical tests require interpretation, factoring in patient risks, pretest probability, and disease prevalence.
- Statistics and formal testing do not remove all uncertainty.
- Informal rules, folklore, and aphorisms provide valuable lessons from the past.
- Individual expertise in diagnosis requires critical thinking skills, reflection, and commitment to personal growth.

References

1. Rejon Altable C. Logic structure of clinical judgment and its relation to medical and psychiatric semiology. *Psychopathology*. 2012;45(6):344–51.
2. Elstein AS, Kagan N, Shulman LS, Jason H, Loupe MJ. Methods and theory in the study of medical inquiry. *J Med Educ*. 1972 Feb;47(2):85–92.
3. Kassirer JP, Kopelman RI. *Learning Clinical Reasoning*. Philadelphia: Lippincott Williams & Wilkins; 1991.
4. Groopman J. *How Doctors Think*. Boston: Houghton Mifflin Company; 2007.
5. Norman GR, Brooks LE. The non-analytical basis of clinical reasoning. *Adv Health Sci Educ Theory Pract*. 1997;2(2):173–84.

6. Dhaliwal G. Going with your gut. *J Gen Intern Med*. 2011 Feb;26(2):107–9.
7. Mamede S, Schmidt HG, Penaforte JC. Effects of reflective practice on the accuracy of medical diagnoses. *Med Educ*. 2008 May;42(5):468–75.
8. Mamede S, Schmidt HG, Rikers RM, Penaforte JC, Coelho-Filho JM. Influence of perceived difficulty of cases on physicians' diagnostic reasoning. *Acad Med*. 2008 Dec;83(12):1210–16.
9. Eva KW. What every teacher needs to know about clinical reasoning. *Med Educ*. 2005 Jan;39(1):98–106.
10. Croskerry P. A universal model of diagnostic reasoning. *Acad Med*. 2009 Aug;84(8):1022–8.
11. Croskerry P. Clinical cognition and diagnostic error: Applications of a dual process model of reasoning. *Adv Health Sci Educ Theory Pract*. 2009 Sept;14 Suppl 1:27–35.
12. Thammasitboon S, Cutrer WB. Diagnostic decision-making and strategies to improve diagnosis. *Curr Probl Pediatr Adolesc Health Care*. 2013 Oct;43(9):232–41.
13. Kahneman D. *Thinking, Fast and Slow*. New York: Farrar, Straus and Giroux; 2011.
14. Gladwell M. *Blink: The Power of Thinking without Thinking*. New York: Little Brown and Company; 2005.
15. Legault MR. *Think!: Why Crucial Decisions Can't Be Made in the Blink of an Eye*. New York: Threshold Edition; 2006.
16. Stanley DE, Campos DG. The logic of medical diagnosis. *Perspect Biol Med*. 2013 Spring;56(2):300–15.
17. Zarefsky D. *Argumentation: The Study of Effective Reasoning*. 2nd ed. Virginia: The Teaching Company; 2005.
18. Brush JE. *The Science of the Art of Medicine: A Guide to Medical Reasoning*. Virginia: Dementi Milestone Publishing; 2015.
19. Stanley DE, Campos DG. Selecting clinical diagnoses: Logical strategies informed by experience. *J Eval Clin Pract*. 2016 Aug;22(4):588–97.
20. Gilbert K, Whyte G. *Argument and Medicine: A Model of Reasoning for Clinical Practice*. June 3, 2009. OSSA Conference Archive. Paper 51. Available at: http://scholar.uwindsor.ca/ossaarchive/OSSA8/papersandcommentaries/51. Accessed February 7, 2016.
21. Haig BD. An abductive theory of scientific method. *Psychol Methods*. 2005 Dec;10(4):371–88.
22. Mirza NA, Akhtar-Danesh N, Noesgaard C, Martin L, Staples E. A concept analysis of abductive reasoning. *J Adv Nurs*. 2014 Sep;70(9):1980–94.
23. Walton DN. Abductive, presumptive and plausible arguments. *Informal Logic*. 2001;21(2):141–69.
24. Davis FD. Phronesis, clinical reasoning, and Pellegrino's philosophy of medicine. *Theor Med*. 1997 Mar–Jun;18(1–2):173–95.
25. Montgomery K. *How Doctors Think. Clinical Judgment and the Practice of Medicine*. Oxford: Oxford University Press; 2006.
26. Gatens-Robinson E. Clinical judgment and the rationality of the human sciences. *J Med Philos*. 1986 May;11(2):167–78.
27. Schleifer R, Vannatta J. The logic of diagnosis: Peirce, literary narrative, and the history of present illness. *J Med Philos*. 2006 Aug;31(4):363–84.

28. Rapezzi C, Ferrari R, Branzi A. White coats and fingerprints: Diagnostic reasoning in medicine and investigative methods of fictional detectives. *BMJ*. 2005 Dec 24;331(7531):1491–4.

29. Mamede S, van Gog T, van den Berge K, van Saase JL, Schmidt HG. Why do doctors make mistakes? A study of the role of salient distracting clinical features. *Acad Med*. 2014 Jan;89(1):114–20.

30. van Ryn D, van Ryn S, Colleen N, Cerak W. *Mistaken Identities: Two Families, One Survivor, Unwavering Hope*. New York: Howard Books; 2009.

31. Sacks O. *The Man Who Mistook His Wife for a Hat and Other Clinical Tales*. New York: Touchstone; 1998.

32. Sanders L. *Every Patient Tells a Story. Medical Mysteries and the Art of Diagnosis*. New York: Broadway Books; 2009.

33. Sackett DL, Haynes RB, Tugwell P. *Clinical Epidemiology: A Basic Science for Clinical Medicine*. Boston: Little Brown; 1985.

34. Riegelman RK, Hirsch RP. *Studying a Study and Testing a Test: How to Read the Medical Literature*. 2nd ed. Boston: Little Brown; 1989.

35. Reid MC, Lachs MS, Feinstein AR. Use of methodological standards in diagnostic test research. Getting better but still not good. *JAMA*. 1995 Aug 23–30;274(8):645–51.

36. Ransohoff DF. Challenges and opportunities in evaluating diagnostic tests. *J Clin Epidemiol*. 2002 Dec;55(12):1178–82.

37. McGee S. Simplifying likelihood ratios. *J Gen Intern Med*. 2002 Aug;17(8):646–9.

38. Grimes DA, Schulz KF. Refining clinical diagnosis with likelihood ratios. *Lancet*. 2005 Apr 23–29;365(9469):1500–5.

39. Wagner JM, McKinney WP, Carpenter JL. Does this patient have appendicitis? *JAMA*. 1996 Nov 20;276(19):1589–94.

40. Diamond GA, Forrester JS. Analysis of probability as an aid in the clinical diagnosis of coronary-artery disease. *N Engl J Med*. 1979 Jun 14;300(24):1350–8.

41. Sackett DL, Straus SE, Richardson WS, Rosenberg W, Haynes RB. *Evidence-Based Medicine. How to Practice and Teach EBM*. 2nd ed. Edinburgh: Churchill Livingstone; 2000.

42. Whiting PF, Davenport C, Jameson C, Burke M, Sterne JA, Hyde C, Ben-Shlomo Y. How well do health professionals interpret diagnostic information? A systematic review. *BMJ Open*. 2015 Jul 28;5(7):e008155.

43. Reid MC, Lane DA, Feinstein AR. Academic calculations versus clinical judgments: Practicing physicians' use of quantitative measures of test accuracy. *Am J Med*. 1998 Apr;104(4):374–80.

44. Richardson WS. Five uneasy pieces about pre-test probability. *J Gen Intern Med*. 2002 Nov;17(11):891–2.

45. Phelps MA, Levitt MA. Pretest probability estimates: A pitfall to the clinical utility of evidence-based medicine? *Acad Emerg Med*. 2004 Jun;11(6):692–4.

46. Croskerry P. The feedback sanction. *Acad Emerg Med*. 2000 Nov;7(11):1232–8.

47. Schiff GD. Minimizing diagnostic error: The importance of follow-up and feedback. *Am J Med*. 2008 May;121(5 Suppl):S38–42.

48. Pauker SG, Kassirer JP. The threshold approach to clinical decision making. *N Engl J Med*. 1980 May 15;302(20):1109–17.

49. Barnes E. Inference to the loveliest explanation. *Synthese*. 1995 May;103(2):251–77.

50. Levine D, Bleakley A. Maximising medicine through aphorisms. *Med Educ*. 2012 Feb;46(2):153–62.

51. Hunter K. "Don't think zebras": Uncertainty, interpretation, and the place of paradox in clinical education. *Theor Med*. 1996 Sep;17(3):225–41.
52. Critical Thinking as Defined by the National Council for Excellence in Critical Thinking, 1987. A statement by Michael Scriven and Richard Paul, presented at the 8th Annual International Conference on Critical Thinking and Education Reform, Summer 1987. Available at: http://www.criticalthinking.org/pages/defining-critical-thinking/766. Accessed February 7, 2016.
53. Defining Critical Thinking. Available at: http://www.criticalthinking.org. Accessed February 7, 2016.
54. Schmidt HG, Boshuizen HP. On acquiring expertise in medicine. *Educational Psychology Review*. 1993;5(3):205–21.
55. Schmidt HG, Norman GR, Boshuizen HP. A cognitive perspective on medical expertise: Theory and implication. *Acad Med*. 1990 Oct;65(10):611–21.
56. Rudolph JW, Morrison JB. Sidestepping superstitious learning, ambiguity, and other roadblocks: A feedback model of diagnostic problem solving. *Am J Med*. 2008 May;121(5 Suppl):S34–7.
57. Stark M, Fins JJ. The ethical imperative to think about thinking: Diagnostics, metacognition, and medical professionalism. *Camb Q Healthc Ethics*. 2014 Oct;23(4):386–96.
58. Meyer AN, Payne VL, Meeks DW, Rao R, Singh H. Physicians' diagnostic accuracy, confidence, and resource requests: A vignette study. *JAMA Intern Med*. 2013 Nov 25;173(21):1952–8.
59. Croskerry P, Norman G. Overconfidence in clinical decision making. *Am J Med*. 2008 May;121(5 Suppl):S24–9.
60. Berner ES, Graber ML. Overconfidence as a cause of diagnostic error in medicine. *Am J Med*. 2008 May;121(5 Suppl):S2–23.

3

Modern Cognitive Approaches to the Diagnostic Process

Pat Croskerry

CONTENTS

Introduction

If people understood their symptoms and signs when they are ill, there would be no need for diagnosticians; we could all self-diagnose. For example, if chest pain always meant a heart problem, then immediate action could be taken to ameliorate the problem, or if headache always meant a migraine attack, then the treatment would be straightforward. Unfortunately, symptoms are rarely specific or pathognomonic. There are at least 25 different causes of chest pain and literally hundreds of conditions associated with headache, and someone has to work out which one it is. Patients, themselves, do not have the knowledge or expertise and often turn to family, friends, books, the Internet, and other sources before consulting a physician for advice (see Chapter 4: Alternatives to Conventional Medical Diagnosis).

Two basic factors influence a physician's ability to make a diagnosis: a sufficient knowledge base and the ability to think, reason, and decide effectively. Those who perform well are referred to as *well calibrated*. *Calibration* is a term taken from quality engineering, used to refer to the operating characteristics of measuring instruments—specifically, the relationship between the measuring device and what is being measured. Cognitive psychologists have adopted the term to describe the process by which an individual makes sound judgments that are reasonably free from favoritism, bias, stereotyping and other factors that can distort reasoning. In this case, the measuring device is the brain, and what is being measured is the quality of judgment. A good knowledge base, and the ability to learn from feedback and experience, leads to a well-calibrated decision maker.

Cognitive science is the discipline that studies human reasoning and decision making. Over the last 40 years, there has been a major research endeavor in this area, and consensus has emerged about how we make decisions; as noted in the previous chapter, we do it in one of two ways. We may use Type 1 processes, which are extremely fast and reflexive, so that we are not aware of making them. These processes, collectively referred to as *intuition*, are how we accomplish most of everyday living (intuition is discussed more fully in Chapter 8: The Rational Diagnostician). Type 2 processes, in contrast, are slow, deliberate, and conscious. Collectively, they refer to *analytical reasoning*. More characteristics of the two types of thinking are given in Table 3.1 [1–3]. This approach, described as *dual process theory* (DPT), has been reviewed recently [1] along with descriptions of its various operating characteristics [4].

These ways of thinking are considered *universal* [5], which means that while the content of the two systems may vary according to prevailing cultural and contextual factors, the basic decision-making processes of a policeman, scientist, mechanic, teacher, factory worker, or any other occupation are similar. The same holds across cultures and races. The decision-making strategies of Italians compare with those of Zulus, Tibetans, or those from Tierra del Fuego. The universality is based on anthropological studies and cognitive evolutionary psychology. Essentially, this holds that our modern brains are the product of Darwinian evolution and that our current thinking is controlled, in part, by hardwired cognitive modules that were selected in our ancient environments many thousands of years ago (see Chapter 6: Stone Age Minds in Modern Medicine) [6].

Dual process decision making is an increasingly well-substantiated theory with converging anatomical, neurophysiological, psychological, and genetic evidence. Specific neuroanatomic loci of Type 1 and Type 2 processes have been identified. Type 1 are located in the older parts of the brain, and importantly, involve the amygdala and parts of the limbic system that process emotion, whereas Type 2 processes are in the newer parts of the brain [7,8]. In experimental studies, neurophysiological substrates for the different thinking states have been described. Neuronal groups underlying Type 1 fire at fast rates of approximately 50 cycles per second (cps) compared with the much slower Type 2 neuronal groups, which fire at only approximately 25 cps or half that

TABLE 3.1

Principal Characteristics of Type 1 and Type 2 Decision-Making Processes

Property	System 1	System 2
Reasoning style	*Intuitive* Heuristic, associative, concrete	*Analytical* Normative, deductive, abstract
Awareness	Low	High
Reliance on language	No	Yes
Prototypical	Yes	No, based on sets
Action	Reflexive, skilled	Deliberate, rule-based
Automaticity	High	Low
Speed	Fast	Slow
Channels	Multiple, parallel	Single, linear
Propensities	Causal	Statistical
Effort	Minimal	Considerable
Cost	Low	High
Vulnerability to bias	Yes	Less so
Reliability	Low, variable	High, consistent
Errors	Common	Few
Affective valence	Often	Rarely
Predictive power	Low	High
Hardwired	Maybe	No
Scientific rigor	Low	High
Context importance	High	Low

Source: Adapted from Evans, J.S, *Annu. Rev. Psychol.*, 59, 255–78, 2008; Dawson, N. V., *Clin. Chem.*, 39, 7, 1468–80, 1993; Croskerry, P, *Can. J. Anesth.*, 52, Suppl 1, R1–R8, 2005.

rate [9]. This argument is further supported by the fact that patients known to have specific areas of brain damage show corresponding functional decrements in Type 1 performance [10]. There is additional supportive evidence from the field of psychology. Personality is very likely associated with reasoning. Personality tests have been developed that distinguish people inclined toward intuitive decision making from those who are predominantly analytical [11]. Differences in personality may in part be based in one's genes. Half the variance associated with personality is said to be genetic [12], and studies have shown that certain behavioral disorders associated with impulsive reasoning in children can be identified with DNA markers [13]. Thus, there appears to be converging evidence from a variety of sources (anatomical, neurophysiological, psychological, and genetic) that supports the dual process model.

These developments in cognitive psychology clearly have application in medicine [2]. The case has been made more explicit with the description of a schematic model (Figure 3.1) that incorporates the major operating characteristics of DPT in diagnostic reasoning [14–16].

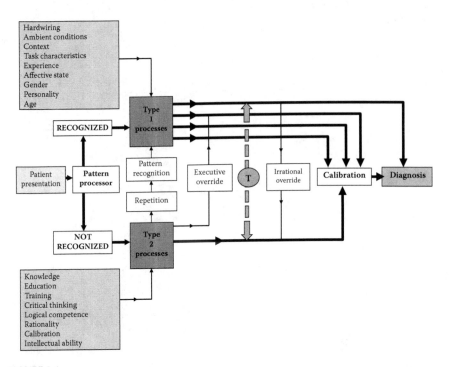

FIGURE 3.1
Schematic model for diagnostic reasoning. (Reproduced with minor modifications from Croskerry, P., Singhal, G., Mamede, S., *BMJ Qual Saf*, 22(Suppl 2): ii58–ii64, 2013; with permission of BMJ Publishing Group Ltd. [15].)

Operating Characteristics of the Model

Type 1 processes may work extremely quickly, such that no active reasoning has time to occur. If Figure 3.2 is shown to a physician, the diagnosis will be made instantly. The rash of herpes zoster is vivid, pathognomonic, and almost impossible to confuse with other rashes. The remarkable feature of this type of thinking is that it is made reflexively. We cannot stop ourselves from reaching the correct diagnosis. Equally, having reached this diagnostic point, we have great difficulty *undoing* it; that is, considering the possibility that it might be something else. We anchor on the salient features at the outset, and cannot easily overcome or adjust our first impressions.

The recognition of the pattern and its processing and matching to discrete knowledge templates are autonomous, in that they happen with no conscious effort. It is accepted that this autonomy of thought characterizes the "pattern recognition" specialties (Dermatology, Radiology, and Anatomic Pathology), although all specialties use pattern matching to a lesser or greater degree.

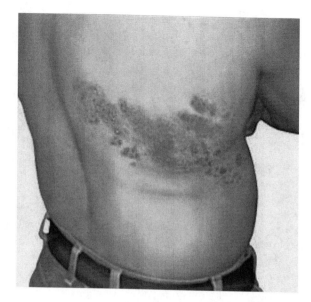

FIGURE 3.2
Typical rash of herpes zoster, identifiable by pattern recognition.

Thus, an internist may react with Type 1 processing to stigmata of particular diseases such as alcoholism, endocrine disorders, cardiovascular disorders, or recognized combinations of salient symptoms or findings (syndromes, toxidromes), also referred to as *illness scripts*. Similarly, a surgeon may have a Type 1 response to a particular combination of symptoms and physical findings. The more pathognomonic the disease, and the more the patients' symptoms and signs match, the greater the diagnostic accuracy. Another feature of Type 1 thinking is that it is multichanneled [4]. One channel may trigger pattern recognition; another may be an emotional or affective disposition toward a particular patient (for example, "this child reminds me of one I saw during my training who died"); another might be activated by the particular context in which the patient presents, and another by a patient seen a week ago who had a stormy course. Type 1 processes are fast and frugal, and often work effectively, but they may occasionally fail.

Type 2 processes, in contrast, are single-channeled, slower, and deliberate. If a patient presents with symptoms that do not immediately match a prototype, or appear ambiguous or inconsistent, a methodical workup may be called for to systematically work through various possibilities in an analytic fashion. Unlike in Type 1 thinking, steps along the way are clearly specified, there is little or no affective component, short-cuts and heuristics are generally not used, reliability is higher, and error will be rare providing input to the process was accurate. This is referred to as *rational* or *analytical decision making*.

This basic two-process model may seem overly simplistic and straight-forward, but in the course of diagnostic reasoning, there are a number of modifiers of these two processes that create a dynamic state of more complex decision making.

1. In practice, diagnosticians do not consciously choose to stay in one system or the other; in fact, it appears that there is an oscillation between the two types of thinking [17], a "toggling" back and forth—this is depicted in Figure 3.1 as a broken double arrow and a T. In some cases, the diagnosis is what it is, especially in highly pathognomonic cases such as herpes zoster, and it would be wasteful to spend much time in Type 2 thinking through other possibilities. In other cases, for example chest pain that is atypical of an acute coronary syndrome, an initial hunch or intuition in Type 1 may force a Type 2 workup, which, if negative, might throw the thinking back into Type 1, a strategy of *hypothesis-hopping*. Well-calibrated physicians are probably those who achieve the optimum balance of Type 1 and Type 2.

2. Type 1 processes may be overridden by Type 2 in an executive control fashion. For example, if a patient presents to an Emergency Department (ED) with flank pain, nausea and vomiting, and hematuria, Type 1 processing may immediately and reflexively generate the diagnosis of ureteral colic. However, the emergency physician may pull back from this reflexive diagnosis and force a reflection on other possibilities that might present in a similar fashion and that might have greater conse-quence, such as a dissecting abdominal aortic aneurysm.

3. Because Type 1 processes are fast, reflexive, and autonomous, it fol-lows that they must use short-cuts. Many involve heuristics, which are strategies of abbreviated thinking, mental short-cuts, educated guesses, common-sense maxims, or rules of thumb that bypass the more laborious, resource-intensive efforts of Type 2. Heuristics are important frugal ways of thinking that get us through the day, and they are often sufficient for our purpose, but they are vulnerable to error, and most error tends to occur in Type 1 [18].

4. Type 2 processing involves analytical skills that are mostly acquired. The box that leads into Type 2 in Figure 3.1 contains some of the pre-cursors of Type 2 processing. Diagnosticians who are well trained, knowledgeable, logical, rational, with good critical thinking skills, and who have enjoyed good feedback from their clinical decision making over many years, are likely to be well calibrated (see further discussion in Chapter 8: The Rational Diagnostician).

5. Repeated Type 2 processing may allow a default to Type 1. When a medical student first sees the rash of herpes zoster, it will have no recognizable significance. He or she may methodically describe it as "clusters of vesicles, with patches of erythema that follow a

dermatomal distribution and stop at the midline." With each new presentation of the rash, however, especially with the classic features in different dermatomes, the diagnosis is made faster and faster and with more confidence till eventually it is made reflexively, without thinking. With vivid, pathognomonic presentations of disease, very few repetitions may be necessary before they are managed with Type 1 processing.

6. Occasionally, even if a Type 2 workup has been done, a Type 1 override may occur. Thus, despite knowing the best thing to do, the diagnostician chooses to do something else. This might be irrational behavior. For example, a clinical decision rule developed in the cold light of day by experts who are well rested and well slept, who have used highly sophisticated, valid, and reliable statistical methods, may be overridden by a busy, fatigued physician. Despite knowing the rule and its efficacy, the physician chooses instead to follow his own intuitions. Here, the physician hasn't engaged in Type 2 processing for this patient, but instead, has had the option of accessing Type 2-by-proxy, and chose to override it. While occasionally there may be virtue in 'situational awareness' and determining an action plan specific to a particular situation [19], this may not be the underlying cause of Type 1 override of Type 2. More often than not, the override reflects an ego bias, cognitive indolence, fatigue, idiosyncratic decision making, other cognitive biases, personality factors, overconfidence, affective state, self-deception, or other quirks in the diagnostician. Or, it may be due to a variety of ambient factors—how many patients are waiting to be seen, cognitive load, team factors, systemic failures, and others. Some of these factors are annotated in the box that connects to Type 1 in Figure 3.1.

A further explanation of Type 1 prevailing when it should not resides in the *cognitive miser function* [20]. This reflects not cognitive laziness, but an overall tendency for thinking to default to Type 1 processing to preserve cognitive energy and resources, a strategy that would have had considerable survival value in our ancient past.

What Does Dual Process Theory Offer?

Universality

A major appeal of DPT is its universality. All other approaches to medical decision making, and all known features of the diagnostic process in particular, can be incorporated into DPT. This means that we can study decision making in the primary areas in which diagnoses are made. Although physicians and dentists have traditionally been the arbiters of the diagnostic process, increasingly the skill is being extended to other domains: physician assistants, nurse practitioners, and the field of Emergency Health Services. Paramedics have considerably expanded their scope of practice, and some of their higher delegated acts now involve interventions based on their diagnosis of time-critical conditions such as stroke [21], congestive heart failure [22], acute myocardial infarction [23], and others. There are, too, a variety of other "alternative" practitioners who purport to provide diagnoses, but as their approach is largely intuitive and mostly lacking an evidentiary basis [24,25], they are unlikely to seek any scientific support (see Chapter 5: Complementary and Alternative Medicine).

Teaching

The expanding number of stakeholders who are required to make diagnoses need to be taught the clinical decision making theory that underlies this skill. While all human minds do engage in dual process thinking, characteristics of the decision maker and ambient conditions will determine the thinking that is appropriate. The good news is that we can teach everyone the essential elements and operating characteristics of DPT such that they have an overall framework and understanding of the process.

Thus, it is possible to accommodate deliberation-without-attention approaches [26], in which decisions can be made without reaching a conscious level, alongside systematic hypothetico-deductive reasoning, which requires an active, deliberate, analytical approach. Both may be appropriate under particular conditions. If we are choosing wallpaper for the dining room, an intuitive approach will be quite sufficient and probably optimal, whereas if we need to build a spacecraft to go to Mars, nothing but an analytical approach will do. Similarly, if a multitrauma victim crashes into an ED, at the outset we want fast, intuitive, shoot-from-the-hip decision making and diagnosis, whereas we would want an oncologist dealing with a complex malignancy to be very analytical both in their diagnosis and in staging of the disease.

Research

The advantage of having a clear, universal approach toward the diagnostic process is that we can identify foci where things can go wrong and research their characteristics. For example, we have discussed Type 2 monitoring of Type 1, the executive control function, anatomically located in the prefrontal cortex. What factors might compromise this control function? Inattentiveness, distraction, fatigue, sleep deprivation, and cognitive indolence may all diminish System 2 surveillance and allow System 1 more laxity than it deserves. With fatigue and sleep deprivation, for example, the diagnostic error rate can increase fivefold [27]. The area of the brain believed to be the neuroanatomical substrate for System 2 reasoning—the anterior cingulate cortex, prefrontal cortex, and medial aspect of the temporal lobe [8]—is the same area that suffers neurocognitive compromise through sleep deprivation [28].

Clinical Practice

An awareness and understanding of the model and its operating character-istics is the starting point toward a more calibrated diagnostic performance and may allow more focused metacognition. Thus, the decision maker can identify which system they are currently using and determine the appropri-ateness and the relative benefits of remaining in that mode versus switching to the other. This insight, coupled with an awareness of the impact of various factors such as excessive cognitive load, fatigue, sleep deprivation, and other adverse ambient conditions, can be a signal to implement protective func-tions; for example, if I am tired and sleep deprived, I am more vulnerable to a Type 1 processing error, and therefore, I should delay making a diagnosis on this patient, or defer to someone who is in better shape; or, if I have a pre-dictable visceral bias (Type 1 response) against a particular group of patients (e.g., drug-dependent patients), I am more likely to make an error with their diagnosis and therefore, should adopt some specific forcing function (always take a good history, do a thorough examination and workup) or refer them to a colleague who does not hold a similar bias.

Another clinical practice strategy that is gaining increasing popular-ity is the use of checklists. In several domains, they have been shown to improve performance [29,30]. Checklists function partly by allowing a moment of reflection and also by forcing a deliberate executive control step. If a diagnosis has been driven largely by Type 1 processes and has inher-ent vulnerability to particular heuristics and biases (anchoring, overconfi-dence, premature diagnostic closure), using a checklist stops the gathering momentum toward a particular diagnosis and forces a consideration of other not-to-be-missed possibilities. A variety of other cognitive bias–mitigating strategies are discussed in Chapter 15: Cognitive Bias Mitigation: Becoming Better Diagnosticians.

A further issue in clinical practice involves individual variability. The Rational-Experiential Inventory (REI) developed by Epstein and his co-workers [11] is a personality test that measures an individual's dispositions toward Type 1 versus Type 2 thinking. It might be useful for diagnosticians to take the test and determine their particular tendencies. If I know that I am predominantly a Type 1 thinker, intuitive and perhaps a little impul-sive in making my diagnoses, perhaps I should deliberately exercise more techniques of reflection in my practice. Also, with aging, Type 2 reasoning tends to be replaced with Type 1 processes [31,32], so perhaps older diag-nosticians should remind themselves of this vulnerability. Gender is also an issue. Females tend to be more intuitive than males and score higher on Type 1 processes and lower on Type 2 [12]. Nursing educators, deliberately or otherwise, have tended to emphasize an intuitive approach in nursing [33]. High-functioning teams should probably embrace a blend of intuitive and analytical approaches. Other aspects of individual variability that may impact the decision making process are discussed in Chapter 9: Individual Variability in Clinical Decision Making and Diagnosis.

SUMMARY POINTS

- Dual process theory has emerged as the dominant model for a universal approach toward diagnostic decision making. Two basic modes of decision making are described: Type 1 (intuitive) and Type 2 (analytical).
- Discrete anatomical and functional areas of the brain underlie each process.
- Type 1 processes operate autonomously through cognitive modules selected in our evolutionary past and from implicitly or explicitly acquired learned material.
- Type 2 processes, in contrast, are conscious and deliberate and depend on the laws of science and rationality.
- A universal model of diagnostic reasoning based on DPT can accommodate previous and current approaches toward diagnostic reasoning.
- The model has several distinct operating characteristics and includes a dynamic interplay between the two systems.
- Well-calibrated diagnosticians achieve an optimal balance between Type 1 and Type 2 reasoning.
- The model has important implications for teaching, research, and clinical practice.

References

1. Evans JS. Dual-processing accounts of reasoning, judgment, and social cognition. *Annu Rev Psychol*. 2008;59:255–78.
2. Dawson NV. Physician judgment in clinical settings: Methodological influences and cognitive performance. *Clin Chem*. 1993 Jul;39(7):1468–80.
3. Croskerry P. The theory and practice of clinical decision making. *Can J Anesth*. 2005 Jan; 52(Suppl 1):R1–R8.
4. Stanovich KE. *The Robot's Rebellion: Finding Meaning in the Age of Darwin.* Chicago: The University of Chicago Press; 2004.
5. Brown DE. *Human Universals*. New York: McGraw-Hill; 1991.

6. Tooby J, Cosmides L. Conceptual foundations of evolutionary psychology. In: Buss DM, editor. *The Handbook of Evolutionary Psychology*. New Jersey: John Wiley; 2005. pp. 5–67.

7. Hardman D. *Judgment and Decision Making: Psychological Perspectives*. Oxford: Blackwell; 2009.

8. Lieberman MD, Jarcho JM, Satpute AB. Evidence-based and intuition-based self-knowledge: An FMRI study. *J Pers Soc Psychol*. 2004 Oct;87(4):421–35.

9. Buschman TJ, Miller EK. Top-down versus bottom-up control of attention in the prefrontal and posterior parietal cortices. *Science*. 2007 Mar 30;315(5820):1860–2.

10. Lieberman MD. Intuition: A social cognitive neuroscience approach. *Psychol Bull*. 2000 Jan;126(1):109–37.

11. Pacini R, Epstein S. The relation of rational and experiential information processing styles to personality, basic beliefs, and the ratio-bias phenomenon. *J Pers Soc Psychol*. 1999 Jun;76(6):972–87.

12. Pinker S. *The Blank Slate: The Modern Denial of Human Nature*. New York: Penguin Putnam; 2002. pp. 45–50.

13. Oades RD, Lasky-Su J, Christiansen H, Faraone SV, Sonuga-Barke EJ, Banaschewski T, Chen W, Anney RJ, Buitelaar JK, Ebstein RP, Franke B, Gill M, Miranda A, Roeyers H, Rothenberger A, Sergeant JA, Steinhausen HC, Taylor EA, Thompson M, Asherson P. The influence of serotonin and other genes on impulsive behavioral aggression and cognitive impulsivity in children with attention-deficit/hyperactivity disorder (ADHD): Findings from a family-based association test (FBAT) analysis. *Behav Brain Funct*. 2008 Oct; 4:48.

14. Croskerry P. A universal model of diagnostic reasoning. *Acad Med*. 2008;84:1022–8.

15. Croskerry P, Singhal G, Mamede S. Cognitive debiasing 1: origins of bias and theory of debiasing. *BMJ Qual Saf*. 2013; 22(Suppl 2): ii58–ii64.

16. Croskerry P. Context is everything *or* how could I have been that stupid? *Healthc Q*. 2009; 12-Spec No Patient: e171–6.

17. Hammond KR. *Human Judgment and Social Policy: Irreducible Uncertainty, Inevitable Error, Unavoidable Injustice*. New York: Oxford University Press; 2000.

18. Gilovich T, Griffin D, Kahneman D. *Heuristics and Biases: The Psychology of Intuitive Judgment*. New York: Cambridge University Press; 2002.

19. Wiggins D. *Deliberation and Practical Reasoning*. Essays on Aristotle's Ethics. Berkeley: University of California Press; 1980. pp. 221–40.

20. Krueger JI, Funder DC. Towards a balanced social psychology. Causes, consequences and cures for the problem-seeking approach to social cognition and behavior. *J Behav Brain Sci*. 2004 Jun;27(3):313–76.

21. Zweifler RM, York D, U TT, Mendizabal JE, Rothrock JF. Accuracy of paramedic diagnosis of stroke. *J Stroke Cerebrovasc Dis*. 1998 Nov–Dec;7(6):446–8.

22. Dobson T, Jensen JL, Karim S, Travers AH. Correlation of paramedic administration of furosemide with emergency physician diagnosis of congestive heart failure. *J Emergency Primary Health Care*. 2009 Jan;7(3), Article 990378.

23. Millar-Craig MW, Joy AV, Adamovicz M, Furber R, Thomas B. Reduction in treatment delay by paramedic ECG diagnosis of myocardial infarction with direct CCU admission. *Heart*. 1997 Nov;78(5):456–61.

24. Bausell RB. *Snake Oil Science: The Truth about Complementary and Alternative Medicine*. New York: Oxford University Press; 2007.

25. Ernst E, Singh S. *Trick or Treatment: The Undeniable Facts about Alternative Medicine.* New York: WW Norton & Co; 2008.

26. Dijksterhuis A, Bos MW, Nordgren LF, von Baaren RB. On making the right choice: The deliberation-without-attention-effect. *Science.* 2006 Feb;311(5763):1005–7.

27. Landrigan CP, Rothschild JM, Cronin JW, Kaushal R, Burdick E, Katz JT, Lilly CM, Stone PH, Lockley SW, Bates DW, Czeisler CA. Effect of reducing interns' work hours on serious medical errors in intensive care units. *N Engl J Med.* 2004 Oct 28;351(18):1838–48.

28. Durmer JS, Dinges DF. Neurocognitive consequences of sleep deprivation. *Semin Neurol.* 2005 Mar;25(1):117–29.

29. Karl R. Briefings, checklists, geese, and surgical safety. *Ann Surg Oncol.* 2009 Jan;17(1):8–11.

30. Reason J. *Human Error.* Cambridge, England: Cambridge University Press; 1990.

31. Jacoby LL. Deceiving the elderly: Effects of accessibility bias in cued-recall performance. *Cogn Neuropsychol.* 1999;16(3/4/5):417–36.

32. Peters E, Finucane M, MacGregor D, Slovic P. The Bearable Lightness of Aging: Judgment and Decision Processes in Older Adults. In: Stern P, Carstensen L, editors. *The Aging Mind: Opportunities in Cognitive Research.* Washington, DC: National Academy Press; 2000. pp. 144–65.

33. Thompson C, Yang H. Nurses' decisions, irreducible uncertainty and maximizing nurses' contribution to patient safety. *Healthcare Q.* 2009;12 Spec No Patient:e176–85.

Section II

Informal and Alternative Approaches to Diagnosis

4

Alternatives to Conventional Medical Diagnosis

Pat Croskerry

CONTENTS

Introduction

Very few of us will live our lives free of injury, illness, or disease. There are estimated to be about 13,000 diseases or health conditions at present, and it is highly unlikely we will complete our lives without being diagnosed with at least a few of them, especially in our later years. On many occasions, the task of diagnosing an illness when we are sick falls to either ourselves or others who may or may not be trained in conventional medicine (Table 4.1). This list does not include a variety of other therapeutic modalities in which

TABLE 4.1

Sources of Diagnosis

- Self
- Family, friends, and acquaintances
- Telephone triage
- Complementary and alternative medicine
- Conventional medicine

treatment, but not necessarily diagnosis, is predominantly offered (faith healing, magnet therapy, spiritual healing, reflexology, guided imagery, and many others). In this chapter, we will discuss the usual alternatives to conventional medical diagnosis, with the exception of complementary alternative medicine (CAM), which will be reviewed separately in the next chapter.

Self-Diagnosis

Little formal study has been done on the subject of self-diagnosis, although it is, by far, the most common type of diagnosis. Essentially, it is the process of diagnosing, or identifying, medical conditions in oneself. The huge industry of over-the-counter (OTC) medications depends upon self-diagnosis or lay diagnosis, the basic assumptions being that people can diagnose their condition as serious or unlikely to be serious, and that OTC medications are relatively inexpensive and mostly harmless. Additionally, harm caused by self-administering an incorrect OTC medication is generally minimal. Self-diagnosis may be assisted by home medical guides, resources on the Internet, past personal experiences, and recognizing symptoms or medical signs of a condition that has been experienced before by self, a friend, or a family member [1]. Self-diagnosis is typically intuitive and usually made with incomplete knowledge. It may be effective or at least harmless much of the time, but is generally considered error-prone and may have a significant impact on the course of a person's illness. Self-diagnosis and treatment by physicians is a special case of self-diagnosis that is fraught with error. As Osler noted, "A physician who treats himself has a fool for a patient."

The human body's ability to repair and heal itself allows us to diagnose and manage many straightforward conditions. Given an estimated 85% of all illnesses are self-limiting and will resolve spontaneously, recovery would be expected to occur without outside medical help in the clear majority of cases. Thus, for the most part, we successfully self-diagnose muscle and joint sprains and strains, abrasions and minor lacerations, menstrual cramps, colds and infections, stomach upset, headaches, and a variety of other minor conditions. Much of this is accomplished by simple cause and effect reasoning that finds ready explanations: my headache is due to over-imbibing last night, my stomach upset to the rich food I ate, my muscle aches to the recent work-out at the gym, my sneezing to an allergy, and so on. This is often reinforced by early resolution of symptoms.

With the increased availability of medical information on the Internet, the general public can now access information that previously was only available in medical libraries. In Australia, it is estimated that 27% of the population now access medical information on the Internet [2], in Canada about 40% [3], and in the United States, a much higher 60% choose the Internet as their first source of health information [4]. However, there is a downside to this, familiar to many physicians. *Cyberchondria*, or *cyberchondriasis*, refers to "the unfounded escalation of concerns about common symptomatology based on a review of search results and literature online" [3,4] that may result in undue anxiety and a further risk of harm from self-diagnosis and self-treatment [5]. This phenomenon is similar to *medical students' disease*, in which medical students tend to diagnose themselves or others with the diseases they are currently studying. In Jerome's comedy classic *Three Men in a Boat (To Say Nothing of the Dog)* [6], published in 1889, George the medical student falls into this trap:

> I remember going to the British Museum one day to read up the treatment for some slight ailment of which I had a touch—hay fever, I fancy it was. I got down the book, and read all I came to read; and then, in an unthinking moment, I idly turned the leaves, and began to indolently study diseases, generally. I forget which was the first distemper I plunged into—some fearful, devastating scourge, I know—and, before I had glanced half down the list of "premonitory symptoms," it was borne in upon me that I had fairly got it.
>
> I sat for a while, frozen with horror; and then, in the listlessness of despair, I again turned over the pages. I came to typhoid fever—read the symptoms—discovered that I had typhoid fever, must have had it for months without knowing it—wondered what else I had got; turned up St. Vitus's Dance—found, as I expected, that I had that too,—began to get interested in my case, and determined to sift it to the bottom, and so started alphabetically—read up ague, and learnt that I was sickening for it, and that the acute stage would commence in about another fortnight.

The prevalence of this phenomenon in medical school is estimated to be as high as 80%; some have seen it as lying along the continuum of the psychiatric syndrome of hypochondriasis, whereas others view it more as a normal perceptual process. In the process of learning about a disease, the learner "creates a mental schema or representation of the illness which includes the label of the illness and the symptoms associated with the condition. Once this representation is formed, symptoms or bodily sensations that the individual is currently experiencing which are consistent with the schema may be noticed, while inconsistent symptoms are ignored" [7]. It is a form of intuitive reverse pattern recognition confounded by confirmation bias (see Appendix A: List of Common Biases in Medicine). Rather than giving it a psychiatric label, which could result in unwanted implications, it has been suggested the condition simply be termed *nosophobia* [8].

The wide variation in human behavior surrounding self-diagnosis may affect the outcome of a given condition. For example, those with an anxious disposition may be especially vulnerable to an Internet-facilitated medical escalation of what may, in the final analysis, prove to be something inconsequential. However, patients with psychiatric conditions, for example somatization disorder (the tendency to persistently complain of a variety of physical symptoms that have no identifiable physical origin), may suffer exacerbation of their symptoms through Web searching (Box 4.1). Conversely, patients with atypical symptoms may be misdiagnosed as having a somatoform disorder (Box 4.2), whereas patients who are stoical and deny symptoms may wait too long before seeking medical help (Box 4.3). What the individual thinks about their potential illness (their cognitive representation of it) is very important. It will influence whether or not they will search out information that will elaborate on the illness, and to what lengths they will go. Aside from the potential for cyberchondriasis noted earlier, there is considerable risk of misinterpretation and misunderstanding of the actual problem when lay people struggle with self-diagnosis [9].

BOX 4.1   RACHEL: A CASE OF SOMATIZATION

Rachel, a well-presented, intelligent woman in her late twenties, is very well known to the four emergency departments of a major city. On occasions she has visited them all on the same day. She is known to the emergency staff as a "frequent flyer." She spends a lot of time on the Internet "researching" her symptoms, and is convinced that she may have one or more of the more rare conditions that are being missed. She usually responds well to reassurance.

Typically, she presents with lower abdominal complaints, more of a urogenital than a gastrointestinal nature. She has had innumerable urinalyses, pelvic examinations, abdominal x-rays, ultrasounds, a CT of her abdomen, referrals to urologists, gynecologists, and, more recently to a psychiatrist who has diagnosed somatization disorder. The head of the emergency department, together with her family doctor and psychiatrist, is in the process of developing a much-needed management plan for her.

BOX 4.2   ROBERT: A CASE OF NOT SO TYPICAL SEIZURES

Robert is a quiet, healthy-looking male in his twenties, who has presented on a number of occasions to his family doctor and to the emergency department with what he has labeled as "seizures." His physical examination, including a neurological exam, electrocardiogram and blood-work, is always within normal limits, and he recovers completely from each episode. He sometimes describes feeling very fearful and appears to have a short post-ictal period. He has been referred to a neurologist and has had an electroencephalogram, which was normal. No other medical abnormalities have been revealed on numerous physical examinations. He has no history of alcohol or other drug use and is otherwise well. On several occasions, he has had a "seizure" in the emergency department with no apparent loss of consciousness or convulsive activity. He has lost considerable time from work, and frequently asks for "sick-notes" to give his employer. Physicians and nurses have arrived at the conclusion that he is having pseudo-seizures or is malingering. A referral to psychiatry resulted in a diagnosis of somatization disorder.

On a recent visit to the emergency department, he presents with his usual symptoms, and after an examination and brief period of observation, he is discharged and advised to follow up with his family doctor. Shortly afterward, he is found collapsed on the sidewalk outside the emergency department. Resuscitation efforts are unsuccessful.

At autopsy, examination was normal other than sclerotic and atrophic changes in his hippocampus. The coroner attributed his death to temporal lobe epilepsy. Later, in the same year as his death, his younger brother presents to the emergency department with a similar history and similar symptoms. He is referred to neurology, who arrange an MRI, which shows changes in the hippocampus consistent with mesial temporal lobe epilepsy.

The Lay Referral System

For infants and children, the first diagnosticians are often immediate caregivers, usually their parents. The symptoms and signs of teething, colic, and simple colds are well known, and seeking a medical opinion is often considered unnecessary. If there is no resolution, however, or if symptoms worsen, lay expertise might be sought from grandparents or other family members, other parents, or neighbors. Grandparents have often had firsthand experience with a variety of illnesses and diseases (the common cold, pinworm,

---

BOX 4.3   JIM: A NOT SO SIMPLE CASE OF CONSTIPATION

Jim is a 55-year-old male who has been experiencing abdominal discomfort for about a week, which he attributes to constipation, although
his bowel movements are usually regular. Over-the-counter laxatives
have been ineffective, and by Saturday, it has worsened to the point
where his wife eventually persuades him to seek medical attention, but
his doctor's office is closed for the weekend. He comes to a busy emergency department, where he is triaged as "constipation" and assigned
a low level of acuity.

After 4 hours elapse, he is seen by the emergency physician. He initially apologizes for wasting the physician's time and says he simply
needs a prescription for a strong laxative. On examination, he is found
to have a tender pulsatile mass in his midline. He adds that he has
been experiencing some back pain in addition to his abdominal discomfort. An ultrasound reveals an abdominal aortic aneurysm measuring 5.5 cm. He is immediately consulted to cardiovascular surgery
and undergoes emergent surgical repair, from which he makes a full
recovery.

---

chickenpox, cold sores, head lice, scabies, and many others) and, besides
providing an accurate diagnosis, may also know of effective treatments and
remedies. Most people will have access to such an informal network of potential consultants, what Friedson calls "the lay referral system" [10]. Given the
pathognomonicity of many illnesses, the benign course they frequently take,
and the general tendency for most conditions to resolve spontaneously, this
referral system mainly works well and spares the resources of the healthcare system. In fact, when young, first-time mothers present to emergency
departments with children showing self-evident, benign conditions, it is a
common lament among staff that the lay referral system has failed somewhere—nurses refer to it as the "absent grandmother syndrome."

Establishing Contact with the Healthcare System

For the majority, if the problem persists or worsens, standard medical care
is then usually sought—this is the point at which the lay and the professional systems meet. The primary care physician ultimately is the "hinge"
between the lay and the traditional referral systems [10], although a non-
physician contact may first be made through telehealth or other means, such
as retail clinics in the United States. Ultimately, the decision to seek medical

help requires some measure of self-triage. Individuals need to make decisions about their symptoms and signs, and delays may occur for a variety of reasons, including novelty of symptoms, their severity, past medical history, socioeconomic status, sociodemographic factors, psychological variables, age, gender, race, and other ambient and contextual variables [11,12].

In terms of the urgency with which the individual will follow up and seek care, two major factors were found to be important: severity of the illness presented and certainty of diagnosis. In one study, for low-severity illness, if the diagnostic certainty was high, care would be sought less urgently. In contrast, for high-severity illness and greater diagnostic certainty, care would be sought more urgently. Importantly, for the high-severity illness case, the *degree of certainty* attached to the diagnosis was found to be critical. For example, telling individuals their symptoms were "consistent with" a brain aneurysm produced less urgency in seeking medical care than if they were told that they "had" a brain aneurysm [12]. Literally, it seems that this uncertainty could be potentially harmful to patients. These findings were replicated and extended in a second study, in which an added feature was to offer a differential diagnosis. This had the effect of attenuating urgency away from a high-severity option [13]. Thus, subtleties in the ways in which degree of certainty is expressed, as well as framing options in a differential diagnosis manner, might have important implications for individuals seeking advice about self-triage from Internet-based resources and from telehealth.

Curbside and Corridor Consultations

Another type of problem occurs when a medical professional is accessed early on, but the contact is informal—these are colloquially known as *curbside* or *corridor* consultations (Box 4.4). For example, a nurse seeks the opinion of a trusted physician colleague in the workplace, a secretary to a medical specialist asks about the likely diagnosis of a family member, or someone asks a physician on the golf course about some recent symptoms they have been experiencing. The main problem in these scenarios is that the formal rules of a patient–physician encounter are usually violated. The interaction typically is not private, it is often hurried, a formal history of presenting illness is not done, past medical history may be ignored, physical exam is often not done or inadequate, accountability is uncertain, and follow-up is incomplete. Often, the physician being approached will ignore their own structured, tried and tested approach and settle for something considerably less. Golub [14] described the process thus: "A physician, who is in a noisy, crowded hallway en route elsewhere, or is button-holed outside his office, that is, 'on the curb,' may be distracted from offering the kind of

BOX 4.4   A SERIES OF CORRIDOR CONSULTATIONS IN THE EMERGENCY DEPARTMENT

A young emergency nurse has dryness and itching of her eyes and photophobia. She asks an emergency physician in the emergency department if he could prescribe something for her symptoms. He readily obliges, and suggests natural tears. After several days, she has experienced no relief and on her next shift asks a second emergency physician if he has any further suggestions. He, too, obliges with a suggestion of continuing the natural tears but adding a topical decongestant. A week later she still feels no significant relief of her symptoms and asks a third emergency physician for his suggestions. He examines her in the department and refers her to the "red eye" clinic at another hospital.

She is seen the following day and diagnosed with bilateral retinal detachment. During the course of her assessment she reveals that her mother was diagnosed with retinal detachment at age 33, and her father at 52. The successive corridor consultations resulted in a delay of about 3 weeks in getting to her correct diagnosis.

TABLE 4.2

Checklist for Physicians Diagnosing Physicians

- Discuss confidentiality and clarify the physician–patient relationship as early as possible.
- Perform thorough examinations in formal circumstances.
- Ask about self-treatment and self-diagnosis, and discourage those practices.
- Discuss diagnostic and treatment plans in detail; do not assume that a physician's professional knowledge makes such discussion unnecessary.
- Avoid engaging in corridor consultations, but do not refuse to help a colleague who is ill. Instead, encourage the colleague to seek appropriate help.

Source: Reprinted with permission from Rosvold, E.O., AHRQ Patient Safety Network. AHRQ WebM&M [serial online]. September 2004. Available at: https://psnet.ahrq.gov/webmm/case/71

thoughtful opinion that may come from a formal consultation or thorough discussion." Several cognitive and affective biases may also creep in (see Chapter 7, Cognitive and Affective Biases and Logical Failures). The problem is compounded further when the interaction is between two physicians. Rosvold [15] has suggested some overall rules of engagement for physicians diagnosing and treating other physicians (Table 4.2).

Physicians Diagnosing Family Members

A related problem arises when physicians diagnose their own family members. These encounters again suffer from many of the problems associated with curbside consultations. Few physicians would hesitate to assist family members with their medical problems, but it is, nevertheless, a path fraught with difficulties, and it is generally considered unwise to diagnose a family member. Nevertheless, it is very common. In one survey, 80% of physicians admitted to diagnosing family members. Interestingly, 33% reported being aware of another physician "inappropriately involved" in a family member's care [16]. There are few justifications for a physician involving themselves with family members other than in situations requiring expediency, in an urgent/emergent situation, or if the physician felt strongly that a misdiagnosis had occurred.

The code of ethics of the American Medical Association in 1901 noted that a family member's illness tends to "obscure [the physician's] judgment and produce timidity and irresolution in his practice." The current view is very clear on the issue: other than in emergency situations or isolated settings where other professional help may not be accessible,

> Physicians generally should not treat themselves or members of their immediate families. Professional objectivity may be compromised when an immediate family member or the physician is the patient; the physician's personal feelings may unduly influence his or her professional medical judgment, thereby interfering with the care being delivered. Physicians may fail to probe sensitive areas when taking the medical history or may fail to perform intimate parts of the physical examination. Similarly, patients may feel uncomfortable disclosing sensitive information or undergoing an intimate examination when the physician is an immediate family member. This discomfort is particularly the case when the patient is a minor child, and sensitive or intimate care should especially be avoided for such patients. When treating themselves or immediate family members, physicians may be inclined to treat problems that are beyond their expertise or training. If tensions develop in a physician's professional relationship with a family member, perhaps as a result of a negative medical outcome, such difficulties may be carried over into the family member's personal relationship with the physician [17].

Drive-By Diagnosis

Finally, there is the problem of what is known colloquially as *drive-by diagnosis*, when someone makes an instant diagnosis after a very brief assessment. It is also known as an *Augenblick* (blink of an eye) diagnosis [18]. It may occur in all of the informal diagnostic modes discussed: in the lay system, in curbside or corridor consultations, in physicians diagnosing family members, in pharmacists diagnosing customers, and in medical environments such as the emergency department. It is characterized by a flash intuitive diagnosis with little or no history or examination (Box 4.5). It may

BOX 4.5 OLIVER PRESENTS WITH A TYPICAL RASH

Oliver is a 35-year-old healthy-looking male who presents to the emergency department with a rash on his neck. He has no other symptoms or complaints, and his vital signs are stable. He is triaged to a nonurgent area. The nurse who is covering that area quickly looks at the rash and tells the patient he has shingles. She later intercepts the emergency physician and says, "There is a quickie in bed 12, a guy with shingles."

The physician, who is bogged down in several complex cases, welcomes the opportunity to see and discharge a patient quickly to keep the waiting room moving, and goes in to see the patient. He is a tanned, healthy-looking man with a rash on his neck.

The rash is a cluster of vesicles with surrounding erythema that appears to follow a C5 distribution on the right side of his neck. On closer examination, however, it appears that the rash crosses the midline. The physician takes further history from the patient. He works as a landscape gardener, and the day previously, had been cleaning up a property. He recalled pulling a vine from a tree and that it caught around his neck. He developed some itching on his neck soon afterward, and thought the rash was similar to something he had experienced a year or two earlier. The physician diagnoses the rash as contact dermatitis from poison oak.

certainly be part of the standard diagnostic process. Emergency physicians, for example, can experience an initial diagnostic impression (see "anchoring" in Appendix A) within seconds of seeing a patient, but their training and discipline usually forces a systematic approach that will include the core requirements—presenting complaint, history of presenting illness, past medical history, physical examination, and laboratory testing and diagnostic imaging if indicated.

Telephone Triage

In 1876, the famous words from Alexander Graham Bell, inventor of the telephone, "Mr. Watson, come here, I want you," was actually a call for medical help, as he had just spilled battery acid on himself [19]. One of the earliest uses of the telephone for medical diagnosis is recorded in a letter published in *The Lancet* in 1879, in which a physician is contacted in the middle of the night by a mother concerned that her child might have croup. The physician instructed her to "Lift the child to the telephone and let me hear it cough."

Mother and child complied, and the physician diagnosed the cough as not being croup, which apparently alleviated the mother's anxiety and ensured a good night's sleep for child, mother, and doctor [20]. The first widespread formal use of the telephone was over 50 years ago in crisis intervention—used by a suicide prevention center in London in 1953 [19]. A variety of options are now available for the delivery of healthcare using telecommunications; *tele-health* is the umbrella term for them.

Physicians' use of the telephone in their clinical practice has become more widespread over the last few decades. While it provides a considerable convenience in communicating with patients, it is not without a downside. In a study of closed malpractice claims, the leading allegation was diagnostic failure in 68% of cases [21]. The largest national insurer of physician and surgeon medical liability in the United States has now issued guidelines to physicians to manage their risk when communicating with patients by telephone [22].

The mode that concerns us mostly here is telephone triage, which, ostensibly, has the purpose of arranging appropriate dispositions for callers with healthcare problems. Underlying the disposition is the goal of detecting, and to some extent diagnosing, conditions that may or may not require urgent care. Typically, nurses and occasionally physicians are employed in these services and sort medical problems according to algorithms and protocols. The process is fraught with difficulty for a variety of reasons. First, many of the signals normally used by nurses and physicians to interpret and diagnose a patient's condition are missing on the telephone. It is estimated that over 65% of social meaning between individuals occurs through non-verbal cues such as motions and gestures [23]. In the absence of such cues, significant meaning is inevitably lost, although one might expect that the non-lexical aspects of speech (variation in loudness, pitch, stress, spacing, rhythm) would assume more importance under these circumstances. Second, the nurse has no vital signs on the patient and no objective means of determining their stability. To put this in some perspective, in a study of triage efficacy, experienced triage nurses who could see the patient in front of them and were able to interact with the patient, had a complete set of vital signs (heart rate, blood pressure, temperature, respiratory rate, pupils, oxygen saturation, and blood sugar), and were using a well-developed, widely used triage system, undertriaged in 12% of pediatric cases and overtriaged in 54% [24]. Overall, it seems very likely that the accuracy of telephone triage would be considerably less than this, yet few studies have directly looked at the efficacy of telephone triage and patient safety.

An Australian study examined telephone triage in a variety of settings using actors to simulate seven medical problems of varying severity on multiple occasions. The triage decision was recorded as Appropriate, Overtriage, or Undertriage according to a predetermined consensus on the appropriate decision for each scenario, all of which were clear-cut. Generally, undertriage was more likely than overtriage. It was particularly worrisome that for two cases with high clinical implications, gastroenteritis with dehydration in a child and presumed meningococcal meningitis in another child, the triage responses were considered inappropriate 60% of the time [25]. Another study looked specifically at the triage process itself and whether or not obligatory questions had, in fact, been asked: that is, questions deemed to be essential for proper evaluation. They also checked on the home management and safety advice given: that is, circumstances in which the patient should call back. The results were not impressive. Appropriate triage outcome was considered to have been achieved in only 58% of cases, and only 21% of obligatory questions had actually been asked. The quality of safety management was assessed as "consistently poor" [26].

Overall, the conclusion from these and other studies is that anything less than the traditional medical hands-on evaluation (presenting complaint, past medical history, evaluation of symptoms and signs, and physical examination) will likely result in an assessment of lower quality and increase the likelihood of diagnostic failure. The very nature of a telephone consultation is error-prone and may lead to delayed or missed diagnoses of serious conditions. Important "contextual" and "intuitive" elements are missing from such evaluations, and they become a type of analytic process by proxy (see Figure 3.1 in Chapter 3: Cognitive Approaches to the Diagnostic Process). However, since many telephone triage systems depend on people following protocols that are designed to elicit responses to critical questions, might a more sophisticated process prove more effective? This may well be possible. Computer systems are currently undergoing development that have the capability to effectively ask and answer natural-language questions over an open and wide range of medical problems. The IBM Deep QA project provided a demonstration of such artificial intelligence, powering Watson to a victory over experts on the television game show *Jeopardy* in February, 2011 [27]. More recently, IBM Watson has formed medical imaging collaboratives with a number of leading healthcare organizations in the United States. Imaging data will be combined with a variety of other medical data, including the electronic health record, reports from radiology and pathology, laboratory results, progress reports, the medical literature, clinical care guidelines, and other sources [28].

Complementary and Alternative Medicine

Those seeking a diagnosis for a range of mild to severe symptoms of injury or disease have a variety of alternatives to orthodox medical diagnosis that are readily available on the Internet. These are reviewed in Chapter 5: Complementary and Alternate Medicine.

Conclusion

People have a variety of alternatives along a continuum before turning to conventional medical diagnosis. It begins with the individual and self-diagnosis, and may proceed through a lay referral system, perhaps via telehealth, to the standard medical system. While conventional medicine does not always make the correct diagnosis, and while medical science still has much to learn about the basis of some diseases, it nevertheless remains the safest option by far. The imperative for all diagnostic strategies is that, insofar as resources will allow, they stand up to scientific scrutiny and are delivered in a safe and ethical manner. In the future, it seems likely that sophisticated computerized systems might well be used alone or in conjunction with human operators to improve the overall reliability of telephone triage, as well as the implicit diagnostic component built into it.

SUMMARY POINTS

- Generally, people have two options for diagnosis: either self or the opinion of another.
- Self-diagnosis is intuitive and often effective for a wide variety of minor complaints and illnesses but may prove harmful in some cases.
- In all societies, ancient and modern, a *lay referral system* has been developed and is usually available to diagnose common disorders. It is based on a network of family, friends, and acquaintances.
- *Cyberchondriasis*, an unfounded escalation of concern over common symptoms, may occur through using the Internet. It appears to be an extension of *medical students' syndrome*—the tendency to identify and self-diagnose with diseases that are of personal interest.
- Curbside or corridor consultations are informal exchanges between professionals that usually result in diagnostic speculation and recommendations for treatment. Unless the purpose is restricted to advice about where to get appropriate help, they should be strongly discouraged.
- Physicians and other healthcare professionals should not engage in drive-by diagnosis, or diagnose themselves or their families except in emergency situations or where they feel strongly that a misdiagnosis might have occurred.
- Telehealth is the provision of healthcare service using telecommunications. Telephone triage attempts to provide appropriate disposition and advice for people by telephone. There are a variety of challenges to present systems that may result in delayed diagnosis.

However, advanced computing systems that bring natural-language processing, information retrieval, knowledge representation and reasoning, and machine learning technologies to the field of open-domain questions hold major promise for the future.

- The hands-on primary care physician is the hinge between the lay and the professional (traditional) referral systems.

References

1. Anonymous. Self-diagnosis. Wikipedia [website]. Accessed September 12, 2016. http://en.wikipedia.org/wiki/Self-diagnosis

2. Better Health Channel [website]. Health information on the Internet. Accessed September 15, 2016. https://hnb.dhs.vic.gov.au/bhcv2/bhcpdf.nsf/ByPDF/Health_information_on_the_Internet/$File/Health_information_on_the_Internet.pdf

3. CBC News [website]. Online health advice sought by more Canadians. Accessed September 15, 2016. www.cbc.ca/news/online-health-advice-sought-by-more-canadians-1.982301

4. White RW, Horvitz E. Experiences with web search on medical concerns and self diagnosis. *AMIA Annu Symp Proc* 2009:696–700. Accessed September 12, 2016. http://research.microsoft.com/en-us/um/people/ryenw/papers/WhiteAMIA2009.pdf.

5. Bengeri M, Pluye P. Shortcomings of health related information on the Internet. *Health Promot Int*. 2003 Dec;18(4):381–6.

6. Jerome JK. *Three Men in a Boat (To Say Nothing of the Dog)*. UK: JW Arrowsmith; 1889.

7. Moss-Morris R, Petrie KJ. Redefining medical students' disease to reduce morbidity. *Med Educ*. 2001 Sept;35(8):724–8.

8. Hunter RCA, Lohrenz JG, Schwartzman AE. Nosophobia and hypochondriasis in medical students. *J Nerv Ment Dis*. 1964 Aug;139:147–52.

9. Cooper AA, Humphreys KR. The uncertainty is killing me: Self-triage decision-making and information availability. *Electronic J Appl Psychol*. 2008:4(1):1–6.

10. Friedson E. Client control and medical practice. *Am J Sociol*. 1960 Jan;65(4):374–82.

11. Dracup K, Moser DK, Eisenberg M, Meischke H, Alonzo AA, Braslow A. Causes of delay in seeking treatment for heart-attack symptoms. *Soc Sci Med*. 1995 Feb;40(3):379–92.

12. Safer MA, Tharps QJ, Jackson TC, Leventhal H. Determinants of three stages of delay in seeking care at a medical clinic. *Med Care*. 1979 Jan;17(1):11–29.

13. Hall EC, Cooper AA, Watter S, Humphreys KR. The role of differential diagnoses in self-triage decision making. *Appl Psychol Health Well Being*. 2010 Mar;2(1):35–51.

14. Golub RM. Curbside consultations and the viaduct effect. *JAMA*. 1998 Sep;280(10):929–30.

15. Rosvold EO. Doctor, don't treat thyself. U.S. Department of Health and Human Services, Agency for Healthcare Research and Quality. Patient Safety Network (PSNet). Accessed September 15, 2016. www.webmm.ahrq.gov/case.aspx?caseI D=71&searchStr=saline

16. La Puma J, Stocking CB, La Voie D, Darling CA. When physicians treat members of their own families. *N Engl J Med*. 1991 Oct 31;325(18):1290–4.

17. American Medical Association, Code of Medical Ethics: Opinion 8.19—Self-Treatment or Treatment of Immediate Family Members. Accessed September 15, 2016. www.doh.wa.gov/Portals/1/Documents/3000/MD2013-03Self-treatmentorTrtmntofFamilyMbrs.pdf

18. Campbell WW. Augenblickdiagnose. *Semin Neurol*. 1998;18(2):169–76.

19. Grumet GW. Telephone therapy: A review and case report. *Am J Orthopsychiatry*. 1979 Oct;49(4):574–84.

20. Aronson S. The Lancet on the telephone 1876–1975. *Med Hist*. 1977 Jan;21(1):69–87.

21. Katz HP, Kaltsounis D, Halloran L, Mondor M. Patient safety and telephone medicine: Some lessons from closed claim case review. *J Gen Intern Med*. 2008 May;23(5):517–22.

22. http://www.thedoctors.com/KnowledgeCenter/PatientSafety/articles/Telephone-Triage-and-Medical-Advice

23. Birdwhistell R. The language of the body: The natural environment of words. In: Silverstein A, editor. *Human Communication: Theoretical Explorations*. New York: Wiley; 1974. As cited by Grumet [19].

24. van Ven M, Steyerberg EW, Ruige M, van Meurs AHJ, Roukema J, van der Lei J, Moll HA. Manchester triage system in paediatric emergency care: Prospective observational study. *BMJ*. 2008 Sept 22; 337:a1501. Accessed September 15, 2016. www.bmj.com/content/337/bmj.a1501

25. Montalto M, Dunt DR, Day SE, Kelaher MA. Testing the safety of after-hours telephone triage: Patient simulations with validated scenarios. *Australas Emerg Nurs J*. 2010 May;13(1–2):7–16.

26. Derkx HP, Rethans JJE, Muijtjens AM, Maiburg BH, Winkens R, van Rooij HG, Knottnerus JA. Quality of clinical aspects of call handling at Dutch out of hours centres: Cross sectional national study. *BMJ*. 2008 Sept 12;337:a1264.

27. IBM Research [website]. The Deep QA Project. Accessed September 15, 2016. www.research.ibm.com/deepqa/deepqa.shtml

28. Monegain B. IBM Watson aligns with 16 health systems and imaging firms to apply cognitive computing to battle cancer, diabetes, heart disease. Healthcare IT News [website]. June 22, 2016. Accessed September 15, 2016. www.healthcareitnews.com/news/ibm-watson-aligns-16-health-systems-and-imaging-technology-apply-cognitive-computing-battle

5

Complementary and Alternative Medicine

Pat Croskerry

CONTENTS

Introduction

When something is given the label of "medicine," the implication is that it provides *both* diagnosis and treatment, but, in fact, a considerable emphasis in complementary and alternative medicine (CAM) is on treatment and less on diagnosis. It may well be that many people accept the diagnosis made through traditional methods but choose to be treated by alternate means. Nevertheless, many alternate medical sources profess expertise in diagnosis, and because patients might pursue CAM for certain symptoms on their own and delay seeking traditional medical opinion for their correct diagnosis, CAM warrants some discussion.

Use of CAM Therapies

Complementary medicine has been defined as the "diagnosis, treatment and/ or prevention which complements mainstream medicine by contributing to a common whole, satisfying a demand not met by orthodoxy, or diversifying the conceptual framework of medicine" [1]. There are many different CAM therapies; some of the main ones are listed in Table 5.1 from Bausell [2]. In 2002, a government survey found that 36% of Americans had used some type of CAM therapy in that year [3]. The Centers for Disease Control and Prevention's National Center for Health Statistics reported that in 2007, Americans spent almost $34 billion on complementary and alternative medicine: $22 billion was spent on alternative medicine products and classes, and $12 billion on the provision of services from professionals [4]. The National Institutes of Health's National Center for Complementary and Alternative Medicine now has an annual budget of more than $121 million.

There has been a noticeable increase of public interest in CAM over the last 20 years. Ironically, some of this may have been due to the emergence of the patient safety movement and the revelation in the Harvard study published in 1991 that approximately 90,000 deaths occurred annually in the United States as a result of adverse events at the hands of the healthcare system,

TABLE 5.1

Use of CAM Therapies in the United States in 2002

CAM Therapy	% of U.S. Population Using Therapy	Number of U.S. Adults Using Therapy
Natural products (e.g., herbs)	18.9	38,183,000
Deep breathing exercises	11.6	23,457,000
Meditation	7.6	15,336,000
Chiropractic therapy	7.5	15,226,000
Yoga	5.1	10,386,000
Massage therapy	5.0	10,052,000
Progressive relaxation	3.0	6,185,000
Megavitamin therapy	2.8	5,749,000
Guided imagery	2.1	4,194,000
Homeopathy	1.7	3,433,000
Tai chi	1.3	2,565,000
Acupuncture	1.1	2,136,000
Energy healing (e.g., Reiki)	0.5	1,080,000
Qi gong	0.3	527,000
Hypnosis	0.2	505,000
Naturopathy	0.2	498,000
Biofeedback	0.1	278,000
Folk medicine	0.1	233,000
Ayurvedic medicine	0.1	154,000
Chelation therapy	0.01	66,000

Source: Reprinted from Bausell, R.B., *Snake Oil Science: The Truth about Complementary and Alternative Science*, 2007, by permission of Oxford University Press.

and that diagnostic error was a significant factor [5]. It became an easy argument to make that if orthodox medicine was failing at this rate, then alternatives might be safer; this was facilitated further by the growth of the Internet through the last decade of the century. Increasing numbers of people could now readily access websites on "alternative medicine" without rummaging around for hours at the local library, not knowing quite what they were looking for.

Although "complementary" and "alternative" medicine are now lumped together in the term *Complementary and Alternative Medicine*, for some, they do not mean the same thing and may be quite distinct from each other. In the United States, the National Institutes of Health initially entered this arena in 1991 with an Office of Alternative Medicine (OAM), which later evolved into the National Center for Complementary and Alternative Medicine (NCCAM) in 1998. In 2014, the name was again changed to the National Centre for Complementary and Integrative Health (NICCIH) apparently in an effort to avoid the term 'alternative', although the strategic plan of NCCAM continues to guide the work of the NICCH [6]. Other countries, such as the United Kingdom and Canada continue to use 'alternative' [6].

In the sense that complementary usually means something that is added to the whole, the inclusion of a healthy diet, routine exercise, a good night's sleep, and avoiding harmful substances, for example, might be considered complementary to the treatment of any disease, and few would argue with these potentially healthy lifestyle changes. Others might consider adding to the orthodox treatment of a disease in the expectation of complementing it; for example, a mother of a child already diagnosed with streptococcal pharyngitis might attempt to complement the treatment by adding *Echinacea* to the already prescribed antibiotic (although *Echinacea* has no proven benefit) [7]. Alternative medicine, in contrast, involves something different from what standard, orthodox medicine has to offer. Although there is a major emphasis on healing rather than curing, deliberate efforts may be made to establish a *de nouveau* diagnosis. Traditional Chinese Medicine, Ayurvedic Medicine, Intuitive Diagnostics, Iridology, Applied Kinesiology, Chiropractic, Naturopathic Medicine, Tibetan Medicine, Homeopathy, and others all have approaches that are inclusive of both diagnosis and treatment.

Challenges to the Claims of CAM

A number of qualified researchers have challenged the unscientific and unsupported claims of CAM: Beyerstein, Ernst, Bausell, Singh, and others. Several detailed works have investigated the efficacy of CAM diagnoses and therapies [2,8–12], with a consensus that they are largely ineffective outside a placebo effect [13]. Of itself, this might mean no more than wasted time and effort in terms of a cure for a disease. However, when CAM diagnoses and therapies lead to a delay, or supplant standard medical diagnosis and treatment, then harm may be done on an individual (see Boxes 5.1 and 5.2) or grander scale. Baum and Ernst [14] note that over a 3-year period from 2003–2006, the National Health Service of the United Kingdom spent £20

BOX 5.1   EDITH'S BACK PAIN

Edith MacDougal is a pleasant 75-year-old woman who presents to the emergency department with lower back pain. Her family doctor has recently retired, and she is waiting to see another doctor who will be taking over the practice. She is apologetic for coming to the emergency department but felt she should get something for pain.

She had a minor slip several months earlier and is unsure whether the pain is related to that event. She was told by a friend that chiropractors could fix back problems, and has been seeing one since. He has had x-rays done of her lower back and performed a series of manipulations for "misalignment." These have been ineffective thus far and fairly costly to her. She has had 12 treatments at a cost of $110 each. Yesterday, she was told that her back continued to be out of alignment and that she would need a further course of treatment.

The emergency physician takes some further history. The patient leads an active life and is generally fairly healthy. She has had only minor surgeries in the past and is currently being treated for hypertension and osteoporosis. Her bowel function is normal, but she reports increased frequency of urination over the last month or two, which she attributes to getting old. He performs an examination of her lower back. There is no localizing tenderness over the lumbar or sacral spine and no evidence of back spasm, and her pelvis is stable. She has good range of motion in both hips, her straight leg raising is about 80° bilaterally, and she shows no signs of sciatic nerve irritation. Power is equal and strong bilaterally. Sensation in her legs is normal, reflexes 2+ bilaterally, and pulses equal and strong.

He reviews her back lumbar-sacral x-rays that had been ordered by the chiropractor, which showed normal alignment and only mild degenerative changes. At triage, she was asked to give a urine specimen. This has now been reported and shows 50–100 WBCs per high power field, significant nitrites, trace of blood, and moderate bacteria.

He starts her on an antibiotic and advises her to return to a followup clinic at the hospital for reassessment in 5 days. She asks whether or not she should start her second round of back manipulations and is advised to hold off till reassessment. She is seen in the clinic for the planned follow-up at 5 days. At that time, her back pain has completely resolved, and a repeat urinalysis is clear.

BOX 5.2   SAM'S NEW DOCTOR

Sam McAvoy is a 60-year-old construction worker who lives in a small rural town. He presents at a walk-in clinic requesting some insurance forms be completed for a back injury he sustained several months ago. His regular family doctor has retired, but he now has a new doctor.

The physician at the clinic asks why he has not asked his new doctor to complete the forms for him, as this would be more appropriate. He says he did, and she sent the forms in for him, but the insurance company has returned them asking that a "real" doctor complete them. Somebody suggested he attend the medical clinic. He appears a little confused about why his doctor's forms were rejected, because she told him she was a doctor. The physician inquires about the patient's new doctor and establishes that she is a naturopathic doctor. Apparently, the patient thought that all doctors were medical doctors. The physician explains that he is willing to complete the forms but will need to complete a history and physical examination. The patient is agreeable to having this done, especially as his back pain has been worsening over the last month. His new doctor has recently performed several spine manipulations, but he feels he is worsening.

He relates that he injured his back about 6 months ago lifting some heavy wood. Because of his discomfort, he has been doing less at work, and attributes this to some loss of conditioning. Both legs have felt weaker over the last few months. He has also been experiencing some problems with urination, which he attributes to his age, as his father had a similar problem. On examination, he had equal straight leg raising bilaterally at about 45°, which was limited by central lumbar back pain. He appeared to have bilateral weakness for his age. He was hyporeflexive, but sensation was grossly normal in his legs. He was reluctant to have a rectal examination, not appreciating any relevance it might have to a back "strain," but did eventually agree. There appeared to be some loss of sphincter tone and sensory loss in the sacral area. He was immediately referred to a nearby hospital for further assessment. An MRI confirmed the suspected diagnosis of cauda equina syndrome, and he was taken to the operating room later that day.

million (then about U.S. $26 million) of British taxpayers' money refurbishing the Royal London Homeopathic Hospital. They estimated that, at this time, this funding would have been sufficient to save the lives of 600 breast cancer patients over the same period.

Many will be puzzled why CAM diagnostics and therapeutics enjoy such widespread support, and why people remain willing to accept a wide variety of unscientific diagnoses and treatments that have no proven efficacy and that may be costly and occasionally even fatal. Given the challenges and complexities of modern medicine, conventional physicians might well marvel at CAM practitioners' fortitude and optimism in pursuing their diagnoses and therapies. Typically, there is no shortage of either confidence or optimism. What helps the CAM practitioner, certainly, is that humans normally tend to be believers rather than skeptics: that is, we have a hard-wired disposition toward believing rather than disbelieving. Disbelief and doubting are not our "natural options." Instead, belief is our default option, and there are abundant follies in modern societies that are testimony to our uncritical acceptance of various belief systems.

Why CAM Is Successful

In a thoughtful and systematic analysis, Beyerstein [8,9] has reviewed the major issues that give rise to such behaviors, summarized in Tables 5.2 and 5.3. A wide range of psychological, social, cultural, and other factors are involved.

An additional factor, proposed by Schermer, theorizes that part of evolved brain modularity is a *belief engine*—a means of finding causal con- nections between objects, events, or phenomena in the environment—and that this has been integral to Darwinian evolution. This "patternicity" (finding patterns of apparent causation) was of critical importance for sur- vival. We are inclined toward making false-positive causations because we believe it is very important not to miss a pattern; in fact, our survival as a species probably depended on it (see *search satisficing* in Appendix A and Chapter 6: Stone Age Minds in Modern Medicine). We thus find patterns where none exist—especially when we confuse correlation with causa- tion, which in turn, may lead to the development of superstitious behavior, "magical thinking," and ultimately self-deception [15]. The medical appli- cation of this idea was nicely captured by Oliver Wendell Holmes Senior's valedictory address to the graduating class of Belleview Hospital College over 140 years ago:

> This is the way it happens: Every grown-up person has either been ill himself or had a friend suffer from illness, from which he has recovered. Every sick person has done something or other by somebody's advice, or of his own accord, a little before getting better. There is an irresistible tendency to associate the thing done, and the improvement which fol- lowed it, as cause and effect. This is the great source of fallacy in medi- cal practice. But the physician has some chance of correcting his hasty inference. He thinks his prescription cured a single case of a particular complaint; he tries it in 20 similar cases without effect, and sets down the first as probably nothing more than a coincidence. The unprofessional

TABLE 5.2

Social and Cultural Factors That Underlie the Apparent Success of CAM

Factor	Description
Anti-intellectualism and antiscientific attitude	New Age gurus have generally promoted emotional rather than empirical criteria for deciding what to believe; that is, we can all create our own reality—"validity" is what works for the individual.
Vigorous marketing	Promotion of alternative options has been aggressive and may have involved intense legislative lobbying; extravagant claims are often made. The Internet has facilitated a variety of marketing ploys.
Inadequate media scrutiny or critique	The lack of evidence for CAM claims, other than anecdote and personal testimonials, has not been adequately challenged by the media partly because there is a fear of appearing racist or sexist, and also because their advertising may provide a source of revenue (thus creating a conflict of interest).
Increasing mistrust of traditional medicine	In recent years, there has been a growing disenchantment with the establishment. CAM has exploited an antidoctor backlash.
Disapproval of modern health-care delivery	Modern medicine is seen as becoming more technocratic, bureaucratic, and impersonal. In contrast, CAM therapists can offer a more personalized individual approach that will hold appeal for many.
Presumption of safety	Some CAM approaches play on the "naturalistic" bias; that is, natural products, coming from nature, are less likely to have harmful side-effects.
Will to believe	People generally are believers rather than disbelievers and are vulnerable to a hardwired disposition toward "magical thinking."
Logical errors and lack of control group	The basis of many decision pitfalls is mistaking correlation for causation. Without properly designed research studies, in particular using inadequate control groups, any conclusions are questionable.
Judgmental shortcomings	The lay public usually exercise intuitive judgment about the efficacy of CAM diagnosis and treatments, often confounded by a variety of biases and other frailties of reasoning, and typically are not aware of what is required for scientific proof.
Psychological distortion of reality	Personal belief systems can be very powerful, and even in the absence of evidence of CAM efficacy, people may still convince themselves of benefit for a variety of psychological reasons (cognitive dissonance, reinterpretation, denial, etc.).
Self-serving biases and demand characteristics	Conflict of interest and other biases (e.g., selective recall) distort reality for both CAM practitioners and their clients. Also, people have a natural tendency to respond to social demands of the therapeutic contract; that is, if I diagnose and treat you, you will feel an obligation to reciprocate by acknowledging the success of the treatment.

Source: Adapted from Beyerstein, B.L., *Sci. Rev. Altern. Med.*, 3, 16–29, 1999.

experimenter or observer has no large experience to correct his hasty generalization. He wants to believe that the means he employed effected his cure. He feels grateful to the person who advised it, he loves to praise the pill or potion which helped him, and he has a kind of monumental pride in himself as a living testimony to its efficacy. So it is that you will find the community in which you live, be it in town or country, full

TABLE 5.3

Other Reasons for Mistakenly Concluding That CAM Works

Factor	Description
Disease may have run its natural course	The vast majority of illnesses are self-limiting with full recovery. Thus, any intervention that precedes the natural termination of an illness can be given credit for the cure.
Many diseases are cyclical	An improvement or apparent remission may simply be due to the cyclic nature of some diseases, but may be attributed to any recent intervention.
Placebo effect	Placebo effects may produce improvement through a variety of means; CAM treatments often fail to take placebo effects into account. Both patients and those who treat them should be double-blinded to which treatment is being given.
Psychosomatic illness	Some illnesses have a psychosomatic basis (the worried well) and may be responsive to support, reassurance, and suggestion, especially from an attentive and charismatic therapist who is willing to offer a "medical" diagnosis.
Symptomatic relief versus cure	For a variety of psychological and other reasons, CAM may provide relief of pain and discomfort, which the patient may equate with cure of the disease.
CAM patients may hedge their bets	People may hedge their bets by accessing multiple diagnostic and treatment modalities, especially when the alternative to orthodox medicine is labeled "complementary" or "integrative." If improvement occurs, the CAM option gets disproportionate (or undeserved) credit.
Misdiagnosis by self or physician	Those who self-diagnose themselves with an illness, but do not receive medical confirmation, are more likely to be diagnosed (and cured) with CAM. Conversely, patients medically misdiagnosed with a significant illness who engage CAM may attribute failure of the illness to progress to CAM.
Derivative benefits	Charismatic CAM therapists may enhance a patient's mood, expectations, and lifestyle such that any orthodox treatments they are receiving may be more effective through greater compliance.

Source: Adapted from Beyerstein, B.L., *Sci. Rev. Altern. Med.*, 3, 16–29, 1999.

of brands plucked from the burning, as they believe, by some agency which, with your better training, you feel reasonably confident had nothing to do with it. Their disease went out of itself, and the stream from the medical fire-annihilator had never even touched it. [16]

Integrated Medicine

Integrated or integrative medicine is an approach that attempts to combine conventional medicine with CAM. This can be construed as "hedging the bet" on both sides of the therapeutic exchange. The patient might feel that they should engage CAM because it has intuitive appeal or they have heard some powerful testimony from a friend, but they will also follow the recommendations of conventional medicine just in case, or because they can have the best of both worlds. While CAM practitioners may strongly advocate their approach, some believe that having standard medicine in the background might have benefits. However, many people oppose this view. As Charlton notes, "The dilemma is that in the short-term a modicum of science (or pseudo-science) may serve to increase the status of New Age practitioners and validate their activities. Yet, in the longer term, the attempt to subordinate science to spirituality will lead to a conflict which science will win" [17]. Charlton and others are opposed to any integration of standard medicine with CAM, as is Ernst, who writes: "The message that emerges … seems clear. Integrated medicine is by no means 'the best of both worlds'. Often it means the substitution of demonstrably effective treatments by remedies which are unproved or even disproved. Integrated medicine promotes CAM no matter what the evidence says. This is a cynical violation of the principles of evidence-based medicine at the expense of effective and cost-effective healthcare. The loser in all this, I fear, would be the patient" [12].

Conclusion

In this chapter, we have reviewed alternatives to conventional medical diagnosis. Unlike the lay referral system, which is an important part of conventional medicine (Chapter 4: Alternatives to Conventional Medical Diagnosis), CAM sets itself apart with radically different philosophies and ideas. A variety of entities under the umbrella term of complementary, integrated, and alternative medicine may be engaged by the general public. Generally, these alternative approaches have not been verified by traditional scientific methods. An important corollary of this is that if the approach itself is not verifiable, then the associated diagnostic and therapeutic approaches may be unsafe. It is true that, on occasions, some patients will benefit from CAM through placebo and psychological factors—symptomatic relief, enhanced mood and optimism, increased motivation toward healthier eating, exercise and sleep habits, reduced stress, an increased sense of well-being, and others. This may well be beneficial, providing it does not delay definitive management of significant illness and is not unduly expensive for the patient. While traditional medicine does not always make the correct diagnosis, and while medical science still has much to learn about the basis of some diseases, it remains by far the safest option. Again, the imperative for all diagnostic strategies is that they stand up to scientific scrutiny and are delivered in a safe and ethical manner.

SUMMARY POINTS

- Complementary and alternative medicine (CAM) involves a wide range of diagnostic and therapeutic modalities that mostly have not been supported by formal scientific investigation.

- Those who use CAM, nevertheless, may receive relief from their symptoms through placebo effects as well as through a wide variety of social, cultural, and psychological mechanisms.

- The appeal of CAM may reside, in part, in an evolutionary imperative to believe rather than disbelieve.

- Integrated medicine is the combination of CAM and conventional medicine—both patient and CAM practitioner may feel some benefit from hedging their bets, and perhaps increasing psychological benefits through CAM. Some notable researchers have advocated strongly against such a combination.

References

1. Ernst E, Resch KL, Mills S, Hill R, Mitchell A, Willoughby M, White A. Complementary medicine: A definition. *Brit J Gen Pract*. 1995 Sep;309:107–11.
2. Bausell RB. *Snake Oil Science: The Truth about Complementary and Alternative Medicine*. New York: Oxford University Press; 2007. pp. 1–22.
3. Barnes PM, Powell-Griner E, McFann K, Nahin RL. Complementary and alternative medicine use among adults: United States, 2002. *Adv Data*. 2004 May 27;(343):1–19.
4. Nahin RL, Barnes PM, Stussman BJ, Bloom B. *Costs of Complementary and Alternative Medicine (CAM) and Frequency of Visits to CAM Practitioners: United States, 2007*. National Health Statistics Reports: no 18. Hyattsville, MD: National Center for Health Statistics. 2009. Available at: www.cdc.gov/nchs/data/nhsr/nhsr018.pdf. Accessed January 18, 2011.
5. Brennan TA, Leape LL, Laird NM, Hebert L, Localio AR, Lawthers AG, Newhouse MP, Weller PC, Hiatt HH. Incidence of adverse events and negligence in hospitalized patients. Results of the Harvard Medical Practice Study 1. *N Engl J Med*. 1991 Feb 7;324(6):370–6.
6. NIH complementary and integrative health agency gets new name. https://nccih.nih.gov/news/press/12172014. Accessed August 15, 2016.
7. Taylor JA, Weber W, Standish L, Quinn H, Goesling J, McGann M, Calabrese C. Efficacy and safety of echinacea in treating upper respiratory tract infections in children: A randomized controlled trial. *JAMA*. 2003 Dec 3;290(21):2824–30.
8. Beyerstein BL. Social and judgmental biases that make inert treatments seem to work. *Sci Rev Altern Med*. 1999; 3:16–29.
9. Beyerstein BL. Alternative medicine and common errors of reasoning. *Acad Med*. 2001 Mar;76(3): 230–7.

10. Ernst E, editor. *Healing, Hype or Harm? A Critical Analysis of Complementary or Alternative Medicine*. Exeter, UK: Societas; 2008.

11. Singh S, Ernst E. *Trick or Treatment. The Undeniable Facts about Alternative Medicine*. New York; WW Norton; 2009.

12. Ernst E. Integrated medicine? In: Ernst E, editor. *Healing, Hype or Harm? A Critical Analysis of Complementary or Alternative Medicine*. Exeter, UK: Societas; 2008.

13. Ernst E. Towards a scientific understanding of placebo effects. In: Peters D, editor. *Understanding the Placebo Effect in Complementary Medicine: Theory, Practice, and Research*. Edinburgh: Churchill Livingstone; 2001. p. 246.

14. Baum M, Ernst E. Ethics and complementary or alternative medicine. In: Ernst E, editor. *Healing, Hype or Harm? A Critical Analysis of Complementary or Alternative Medicine*. Exeter, UK: Societas; 2008. pp. 104–11.

15. Schermer M. *Why People Believe Weird Things: Pseudoscience, Superstition and Other Confusions of Our Time*. New York: W.H. Freeman; 1997.

16. Oliver Wendell Holmes Senior: The Young Practitioner [A Valedictory Address delivered to the Graduating Class of the Bellevue Hospital College, March 2, 1871.] Accessed from The Literature Network: www.online-literature.com/oliver-holmes/medical-essays/7/ on September 17, 2016.

17. Charlton BG. Healing but not curing. Alternative medical therapies as valid New Age spiritual healing practices. In: Ernst E, editor. *Healing, Hype, or Harm? A Critical Analysis of Complementary or Alternative Medicine*. Exeter, UK: Societas; 2008.

Section III

The Elements of Reasoning

6

Stone Age Minds in Modern Medicine: Ancient Footprints Everywhere

Pat Croskerry

CONTENTS

Introduction

As we have discussed, much of diagnostic reasoning is about thinking—what is going on in the brain of the decision maker. The human brain, like any organ in the body, has evolved from very humble beginnings over many millions of years since the first primitive neural networks in simple organisms started working together. In the process of natural selection, Darwinian forces have acted on the brain to produce a highly complex organ. And, like the heart or the kidney, the brain performs basic functions that served us well millions of years ago and continue to do so. What this does mean, however, is that certain operations of the brain that were selected millions of years ago to produce specific kinds of behaviors may still produce those behaviors today in modern times. This is the gist of evolutionary psychology, a discipline founded in the latter part of the twentieth century.

Evolutionary Psychology

Evolutionary psychology holds that major selective pressures on human brain development occurred over many thousands of years in our ancestral environment, principally during the nearly two million years that spanned the Pleistocene period. Thus, some modern human thinking, reasoning, decision making and behavior are the product of contemporary influences interacting with our ancestrally designed brains. We might like to think that we are highly sophisticated thinkers and significantly superior in intellect to our ancestors from 50,000 years ago, but with a few minor exceptions, there do not appear to have been any major evolutionary changes in our brains for at least the last 50,000 years [1]. So, if we were able to take an infant born to prehistoric cave people and raise it in our present environment, it would be as able and accomplished as children born today. Living examples of this are the aborigines of Australia. They migrated to that continent about 40,000–50,000 years ago and, with no major selective pressures on the brain, remained as hunter-gatherers till the first Europeans arrived a few centuries ago. Yet, they are just as capable of becoming engineers, pilots, physicians, or any other career in modern society. Given the accomplishments of the modern brain (putting men on the moon, smartphones, Michelangelo's masterpieces, or a Beethoven concerto), it may be a humbling thought for some that we are no "brighter" than our caveman ancestors, but this appears to be the case. As Cartwright has observed, we now "carry Stone Age minds in modern skulls" [2]. While hardwiring provides the substrate for at least some of our brain function, the environment and learning significantly influence what the final product will look like. The question for us here concerns what relevance this might have to clinical decision making, and to diagnostic reasoning in particular. To answer it, we need to take a closer look at the evolution of cognition.

Empiricism and Nativism

The competing view against evolutionary psychology is that our brains are blank slates at birth—the empiricist position. Some still believe this; it was only a few decades ago that behaviorists argued that our entire behavioral repertoire was the result of learning. These reinforcement learning theories

TABLE 6.1

Hardwired Behaviors in Humans

- Sympathetic arousal can improve or impair performance.
- Emotional arousal changes our judgments and decision making.
- Sexual arousal changes moral standards and reduces willpower.
- We inappropriately crave sugars and fats.
- Women's social and sexual behaviors increase at ovulation.
- Women's appetite decreases at ovulation.
- We have phobias (snakes, spiders, heights, strangers).
- We make overly predictive inferences about others' personalities.
- We judge approaching sound to be closer than receding sound.
- We have aversions to diseased or injured persons.
- Men overinterpret the sexual interest of women.
- Women are biased toward underestimating men's commitment.
- Racial and ethnic stereotyping increases in reduced ambient lighting.
- We overrate our own qualities and our degree of control over the environment.
- We perceive members of out-groups as less generous and more dangerous than in-group members.
- We search satisfy in a wide variety of tasks.

lasted a span of about 70 years, from Pavlov at the turn of the century to Skinner in the late 1960s. In some ways, it proved to be a costly diversion of time and effort. Although it never enjoyed complete acceptance, it was the prevailing dogma for many psychologists until the latter part of the twentieth century. However, common sense and nativism have come to prevail—Stephen Pinker has given a thorough exposition of the issues underlying the modern denial of human nature [3], and we now appear to be on the right track, recognizing abundant examples of human prewired behaviors. Ancient footprints (as Bob Dylan noted on a trip to Rome) are everywhere [4]. A variety of human behaviors, posited to be hardwired, have been described in the evolutionary psychology literature (Table 6.1). Once we have accepted the nativist position that our bodies and brains are the products of millions of years of evolution, we can more closely look at the evolution of cognition and decision making.

Evolution of Cognition

With the exception of some basic reflexes, all of our behavior is under some degree of cognitive control, as it is with other animals. Further back on the evolutionary scale, the behavior of simpler forms is driven by *instinct*—a particular stimulus appears in the environment and triggers a fixed-action pattern (Figure 6.1). No conscious mediation is involved. We talk about the

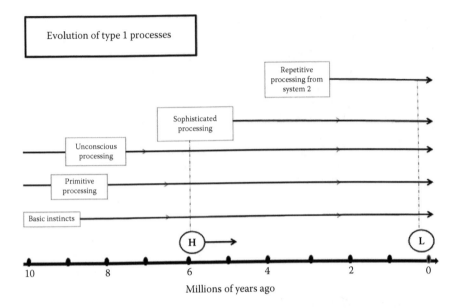

FIGURE 6.1
Developmental stages in human cognition. H indicates when hominids (of which there were at least 20 different species) first began to appear. L indicates the recent emergence of language. We (*Homo sapiens*) have been in existence for about 200,000 years.

instinct of birds to migrate, cats to hunt, and salmon to swim upstream in the breeding season. All of these complex behaviors are instincts—no purposeful thought or decision making is involved. Although a robin's nest may appear to be a thoughtful design, no learning was involved, no deliberate thinking went into it, and the next generation will build one almost exactly the same with no input from parents. Following instinct, Hogarth [5] describes the next level of cognitive function as *primitive processing*—this again involves innate automatic responses but is distinct from instinct in that there is now cognitive involvement that can recognize co-variation of events and frequencies, and some basic inferences might be made about weather, food, shelter, and predators. The next level is *unconscious processing*, which again, is automatic and unconscious and does not require specific attention, but involves tacitly learning about the environment and may involve a memory for important stimuli. *Sophisticated processing* is the next level and involves meaning and affect.

When individuals are able to attach meaning and affect to their experiences, they have a template for building a "personal" logic of inductive experience, and the beginnings of individual differences in decision making become apparent. Sophisticated processing is quintessentially human, says Hogarth, but some would argue that there are examples of other species showing aspects of sophisticated processing (crows come to mind). One can imagine that, as soon as the threshold of sophisticated processing has been reached, then we are not far from learning through practice and the acquisition of important skills.

In evolutionary terms, however, this took a very long time. As early as approximately 2.5 million years ago, humans were making stone tools—a behavior that required relatively sophisticated psychomotor skills. It is perplexing that very little appeared to change over the next 2.45 million years. Overall, it appears that simply being capable of sophisticated processing was not associated with any rapid expansion of motor skills over this period.

Analytic reasoning (Type 2 processes) appears to have emerged relatively recently. One of its intrinsic properties is that it requires abstract thinking around operations that can be verbalized. Thus, it seems highly likely that it underwent significant development in *Homo sapiens sapiens* within the last 100,000 years. Once it did, repetitive processing allowed a variety of learned skills to be relegated to the intuitive mode and become Type 1 processes, as shown in Figure 6.2. (The figure is an expansion of the first part of the schema described in Figure 3.1.) With this overall view of the evolution of the different types of reasoning, we can now look more closely at how ancient thinking might influence modern decision making and the diagnostic process.

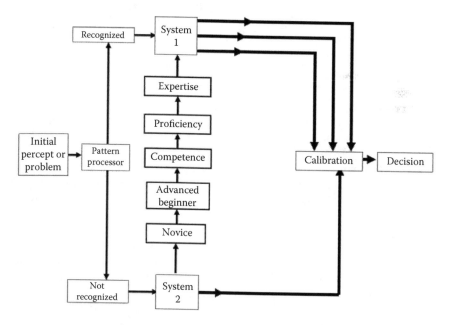

FIGURE 6.2
Movement of a newly experienced percept from System 2 into System 1 through repeated processing, as in the acquisition of a learned skill or habit.

Characteristics of the Intuitive Mode of Decision Making

It is important to understand intuitive decision making of the type characterized by Type 1 processes described in Chapter 3: Cognitive Approaches to the Diagnostic Process (see also Chapter 8: The Rational Diagnostician). It is where we spend most of our time and where most error is believed to occur. The Type 2 processes that underlie analytic thought are extremely important in modern medicine but are less a cause for concern in diagnostic reasoning.

There appear to be two main sources of input into intuitive thinking. One of them is depicted in Figure 6.2. Patterns and combinations of symptoms and signs that are not recognized are processed in System 2, but repeated processing there eventually leads to recognition, familiarity, and competence, such that future presentations may be reliably managed in System 1. The other major sources of input into System 1 are these hardwired responses that have their origins in basic instincts and in primitive, unconscious, and sophisticated processing. An example is given in Box 6.1.

This clinical case describes a "modern thinking failure" of search satisficing. *Search satisficing* was first described by Herbert Simon [6], who won the Nobel Prize for his work in the economics of decision making. Satisficing is an amalgam of *satisfy* and *suffice*. We satisfy ourselves that the search we have done will suffice for our purposes. Evolutionary psychology would argue that, from an evolutionary standpoint, search satisficing conferred some selective advantage; that is, individuals who called off the search earlier would have been more likely to get their genes into the next generation. Consider the example in Box 6.2.

Decision making in ancient times, what the evolutionary psychologists refer to as the *environment of evolutionary adaptedness* (EEA), has been characterized as a form of evolutionary probability gambling. In *Unweaving the Rainbow*, Richard Dawkins discusses how natural selection has operated on intuitive decision making (System 1) by selecting differentially for false positives (believing something is there when it isn't) over false negatives (believing something is not there when it is), and describes some human behaviors as evidence of false positive errors: for example, superstitions, phobias, magical thinking, and so on. Like Cartwright, he concludes that "parts of our brains for doing intuitive statistics are still back in the stone age" [7]. Much of survival depended on pattern recognition and pattern matching, and there was probably some value in erring on the side of false positives rather than false negatives. Search satisficing strategies need not be optimal, only sufficient to improve chances of survival.

In evolutionary psychology, error management theory (EMT) proposes that such "thinking failures" evident in modern environments are the result of evolved, naturally selected patterns of behavior that served us well in our evolutionary past and for which we are now hardwired [8]. Dual process theory, in turn, proposes that many of these failures are examples of cognitive biases and heuristics, and that these occur largely in the intuitive mode.

BOX 6.1   MODERN SEARCH SATISFICING

A 45-year-old woman presents to the emergency department in an agitated state. She is holding an empty bottle of aspirin and says she has taken all of the pills a few hours ago to "end it all." Her breathing and heart rate are fast; she is nauseous and complains of ringing in her ears. Blood work is drawn that includes a toxic screen, intravenous lines are started, and treatment is begun for salicylate poisoning. Within an hour, her salicylate level comes back at a toxic level.

Although her condition initially showed marginal improvement, when she is reassessed by the emergency physician after 2 hours, the impression is that she is not progressing as well as expected. She now appears confused, and her monitor shows a marked tachycardia. While the ED physician is reflecting on her condition, the patient's partner comes to the ED to inquire how she is doing.

The physician tells him that she is not doing as well as expected, but, given that she has taken a major overdose of salicylate, she may take a little time to stabilize. Her partner pulls an empty bottle of a tricyclic antidepressant out of his pocket and says that he found it on the bedroom floor when he got home from work. He wonders if this is important.

Shortly afterward, the patient becomes hypotensive, with the monitor showing an intraventricular conduction delay with wide QRS, first degree block, and a prolonged QT interval; she then seizures. She is intubated and transferred to the ICU.

Comment: the emergency physician and nurses initially anchor on the patient's story, the empty bottle of salicylate, and the signs of acute salicylate toxicity. They do not inquire about any other medication she might have taken. The initial toxic screen measures only salicylate, acetaminophen, and alcohol levels, and the initially high salicylate level confirms their belief that they are dealing with a salicylate toxidrome. However, they have satisfied themselves that they were dealing with only a single toxidrome and called off the search for any additional toxidromes. Had they asked the patient whether she had taken anything else, they might well have discovered the concurrent (and potentially more serious) tricyclic overdose, or might have detected tricyclics in a qualitative urine screen that would have alerted them to the possibility of a concurrent toxidrome.

As noted in Chapter 3, these hardwired characteristics of human behavior are described in the anthropology literature as "universals," meaning they are common to all known cultures on Earth (wariness around snakes is one of them). They are evidence of hardwiring acquired from a common ancestor—*Homo erectus*, originating from Africa about two million years

BOX 6.2   ANCIENT SEARCH SATISFICING

About 75,000 years ago, in the Upper Pleistocene period, two cavemen are walking through low grasslands towards a group of trees, anticipating a meal of fresh fruit. They both hear a rustling sound ahead. One turns immediately away from the sound and makes a wide detour, believing there is danger ahead. The other is less convinced, thinking the sound is more likely due to leaves being blown around by the wind. He continues towards the trees and suffers a bite from a black mamba snake, which (in those days) carried a mortality rate of 100%.

The first caveman satisfied himself quickly that there was danger ahead and took immediate evasive action—his genes went into the next generation. The second was less satisfied and didn't believe there was sufficient evidence of imminent danger—his genes didn't.

Comment: snakes were probably the first serious predators of modern mammals—they evolved about 100 million years ago. Decisions around detection and avoidance of snakes were therefore critical to survival and getting genes into the next generation.

Black Mamba

ago. All human decision making, in all walks of life and in all cultures, has universal properties, although local contingencies, ambient factors, and cultural issues may be expected to exert an influence. For example, Nisbett [9] points out the impact of dialectical reasoning of Eastern cultures on the expression of several major biases (fundamental attribution error, confirmation bias, and contextual biases). Nevertheless, all human decision making, from the inhabitants of Tierra del Fuego at the tip of South America to the Inuit of Northern Canada, has strong fundamental similarities. In 1986, the anthropologist Brown described over 300 "universals" of human behavior, characteristics such as jealousy, parenting, wariness of snakes, recognition of cheats, and many others that are common to all cultures, and his list has been expanded further since [10]. That the Inuit may suffer from snake phobia is fairly convincing proof—the Arctic is no place for a poikilotherm, and fear of them doesn't make much sense. The existence of these collective ways of thinking and behaviors represents a *collective conscious*, the substrate of which is the brain and the genes that control its development.

Implications for Diagnostic Reasoning

Most of us would like to think that our daily thinking is deliberate, rational, objective, and in medicine (preferably) evidence-based. However, it is the rule of thumb among cognitive scientists that probably about 95% of what shapes and structures all decision making is below the surface of conscious awareness [11]. Much of our behavior is simply driven by what our bodies are doing—if we are sitting at the wheel of a car, many of the decisions that are made are automatic and reflexive Type 1 processes. Occasionally, things will come along that require attention—a new road works, or a danger of hydroplaning—but for the most part, driving is a mindless, automatic act. On any particular day, most of us cannot remember driving to work. Similarly, when sitting at a table at mealtime, we do not consciously ask: how small do I need to cut the pieces to get them easily into my mouth, how much chewing should I devote to a particular food to make sure the bolus goes down my esophagus easily? All these questions and decisions, for the most part, can be relegated to a subconscious level. To use an expression from Pink Floyd [12], we become "comfortably numb" with the highly familiar; over-learned skills provide for a great redundancy and inevitably will lead to economy of thought and action, and a state of mindlessness.

As we have noted, the other major substrate for Type 1 processes are those reflexive patterns of behavior that have been acquired through natural selection, and which, in our evolutionary past, have served us well. They may still serve us well in modern environments—if you are skiing down a mountain and round a bend to find an avalanche has blocked most of the path, there is no time for analytic reasoning (System 2). Instead, fast, evasive action (System 1) is required to avoid accident and injury. What saves the day are those Type 1 escape processes that are driven by hardwiring (fear, sympathetic arousal, reflexive movement, search satisficing, and others).

Implications for Medical Education

This discussion has important implications for how we teach medical students and residents, and for continuing medical education. Tooby and

Cosmides have come down strongly against the models that we currently use in education. Historically, the approach toward education in the medical sciences has followed the traditional conceptual framework of the standard social science model (SSSM) developed in the twentieth century. This had a culturally deterministic, blank slate, empiricist view that was largely descriptive and took insufficient account of the biological and evolutionary substrates of human emotions, thoughts, and behavior [13]. While there probably remain few proponents of *blank slate* empiricism nowadays, the legacy of the SSSM approach remains firmly embedded in the orthodoxy of medical education. New approaches are now required that will allow greater understanding of the evolved mechanisms that have led to present-day human nature, not only of patients but also of those charged with their care (see Chapter 14: Medical Education and the Diagnostic Process). In the process, we may come to a deeper understanding of what occurs during the process of diagnostic reasoning and especially what underlies diagnostic failure.

SUMMARY POINTS

- The evolution of thinking is relevant to understanding modern-day decision making.
- Thinking, reasoning, and decision making are not simple or singular processes.
- Intuitive processes of the human brain have evolved over millions of years from basic animal-like instincts to sophisticated processing.
- The existence of "universals" of behavior in all cultures gives testimony to its inheritance from common ancestors.
- Analytic reasoning appears to have evolved fairly recently, probably co-evolving with the emergence of language over the last 100,000 years.
- Most decision making is unconscious and autonomous.
- Unconscious thought and action have four major sources. The two principal ones are: decision making and behaviors that have undergone Darwinian selection, and over-learned thinking and behaviors that have moved from System 2 into System 1.
- Many hardwired decisions and behaviors were selected in the environment of evolutionary adaptedness (EEA) when they had survival value.
- The human EEA is considered largely to have been the Pleistocene period (over the last 1.8 million years).
- Some medical decisions will be influenced by hardwired tendencies in the brain such as search satisficing.
- These concepts need to be acknowledged in medical education and in our approach toward diagnostic reasoning.

This challenge, for all domains of human behavior, is exquisitely articulated by the novelist Lawrence Durrell in the context of his spiritual and philosophical excursion into Taoism: "The greatest delicacy of judgment, the greatest refinement of intention was to replace the brutish automatism with which most of us exist, stuck like prehistoric animals in the sludge of our non-awareness" [14].

References

1. Tooby J, Cosmides L. Conceptual foundations of evolutionary psychology. In: Buss DM, editor. *The Handbook of Evolutionary Psychology*. Hoboken, NJ: Wiley; 2005. pp. 5–67.
2. Cartwright J. *Evolution and Human Behavior: Darwinian Perspectives on Human Nature*. 2nd ed. Cambridge, MA: MIT Press; 2008. p. 151.
3. Pinker S. *The Blank Slate: The Modern Denial of Human Nature*. New York: Penguin Putnam; 2002.
4. Dylan B. When I paint my masterpiece. Big Sky Music, 1971. http://bobdylan.com/songs/when-i-paint-my-masterpiece/
5. Hogarth RM. *Educating Intuition*. Chicago, IL: University of Chicago Press; 2001.
6. Simon HA. Rational choice and the structure of the environment. *Psychol Rev.* 1956 Mar;63(2):129–38.
7. Dawkins R. *Unweaving the Rainbow: Science, Delusion and the Appetite for Wonder*. New York: Houghton Mifflin; 1998. p. 178.
8. Haselton MG, Buss DM. Error management theory: A new perspective on biases in cross-sex mind reading. *J Pers Soc Psychol.* 2000 Jan;78(1):81–91.
9. Nisbett RE. *Mindware: Tools for Smart Thinking*. New York: Farrar, Straus and Giroux; 2015. p. 240.
10. Brown DE. *Human Universals*. New York: McGraw-Hill; 1991.
11. Lakoff G, Johnson M. *Philosophy in the Flesh: The Embodied Mind and Its Challenge to Western Thought*. New York: Basic Books; 1999.
12. Pink Floyd. Comfortably Numb. *The Wall*. Harvest Records, United Kingdom. 1979.
13. Barkow J, Cosmides L, Tooby J. *The Adapted Mind: Evolutionary Psychology and the Generation of Culture*. Oxford: Oxford University Press; 1992.
14. Durrell L. *A Smile in the Mind's Eye*. London: Granada; 1980. p. 19.

7

Cognitive and Affective Biases, and Logical Failures

Pat Croskerry

CONTENTS

Introduction

It would be a simpler and safer world if all human thinking was objective, rational, consistent, and predictable, so that whenever we asked a particular question we would get comparable or similar answers. Unfortunately, this is not the case. For example, if a student in the Bachelor of Arts program at Tennessee Temple University was asked to explain the origins of human existence, his answer would likely be framed in terms of intelligent design, a young earth, and supernatural influences, whereas the same question to a biology major at the University of Copenhagen would likely elicit a response strongly couched in secular, scientific terms and perhaps framed with reference to Darwin and the theory of evolution (and a very much older earth). Despite coming from members of the same hominid species *Homo sapiens*, and from individuals of comparable intellect and age, the two explanations are radically different from each other. Yet, both individuals are strongly convinced that theirs is the true explanation. How is it that biologically similar organs can produce such different and conflicting outputs? The answer is that the human brain is fundamentally biased [1] and flexible. As wonderful and as powerful as the natural computer is that we have as a brain, at times its output may be completely flawed. Despite the anatomical and physiological congruence of human brains, they can come to function entirely differently. There may be some advantages to such variance. The Danish biologist might note that evolution needs variation to increase reproductive fitness, but thus far, alas, it does not appear to have selected for rationality [2].

Nature and Nurture

Two main sources of influence on cognitive aspects of brain function are generally accepted: nature and nurture. From the nature standpoint, just as there is clear evidence for the inheritance of height, habitus, eye color, and many other physical characteristics, there is mounting evidence that parts of our cognitive output are hardwired; that is, some of our decisions and behaviors are genetically determined. As noted in the previous chapter, this is the field of evolutionary psychology [3]. The basic tenet is that some of our present-day behaviors were biologically selected for survival in our distant evolutionary past, and may be transmitted through our DNA in modern brains. The nurture input, in contrast, is extra-genetic. Individuals acquire a variety of behaviors that are learned (tacitly or explicitly) through interaction with the environment. It is generally accepted that there is an ongoing interaction between the biological substrate and the environment leading to behaviors that will not be passed on to subsequent generations genetically.

Heuristics and Biases

Predictable constancies in our environment result in repetitive patterns in our everyday life and lead to specific dispositions to respond to these patterns. These set patterns of responding are referred to as *biases*, which in turn, may be due to heuristics. Both terms need explanation. Heuristics have been defined as "strategies that guide information search and modify problem representations to facilitate solutions" [4]; that is to say, they are information processing rules. Heuristics are usually an abbreviated form of decision making that provides an approximation or short-cut to get to a reasonable conclusion. Thus, instead of the laborious process of reasoning deductively through a series of premises, a heuristic may be employed that saves time and effort; for example, if I have to make a decision about whether or not to take an umbrella with me to work, I can muster everything I know about weather patterns, read some respected meteorological texts, study the cloud formations, review local geographical conditions, and make a prediction about whether it will rain or not. This will take considerable time and effort and may not even be reliable. Alternatively, I could look out into the city street and note how many people are carrying umbrellas. This is a heuristic behavior that relies on an observation that has likely proved reliable in the past. An example of a medical heuristic is: take extra care with elderly patients who appear sick but whose vital signs may be normal. We use a lot of heuristics in our everyday clinical decision making—they save us time and effort and for the most part, provide reasonable approximations of the likelihood that something will happen. When they work, they often go unnoticed, but when they fail, there is a tendency to call them biases.

The word "bias" has a negative connotation. We are aware of racial biases, ageism bias, gender bias, obesity bias, biases against psychiatric patients, biases against those with addictions, and others [5]. They all suggest some unjustified perception in the decision maker that may result in certain patients being disadvantaged or treated unfairly. The various heuristics that underlie these particular biases are not well founded, but those underlying other biases may well be. Interestingly, one of the original meanings of bias was simply to increase the likelihood of a useful outcome; for example, in lawn bowling, a popular sport in England, lawn bowls are deliberately constructed with a weight bias so that they will tend to veer to the right or the left. This feature of the bowl allows an increased variety of shots that can be played; that is, the biased ball can do more things than balls that only go in straight lines. So, under certain circumstances, having a bias might be preferred, because it serves some useful purpose. An example of a useful medical bias is always establishing a differential diagnosis rather than a solitary diagnosis on any patient. An example of a negative bias is to call off a search for further findings on an x-ray once a significant one has been found (search satisficing). The heuristic that underlies this particular bias is: once something acceptable has been found, the likelihood of finding another significant finding is very low and not worth the effort; therefore, call off the search. Many of our biased behaviors are positive and useful and support the well-being of our patients, but some will not. An important feature of heuristics and biases is that they are usually unconscious. They tend to be Type 1 processes; that is, they are reflexive and autonomous (see Chapter 3: Cognitive Approaches to the Diagnostic Process). Therefore, we may not be aware of them. This property makes teaching and educating about biases difficult. People may simply not recognize them in their own behavior, or if they do, may discount their importance, and be less inclined to teach about them.

Experimental research into heuristics and biases began in the mid-1970s with the classic work of Daniel Kahneman and Amos Tversky, two cognitive psychologists at Princeton. At the outset, they described only a handful of heuristics that could lead to biased judgment in decision making [6], but the proverbial cat was out of the bag. The number has risen steadily over the years, and new ones continue to be added as cognitive scientists delve ever deeper into the vagaries of human decision making. New biases are added if they describe a consistent and predictable pattern of departure from "normative decision making." By "normative" reasoning in decision making, psychologists are referring to the best decision that an individual might make if they were rational, capable of computing with a high degree of accuracy, well slept, appropriately motivated, and fully informed. Dobelli describes almost a hundred common biases for the layperson along with suggestions on how to deal with them [7], and Jenicek lists 110 of them in his *Medical Error and Harm: Understanding, Prevention, and Control* in 2010 [8]. Wikipedia currently lists 103 cognitive biases, 27 social biases, and 49 memory biases [9], which do not include many of those cited in other sources. Students new to the area may experience despair as the list continues to grow; however, in a recent chapter in a classic medical text on clinical decision making for medical students, Cooper focuses in on 12 common biases [10] extracted from an original list of 30 that were published in one of the earlier papers [11]. This seems to be a sensible approach to avoid the onus some may feel to memorize them. Knowing a dozen is to be well armed, and further alerts us to what is out there. A list of common cognitive biases is given in Appendix A.

It appears that all biases have some affective component; therefore, any cognitive and affective separation may be arbitrary. Often, however, biases will be characterized in this way for ease of typifying their main characteristics. It may be worth noting that our emotional reactions to patients often are our very first reactions, occurring reflexively and subsequently influencing information processing, judgment, and decision making [12]. Mamede et al. recently demonstrated that the simple insertion of a fragment of text into a clinical case scenario describing a patient's behavior as "difficult" was sufficient to reduce the accuracy of subsequent diagnostic decision making [13]. Extensive research into the nature and extent of biases has been conducted by cognitive scientists in the fields of psychology and sociology over the last 40 years. There is an extensive literature of books and literally hundreds of scientific papers on the topic. It has also captured the interest of the lay public, and a number of books targeted at this audience have appeared over the last 20 years [see 14–25] and continue to do so. At issue is the question of just how irrational the human brain can be. Although the distinguishing feature of our species is its rationality, it is readily apparent that some humans are more rational than others. There are significant individual differences (Chapter 9: Individual Variability in Clinical Decision Making and Diagnosis) and therefore a continuum of decision-making performance from those who are impulsive, quick, and less than fully informed at one end to those who are thoughtful, deliberate, and evidentiary at the other. Stanovich et al. [2] specifically define irrationality as the vulnerability to cognitive bias: "Degrees of rationality can be assessed in terms of the number and severity of such cognitive biases that individuals display. Conversely, failure to display a cognitive bias becomes a measure of rational thought."

Despite the protean and convincing findings from the field of cognitive science, there remains a significant minority who do not accept that the normal brain is biased with the potential to produce significant departures from rationality. There are abundant examples in everyday life: those who read astrology forecasts, the belief in alien abduction, tarot cards, extra-sensory perception, homeopathy, and other magical thinking in a variety of forms and guises. Historically, nonsecular extremist irrationality has led to human suffering and death on a scale that has few rivals in other belief systems. At a more granular level, there are many examples of suboptimal decision making. Stanovich cites numerous studies demonstrating failings in human rationality: "people's responses sometimes deviate from the performance considered normative on many reasoning tasks. For example, people assess probabilities incorrectly, they test hypotheses inefficiently, they violate the axioms of utility theory, they do not properly calibrate degrees of belief, their choices are affected by irrelevant context, they ignore the alternative hypothesis when evaluating data, and they display numerous other information processing biases" [2, p. 345]. The experimental demonstrations of irrationality in the earlier studies in the "heuristics and biases" literature were dismissed by some as the selective study of psychology students and/or a laboratory artifact—a "cartoonish" characterization that now appears to have been laid to rest [26], but not, apparently, to the satisfaction of all [27].

Logical Fallacies

When clinicians communicate with patients and when members of the medical team communicate with each other the exchange of information is mostly verbal. Our perceptions and understanding of what the other person is saying, together with their appearance and body language, are influenced by a number of factors that may bias our interpretation. For example, at a fairly basic level, what we remember about the information that is presented to us is determined by its serial position, such that we recall better what we hear first (primacy) and what we hear last (recency), as illustrated in Figure 7.1 [28]. Further, the juxtaposition of one idea or concept in a message to another idea may influence the receiver's interpretation of the second; the idea that is there are intermessage influences that are not simply based on recall. This is referred to as *verbal priming* and may be employed as a strategy to influence how a person may express their attitude or belief about something; for example, simply asking someone about the weather where they are may influence

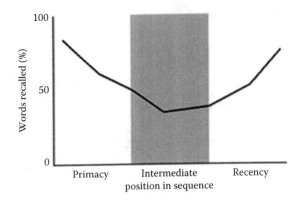

FIGURE 7.1
Recall is a U-shaped function dependent on serial position.

their response to a next, unrelated question [29]. However, there are less subtle biases than these in the context of the information presented that may distort our understanding. A patient might say, for example, that they have strained their left shoulder while mowing the lawn and would like an x-ray of the shoulder (see Box 14.1). The physician, however, may look beyond the patient's correlation and see instead the possibility of a causal relationship between exertion and referred pain to the left arm, and therefore a possible cardiac origin of the shoulder pain. The confusion of correlation with causation is an example of the *post hoc* logical fallacy; that is, the fact that B follows A does not mean A caused B. There are many examples of logical fallacies that lead to distorted reasoning, misunderstandings, and failures of communication (Appendix B) [30]. Some of the fallacies themselves are associated with cognitive and affective biases. For example, if a patient's complaint of pain is not believed because they are currently being treated with methadone for an opiate addiction, the physician may be engaging the *ad hominem* fallacy that discredits the source of the information rather than objectively interpreting the symptom of pain in the context of illness. Opiate-dependent patients on methadone may experience pain just as another patient does. This is an example of a social stigma or bias [5]. As with cognitive and affective biases, many people are simply unaware of them and may be led toward irrational conclusions.

Meliorists and Panglossians

In his discussion of The Great Rationality Debate, Stanovich describes the opposing sides as Meliorists and Panglossians [2]. The Meliorists argue that human rationality is not everything it could be but might be improved (ameliorated) by educational interventions. The Panglossians take an alternate view: that the characteristics described above do not demonstrate irrationality but rather, an optimal information processing adaptation [31–33]. Gigerenzer and his colleagues argue for the general adaptive utility of heuristics in everyday life [34,35], which few would disagree with, as most heuristics serve us well. However, as the philosopher Horton [36] notes, this everyday usefulness is, in turn, "consistent with the claim that such 'heuristics' sometimes lead us astray, and, in particular, with the claim that when we have time to reflect, and the question at hand is an important one, it is sensible to take account of the possibility that this may happen." This is exactly the point. Medical heuristics do have broad utility in clinical medicine, and in diagnostic reasoning in particular, but both vigilance and reflection are needed to make sure they do not lead us astray.

The Panglossian view,* emphasizing the "trusting one's instincts" approach, has undergone some resuscitation recently in the medical literature in an exchange of seemingly opposing views about intuitive versus analytical approaches toward clinical reasoning [38–41], although the polarization here appears to be more imagined than real. Part of the issue has been that different people have differing views of what intuition is (see Chapter 8: The Rational Diagnostician). The pro-intuition camp appears to take the view that intuition has been unduly vilified as an unreliable process that inevitably leads to diagnostic failure, although this claim does not appear to have been explicitly made by medical Meliorists. What the Meliorists have said is that most mistakes occur in the intuitive mode, and that we should expect this, as we spend most of our time there, compared with the analytical mode. Further, heuristics are working approximations and can be expected to fail occasionally, whereas the analytic method, executed deliberately, carefully, and following the rules of science, is less likely to fail. The consensus view of medical Meliorists, rather, is that medical decision making involves an equipoise of intuitive and analytical decision making that will be determined by ambient and contextual influences. The optimal approach appears to lie in promoting a finer tuning of intuition generally, recognizing its limitations and pitfalls and the dangers of affective and cognitive bias, conflict of interest, and logical failures [42], while retaining the option to exercise more deliberation in reasoning and problem solving if required. Thus, the view that clinical reasoning can be improved by cognitive interventions (e.g., critical thinking, rationality training) is an ameliorative one aimed at improving the performance of the intuitive system [43].

---

* The term appears to have been first used by Gould and Lewontin [37] and derives from the Professor of Philosophy Pangloss in Voltaire's *Candide* (1759). Pangloss famously inverted explanations: "Our noses were made to carry spectacles, so we have spectacles." The Panglossian view of rationality defines it as an evolutionary adaptation for optimal information processing, so that whatever human rationality has evolved to be is how rationality should be defined, and not how rationality experts would define it.

All disciplines and several subdisciplines in medicine have now acknowledged the impact of cognitive bias in clinical reasoning (for references, see Chapter 14). The consensus view is that clinical decision making is vulnerable to cognitive and affective biases, as well as logical failures, and is amenable to improvement. Importantly, recent papers have begun to address how bias might be detected and mitigated in specific disciplines such as radiology [44], pediatrics [45], neurology [46], neurosurgery [47], and forensic science [48], and in patient safety generally [49]. Concurrently, there is a growing awareness of the problem in such diverse areas as bryology [50], the business community [51], the judicial system [52], the National Aeronautics and Space Administration [53], U.S. foreign policy [54], the broader scientific community [55], and the Sirius program of the Intelligence Advanced Research Projects Activity (IARPA), an organization within the Office of the Director of U.S. National Intelligence [56]. All are keenly aware of the need to find solutions to the problems of cognitive bias. For medical decision making, too, it is probably fair to say that a consensus now prevails that there is room for improvement in physician performance just as there was when evidence-based medicine (EBM) began to emerge in the 1990s. In fact, Guyatt et al. [57] referred several times to the potential ameliorating effects of EBM strategies in clinical reasoning in those clinicians resistant to change. Interestingly, they also took the view that while EBM should de-emphasize intuition, it should not reject it, which concurs with the current Meliorist approach [43].

Impediments to Medical Meliorism

Although a number of fields of human endeavor have now embraced the Meliorist view [58], uptake has been slow in medicine until fairly recently. This has impeded our understanding of diagnostic failure. While a number of other impediments have been described [1,59], ironically, three particular biases have slowed the uptake of Medical Meliorism: they are the Not Invented Here (NIH) bias, blind spot bias, and myside bias [60].

Not Invented Here Bias: medicine has a fairly long history of following its own path and doing its own thing. Clinicians were making a living practicing medicine long before the Enlightenment and the study of pure sciences emerged. The imperative to care for the sick dates back to the emergence of our "conscious" brains many thousands of years ago. There is evidence, for example, that Neanderthals cared for their physically crippled, which presumably went beyond simply providing food and shelter. Specialists in caring for the sick (the earliest diagnosticians) in the form of shamans eventually emerged. Archaeological evidence of the existence of shamans dates back at least 30,000 years (see Figure 7.2) [61]. A long history of self-sufficiency

FIGURE 7.2
The earliest known depiction of a Siberian shaman, by the Dutch explorer Witsen in 1692, in his account of his travels among the Samoyedic- and Tungusic-speaking peoples. Labeling the figure a "Priest of the Devil," he gave it clawed feet to characterize its demonic nature. (From Wikipedia, https://en.wikipedia.org/wiki/Shamanism [61].)

in medical problem solving within medicine may have accounted for the disinclination of medical practitioners to look outside of the profession for help in understanding how we make decisions. Part of the reluctance to accept the findings from cognitive science on decision theory may have stemmed from the fact that medical scientists did not invent the approach. Although times are changing, a significant number of modern practitioners still believe that cognitive science is not relevant to modern clinical decision making. However, in many areas of medicine now, there is a new openness and willingness to embrace its findings. There is not only an increased vigilance in detecting biases in clinical decision making, but also a growing armamentarium of strategies to mitigate them (see Chapter 15: Cognitive Bias Mitigation: Becoming Better Diagnosticians).

Blind Spot Bias: even though many people can detect bias in the decision making of others, they may not be so vigilant or effective in detecting it in themselves. This metabias, which shows considerable individual variation [62], was originally described by Pronin et al. [63] as the bias blind spot. Our assessments of ourselves are based largely on our thoughts and feelings, whereas our assessments of others are based on their external observable behavior [64]. We have a strong tendency to believe that our own perceptions reflect true reality (naïve realism), while those of others are seen to be biased by self-interest, personal allegiances, an emphasis on dispositional rather than situational explanations (fundamental attribution error), and other factors. In the process, we are less likely to follow the advice of others [65], although it does not appear to make one any less competent at decision making overall [62]. An important question is whether or not vulnerability to biases makes it more or less likely to exhibit blind spot bias. Testing people on six common cognitive biases, West et al. [66] found that cognitive sophistication did not reduce vulnerability to blind spot bias; however, Scopoletti et al. [62] did report that a higher susceptibility to bias blind spot appeared to be a barrier to bias mitigation training for the robust bias fundamental attribution error. An important aspect of bias blind spot is that those who are more susceptible to it are *less* likely to engage in strategies to improve their decision making through taking the advice of others or by corrective training, and may be resistant to bias mitigation training [62].

Myside Bias: this is a tendency by people to "evaluate evidence, generate evidence, and test hypotheses in a manner biased toward their own prior opinions and attitudes" [67]. Thus, people may gather information selectively, and recall information selectively, that supports their own viewpoint. It is a form of ascertainment bias, in that people see what they want to see. It does not appear to be related to intelligence. It is of particular concern for decision making, as a critical feature of critical thinking is the ability of the decision maker to separate out their own beliefs and opinions in their evaluation of the evidence and the arguments around an issue. The bias is an impediment at both the individual and the group level. It is not difficult to see the problems it generates at an individual level, but the effects of myside bias may be exacerbated in a group situation. For example, a football fan may see no failings in his home team but is keenly aware of a multitude of egregious behaviors in the opposing team. It is not hard to imagine that a group of fans will augment the bias through various groupthink [68] phenomena; that is, the strength of an individual's myside bias may be amplified by group membership. This is similar to the biased thinking that may occur in members of an affinity group, known as *identity protective cognition* [69]. Thus, an individual may interpret evidence in such a way that it is consistent with, and aligns with, the values of the particular group to which they perceive themselves as belonging. Lest anyone think that intellectually sophisticated people are above such "groupthink" vulnerabilities, it appears that (just as for the blind spot bias [62]) the more cognitively able members of the group are more likely to interpret evidence in such a biased fashion [70,71]. In the case of the Meliorists and Panglossians, then, individuals may take strength from their respective group and polarize themselves further than they might otherwise have done on their own. Concerns have been expressed that this polarization may lead to deadlock, such that progress in understanding the processes of clinical reasoning may be impaired [72]. Though other contributory factors may have impeded the uptake of cognitive science into medical decision making, the combination of these three specific and independent biases may have been a particularly potent challenge. Those who are reluctant to accept the impact of biases on clinical decision making would have been especially vulnerable.

Conclusions

About 10 years ago, David Eddy, who did much to set the stage for evidence-based medicine, said of medical decision making in the 1970s: "Up until about 40 years ago, medical decisions were doing very well on their own, or so people thought. The complacency was based on a fundamental assumption that through the rigors of medical education, followed by continuing education, journals, individual experiences, and exposure to colleagues, each physician always thought the right thoughts and did the right things. The idea was that when a physician faced a patient, by some fundamentally human process called the 'art of medicine' or 'clinical judgment,' the physician would synthesize all of the important information about the patient, relevant research, and experiences with previous patients to determine the best course of action. 'Medical decision making' as a field worthy of study did not exist" [73].

It might be argued that in some training programs, nothing much has changed, and Eddy's status quo prevails. However, in others, there does now appear to be a growing awareness of the need to study clinical decision making and all the processes that it involves, as part of the general skill domain of the clinician within the complex arena of evidence-based decision making [74]. Interestingly, the resistance from those who see biases as natural and optimizing factors in decision making is reminiscent of the

SUMMARY POINTS

- There is substantial variation in the ways in which different people see things. The normal operating function of the brain is biased.

- The two major influences on the way in which a particular brain functions are genetic (what we inherit from our parents) and environmental (what we learn from the environment as the brain matures).

- Abundant studies demonstrate that heuristics (mental short-cuts) and biases (heuristics and other predictable failures of rationality) have significant impact on our decision making.

- The ways in which we reason about things may be logical or otherwise. Logical failures in the ways in which people handle information are common.

- The Great Rationality Debate refers to a polarization of attitudes toward heuristics and biases. The major opposing groups are referred to as *Meliorists* (reasoning and decision making is often suboptimal but can be improved on) and *Panglossians* (the negative impact of bias is exaggerated, and many biases are useful).

- Three biases in particular have contributed to the polarization between Meliorists and Panglossians: the *not-invented-here bias*, the *blind spot bias*, and the *myside bias*.
- Although progress has been slow, there now appears to be a growing awareness of the imperative to understand the multiple and complex processes that underlie clinical reasoning and decision making.

polarization against scientific enlightenment that came from the Romantics in the early nineteenth century. The last 20 years have seen what might be termed a *Medical Enlightenment*, with a growing understanding that cognitive and affective biases, logical failures, and any other factors that interfere with rationality rightfully belong in the study of clinical reasoning and decision making.

References

1. Croskerry P. Bias: A normal operating characteristic of the diagnosing brain. *Diagnosis*. 2014 Jan;1(1):23–7.
2. Stanovich KE. On the distinction between rationality and intelligence: Implications for understanding individual differences in reasoning. In: Holyoak KJ, Morrison RG, editors. *The Oxford Handbook of Thinking and Reasoning*. New York: Oxford University Press; 2012. pp. 343–65.
3. Barkow JH, Cosmides L, Tooby J. *The Adapted Mind: Evolutionary Psychology and the Generation of Culture*. New York: Oxford University Press; 1992.
4. Goldstein DG, Gigerenzer G. Models of ecological rationality: The recognition heuristic. *Psychol Rev*. 2002 Jan;109(1):75–90.
5. Croskerry P, Nimmo GR. Better clinical decision making and reducing diagnostic error. *J R Coll Physicians Edinb*. 2011 Jun;41(2):155–62.
6. Tversky A, Kahneman D. Judgment under uncertainty: Heuristics and biases. *Science*. 1974 Sep 27;185(4157):1124–31.
7. Dobelli R. *The Art of Thinking Clearly*. New York: HarperCollins; 2013.
8. Jenicek M. *Medical Error and Harm: Understanding, Prevention and Control*. New York: Productivity Press; 2011.
9. Available at https://en.wikipedia.org/wiki/List_of_cognitive_biases. Accessed March 4, 2017.
10. Cooper N. Clinical decision making in medicine. In: *Davidson's Principles and Practice of Medicine*, 23rd ed. (forthcoming).
11. Croskerry P. The importance of cognitive errors in diagnosis and strategies to minimize them. *Acad Med*. 2003 Aug;78(8):775–80.
12. Zajonc RB. Feeling and thinking: Preferences need no inferences. *Am Psychol*. 1980 Feb; 35(2):151–75.

13. Mamede S, Van Gog T, Schuit SC, Van den Berge K, Van Daele PL, Bueving H, Van der Zee T, Van den Broek WW, Van Saase JLCM, Schmidt HG. Why patients' disruptive behaviours impair diagnostic reasoning: A randomised experiment. *BMJ Qual Saf.* 2017; 26: 13–18.

14. Konnikova M. *Mastermind: How to Think Like Sherlock Holmes*. New York: Viking; 2013.

15. Markman A. *Smart Thinking; Three Essential Keys to Solve Problems, Innovate, and Get Things Done*. New York: Perigree; 2012.

16. Chabris C, Simons D. *The Invisible Gorilla. How Our Intuitions Deceive Us*. New York: Crown; 2010.

17. Lehrer J. *How We Decide*. New York: Houghton Mifflin Harcourt; 2009.

18. Hallinan JT. *Why We Make Mistakes: How We Look without Seeing, Forget Things in Seconds, and Are All Pretty Sure We Are Way above Average*. New York: Broadway Books; 2009.

19. Thaler RH, Sunstein CR. *Nudge: Improving Decisions about Health, Wealth, and Happiness*. New York: Penguin; 2008.

20. Brafman O, Brafman R. *Sway: The Irresistible Pull of Irrational Behavior*. New York: Broadway Books; 2008.

21. Ariely D. *Predictably Irrational: The Hidden Forces that Shape Our Decisions*. New York: HarperCollins; 2008.

22. Wolpert L. *Six Impossible Things before Breakfast: The Evolutionary Origins of Belief*. New York: W.W. Norton; 2007.

23. Tavris C, Aronson E. *Mistakes Were Made (But Not by Me): Why We Justify Foolish Beliefs, Bad Decisions and Hurtful Acts*. Orlando: Harcourt Inc.; 2007.

24. Myers DG. *Intuition: Its Powers and Perils*. New Haven, CT: Yale University Press; 2002.

25. Piattelli-Palmarini M. *Inevitable Illusions: How Mistakes of Reason Rule Our Minds*. New York: Wiley; 1994.

26. Kahneman D, Klein G. Conditions for intuitive expertise: A failure to disagree. *Am Psychol.* 2009 Sep; 64(6):515–26.

27. Norman G. Editorial: A bridge too far. *Adv in Health Sci Educ.* 2016; 21:251–6.

28. Colman A. *Dictionary of Psychology*, 2nd ed. New York: Oxford University Press; 2006. p. 688.

29. Schwarz N, Clore GL. Mood, misattribution, and judgments of well-being: Informative and directive functions of affective states. *J Pers Soc Psychol.* 1983;45(3):513–23.

30. Available at: http://changingminds.org/disciplines/argument/fallacies/falla-cies_alpha.htm. Accessed April 24, 2016.

31. Cosmides L, Tooby J. Are humans good intuitive statisticians after all? Rethinking some conclusions from the literature on judgment under uncertainty. *Cognition.* 1996;58:1–73.

32. Gigerenzer G. *Gut Feelings: The Intelligence of the Unconscious*. New York: Viking Penguin; 2007.

33. Marewski JN, Gaissmaie W, Gigerenzer G. Good judgments do not require complex cognition. *Cogn Process.* 2010 May;11(2):103–21.

34. Gigerenzer G, Todd PM, ABC Research Group. *Simple Heuristics That Make Us Smart*. Oxford: Oxford University Press; 1999.

35. Gigerenzer G. *Adaptive Thinking: Rationality in the Real World*. Oxford: Oxford University Press; 2000.

36. Horton K. Aid and bias. Inquiry: An interdisciplinary. *J Phil*. 2004:47(6):545–61.
37. Gould SJ, Lewontin RC. The spandrels of San Marco and the Panglossian paradigm: A critique of the adaptionist programme. *Proc R Soc Lond*. 1979;205(1161):581–98.
38. Norman GR, Eva KW. Diagnostic error and clinical reasoning. *Med Educ*. 2010 Jan;44(1):94–100.
39. lgen JS, Bowen JL, McIntyre LA, Banh KV, Barnes D, Coates WC, Druck J, Fix ML, Rimple D, Yarris LM, Eva KW. Comparing diagnostic performance and the utility of clinical vignette-based assessment under testing conditions designed to encourage either automatic or analytic thought. *Acad Med*. 2013 Oct;88(10):1545–51.
40. McLaughlin K, Eva KW, Norman GR. Reexamining our bias against heuristics. *Adv Health Sci Educ Theory Pract*. 2014 Aug;19(3):457–64.
41. Dhaliwal G. Premature closure? Not so fast. *BMJ Qual Saf*. 2017 Feb;26(2):87–9.
42. Seshia SS, Makhinson M, Phillips DF, Young GB. Evidence-informed person-centered healthcare part I: Do "cognitive biases plus" at organizational levels influence quality of evidence? *J Eval Clin Pract*. 2014 Dec;20(6):734–47.
43. Croskerry P, Petrie DA, Reilly JB, Tait G. Deciding about fast and slow decisions. *Acad Med*. 2014 Feb;89(2):197–200.
44. Bruno MA, Walker EA, Abujudeh HH. Understanding and confronting our mistakes: The epidemiology of error in radiology and strategies for error reduction. *Radiographics*. 2015 Oct;35(6):1668–76.
45. Jenkins MM, Youngstrom EA. A randomized controlled trial of cognitive debiasing improves assessment and treatment selection for pediatric bipolar disorder. *J Consult Clin Psychol*. 2016 Apr;84(4):323–33.
46. Vickrey BG, Samuels MA, Ropper AH. How neurologists think: A cognitive psychology perspective on missed diagnoses. *Ann Neurol*. 2010 Apr;67(4):425–33.
47. Fargen KM, Friedman WA. The science of medical decision making: Neurosurgery, errors, and personal cognitive strategies for improving quality of care. *World Neurosurg*. 2014 Jul–Aug;82(1–2):e21–29.
48. Dror IE, Thompson WC, Meissner CA, Kornfield I, Krane D, Saks M, Risinger M. Letter to the editor—Context management toolbox: A linear sequential unmasking (LSU) approach for minimizing cognitive bias in forensic decision making. *J Forensic Sci*. 2015 Jul;60(4):1111–12.
49. Sibinga EM, Wu AW. Clinician mindfulness and patient safety. *JAMA*. 2010 Dec 8;304(22): 2532–3.
50. Sleath J. The dual process theory as applied to bryological identification. *Field Bryology*. 2011 Nov;105:33–6.
51. Stenner T. Why the average investor is so bad at it. *The Globe and Mail*. July 1, 2016. Available at: www.theglobeandmail.com/globe-investor/investor-education/why-the-average-investor-is-so-bad-at-it/article30728320/. Accessed September 26, 2016.
52. Guthrie C, Rachlinski JJ, Wistrich AJ. Blinking on the Bench: How Judges Decide Cases. 2007; Cornell Law Faculty Publications. Paper 917. Available at: http://scholarship.law.cornell.edu/facpub/917. Accessed September 26, 2016.
53. SIDM wins NASA grant. Available at: www.improvediagnosis.site-ym.com. Accessed October 12, 2016.
54. Yetiv SA. *National Security through a Cockeyed Lens: How Cognitive Bias Impacts U.S. Foreign Policy*. Baltimore, MD: Johns Hopkins University Press; 2013.

55. Editorial. Let's think about cognitive bias. The human brain's habit of finding what it wants to find is a key problem for research. Establishing robust methods to avoid such bias will make results more reproducible. *Nature*. 2015 Oct 7; 526(7572):163.
56. Sirius program of IARPA: www.iarpa.gov/index.php/research-programs/sirius.
57. Guyatt G, Cairns J, Churchill D, et al. Evidence-Based Medicine Working Group. Evidence-Based Medicine. A new approach to teaching the practice of medicine. *JAMA*. 1992 Nov 4;268(17):2420–5.
58. Nuzzo R. How scientists fool themselves—and how they can stop: Humans are remarkably good at self-deception—But growing concern about reproducibility is driving many researchers to seek ways to fight their own worst instincts. *Nature*. 2015 Oct 7;526(7572):182–5.
59. Croskerry P. Perspectives on diagnostic failure and patient safety. *Healthc Q*. 2012;15 Spec No. 50-6.
60. Croskerry P. Our better angels and black boxes. *Emerg Med J*. 2016 Apr;33:242–4.
61. https://en.wikipedia.org/wiki/Shamanism.
62. Scopoletti I, Morewedge CK, McCormick E, Min HL, Lebrecht S, Kassam KS. Bias blind spot: Structure, measurement, and consequences. *Manag Sci*. 2015;61(10):2468–86.
63. Pronin E, Lin DY, Ross L. The bias blind spot: perceptions of bias in self versus others. *Pers Soc Psychol Bull*. 2002 Mar;28(3):369–81.
64. Pronin E. How we see ourselves and how we see others. *Science*. 2008 May 30;320(5880):1177–80.
65. Liberman V, Minson JA, Bryan CJ, Ross L. Naïve realism and capturing the "wisdom of dyads." *J Experiment Soc Psych*. 2012 Mar;48(2):507–12.
66. West RF, Meserve RJ, Stanovich KE. Cognitive sophistication does not attenuate the bias blind spot. *J Pers Soc Psychol*. 2012 Sep;103(3):506–19.
67. Stanovich KE, West RF, Toplak ME. Myside bias, rational thinking, and intelligence. *Curr Dir Psychol Sci*. 2013 Aug;22(4):259–64.
68. Whyte WH. Groupthink. *Fortune Magazine*. 1952.
69. Kahan DM, Braman D, Gastil J, Slovic P, Mertz CK. Culture and identity-protective cognition: Explaining the white male effect in risk perception. *J Empirical Legal Stud*. 2007 Nov;4(3):465–505.
70. Kahan DM. Ideology, motivated reasoning, and cognitive reflection. *Judgm Decis Mak*. 2013 Jul;8(4):407–24.
71. Taber CS, Lodge M. Motivated skepticism in the evaluation of political beliefs. *Am J Pol Sci*. 2006 Jul;50(3):755–69.
72. Rotgans JI, Low-Beer N, Rosby VL. The relevance of neuroscientific research for understanding clinical reasoning. *Health Professions Education*. 2016;2:1–2.
73. Eddy DM. Evidence-based medicine: a unified approach. *Health Aff (Millwood)*. 2005 Jan–Feb;24(1):9–17.
74. Seshia SS, Makhinson M, Young GB. "Cognitive biases plus": Covert subverters of healthcare evidence. *Evid Based Med*. 2016 Apr;21(2):41–5.

8

The Rational Diagnostician

Pat Croskerry

CONTENTS

Introduction

Within the last decade or so, significant advances in cognitive science have increased our understanding of what it is to be rational in medical decision making. A consensus has emerged that rationality is the quintessential characteristic of the well-calibrated decision maker [1]. Rather than decision makers being seen as either rational or not, it appears instead that rationality is normative; that is, we have an understanding of what is generally considered to be the normal or correct way of doing something with reference to an ideal standard or model. Thus, rationality is a variable individual characteristic, such that we can expect some diagnosticians to be more rational than others.

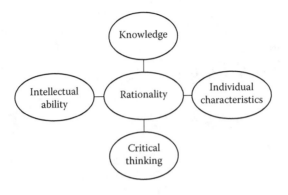

FIGURE 8.1
Components of rationality.

One's capacity for rationality appears to be influenced by several major components: extent of medical knowledge, intellectual abilities, critical thinking, and individual characteristics (Figure 8.1).

Components of Rationality

If a clinician has limited medical knowledge about a disease, then no amount of cognitive effort can yield a rational decision about it. Depth of knowledge is important, and there is little room for maneuver around this in medicine. Inadequate knowledge of a topic will compromise decision making. In terms of the etiology of diagnostic failure, however, several studies have found that physician knowledge deficits are not nearly as significant as some might think. Most diagnostic errors involve conditions that are common, and about which physicians have considerable knowledge and familiarity. This is supported by studies in primary care [2], general medicine [3], emergency medicine [4], hospitalized patients [5], and in the intensive care setting (ICU) [6]. In the ICU, where patients are generally sicker, and where diagnostic assessment would be expected to be more aggressive and thorough, the results of 5863 autopsies over a 45-year period found that of the autopsies identified with a Class I diagnostic error (major missed diagnosis that likely was the cause of death or contributed to the death), pulmonary embolus, myocardial infarction, pneumonia, and aspergillosis were the most common missed diagnoses [6]. There are a number of reasons that might explain how common things can be (and are) missed. Many illnesses are complex, and patients often have multiple problems; signs and symptoms of one condition may well overlap those of another. Diseases may have atypical manifestations. Perhaps our understanding of disease is not completely developed, such that even our science fails to give us the tools and methods we need. But it also appears that it isn't what the clinician doesn't know that leads to diagnostic failure, but rather, the way the clinician thinks. Historically, medical education has done a good job of imparting the necessary knowledge to would-be practitioners but has been less effective at addressing the question of how they should subsequently think about the knowledge they have painstakingly acquired (see Chapter 14: Medical Education and the Diagnostic Process).

Being rational also requires a certain degree of intellectual ability. Generally, a higher level of intelligence is associated with being able to grasp things more readily, having a broader knowledge about the world and how things work, and perhaps finding studying less arduous. However, there are again caveats. Some studies show that intelligent people can exercise poor judgment. While tests that measure intelligence quotient (IQ) are widely regarded by the lay public and by many scientists as tests of all cognitive abilities and generally of "good thinking" [7], there are abundant examples of intelligent people not being so smart. In a survey of members of Canadian Mensa (membership requires an IQ in the 98th percentile or higher), 51% believed in biorhythms, 44% in astrology, and 56% in visitors from outer space [8]. None of these topics is supported by valid evidence. Such failure to think rationally despite a normal intelligence is referred to as *dysrationalia* [9]. As noted in Chapter 9 (Individual Variability in Clinical Decision Making and Diagnosis), those who pursue a career in medicine will score in the upper range on tests that measure IQ, and so deficiencies in thinking are unlikely due to overall intellectual deficiencies. But, as Stanovich notes, IQ does not guarantee rationality, which is considered superordinate to IQ [10] and is a more desirable attribute than scoring well on IQ tests. So, an above average IQ (100 is defined as the average) may be a necessary but insufficient attribute for the level of rationality required to be a good diagnostician.

Many aspects of critical thinking (CT) are desirable and would be expected to support rationality, but CT, too, qualifies as another necessary but not sufficient condition for good rational thinking. Despite someone having acquired good CT habits (e.g., being precise, relevant, accurate, and logical), there is no guarantee that sufficient account has been taken of training other cognitive attributes (probabilistic and statistical reasoning, scientific reasoning skills, rules of logical consistency and validity) that are known to be important for sound decision making [10]. So, critical thinkers may still make erroneous judgments. Thus, it appears that critical thinking is "a subspecies of rational thinking" [10].

The fourth major factor is individual variability. Despite being an intelligent, critical thinker with a sound knowledge base, there is still considerable variability due to individual characteristics that have to do with gender, age, personality factors, traits, cognitive styles, and other variables. These are reviewed in the following chapter (Chapter 9: Individual Variability in Clinical Decision Making and Diagnosis). One in particular, actively open-minded thinking (AOMT), is highly correlated with rational thinking ability [1].

Medical Rationality

Rationality is defined as "the quality or state of being reasonable, based on facts or reason" and implies conformity of one's beliefs with one's reasons to believe, or of one's actions with one's reasons for action. "Rationality" has different specialized meanings in economics, sociology, psychology, evolutionary biology, and political science [11]. In medicine, Hippocrates was responsible for the early transition away from the divine toward "rational" decision making, and over the ensuing two millennia, we have gradually moved closer toward a secular understanding of what it is to make rational medical decisions. But the understanding of "medical rationality" varies considerably within medicine, and many researchers and educators still have differing views of the concept.

Like Stanovich [1], Bornstein and Emler [12] equated irrational decision making with the influence of biases and proposed that improved rationality might be achieved with an increased awareness of bias. Cognitive debiasing training, therefore, might mitigate the influence of biases and improve rationality. They further pointed out that evidence-based medicine (EBM) itself would facilitate improved decision making: "by providing the most relevant and objective empirical information available, and incorporating it with clinical expertise, test results and patient preferences, many of the biases associated with doctors' relying too heavily on intuition and selectively attending to some information while ignoring other relevant information could be avoided." In contrast, the treatment of rationality by Rao in *Rational Medical Decision Making* [13] contains no reference to biases other than those associated with study design: allocation of subjects and bias in the preparation of systematic reviews (publication bias, citation bias). There is no discussion of any of the common cognitive biases anywhere in the book. This is not to disparage Rao's thorough treatment of the important area of clinical epidemiology and biostatistics, that is, quantitative medical decision making, which, as noted in Chapters 2 and 14 (Medical Decision Making; Medical Education and the Diagnostic Process), is a vital aspect of clinical decision making, but it serves to illustrate the widely differing views of what it is to be rational in medicine.

A similar comment can be made about a long-running series in the *Journal of the American Medical Association* (*JAMA*) titled *The Rational Clinical Examination*. Over the years since its inception in 1993, this series has provided an invaluable, up-to-date, evidence-based review of the approach toward the diagnosis and management of a number of common medical conditions, and is available in book form [14]. The focus is on the appropriate interpretation of patient history, symptoms, and signs, emphasizing precision and accuracy. However, the issue of the rationality of the clinician's thinking is not explicitly addressed. All clinical examinations include taking and thinking about the patient's history, forming an impression of the patient, and evaluating his or her reliability and credibility, which may all have a significant influence on the subsequent physical examination. In fact, in the first article of the *JAMA* series, Sackett recognized this problem in a patient referred to an orthopedic clinic on whom he could smell alcohol: "For an item of the clinical history or physical examination to be accurate, it first must be precise. That is, we need to have some confidence that 2 clinicians examining the same, unchanged patient would agree with each other on the presence or absence of the symptom (such as our patient's answer to one of the CAGE questions) or sign (such as the presence of spider nevi on our patient's chest). The precision (often appearing under the name of 'observer variation' in the clinical literature) of such clinical findings can be quantitated" [15].

It seems likely that at least some observer variation will be explained by individual variation in rationality, referred to above. In terms of precision in eliciting the patient's clinical history, much can be won or lost at the very outset of the clinician–patient encounter. For example, if a patient is simply described as "difficult" [16], or even if the patient smiles [17], this can result in a change in attitude toward the patient that may influence diagnostic outcomes. Thus, although it may not seem immediately relevant to the rational physical exam, the physician's rational approach toward taking the patient's clinical history would seem to be very important. Factors that bear on individual rationality, various facilitators and inhibitors (see below) as well as patient characteristics, and ambient factors (cognitive load, time of day, teamwork issues, and ergonomic features of the workplace) will influence patient assessment.

Types of Rationality

Which aspects of rationality, then, are most relevant to the diagnostic process? The view taken in this and other chapters in this book is that while it is important and necessary for clinical practitioners to acquire the special knowledge and expertise of clinical epidemiology and biostatistics, it is also extremely important to train the appropriate skills to overcome the cognitive failures described in Chapter 7 (Cognitive and Affective Biases, and Logical Failures). The sources of expertise for such rationality come from the field of cognitive science. Thus, for well-calibrated decision making in medicine, we need both.

Rationality, one of the most important characteristics of human cognition, comes in two forms: *instrumental* and *epistemic* (or *evidential*) rationality. Instrumental rationality describes thinking behaviors that get us what we most want given the resources available to us [18]. So, in the present context, these would be those thinking behaviors that achieve the most accurate diagnosis for a patient given the resources available to us. Given that our overarching goal is accurate and timely diagnosis, we can say it is the thinking behaviors that optimize that goal. The other aspect of rationality is epistemic rationality, which describes how well our beliefs map into the real world; that is, holding beliefs "commensurate with available evidence" [10]. Thus, to achieve our goal of making an accurate diagnosis, our actions need to be based on beliefs that are properly calibrated to the evidential nature of disease and its manifestations [14]. This assumes, of course, that the quality of evidence (and thus the quality of our science and facts that we accept as true and use to make decisions) has not been compromised (see Chapter 15: Cognitive Bias Mitigation: Becoming Better Diagnosticians) by factors outside the decision maker's control. The two types of rationality can be remembered in practical terms as *what is true* (epistemic) and *what to do* (instrumental), each integrated with the other.

Stanovich has described important elements of rationality that support sound clinical decision making [10]. Attributes of rationality are listed in Table 8.1. In the left column are those that make up Stanovich's conceptual structure of rational thought [10]. They emphasize that rationality is not a single mental construct but is instead heterogeneous. Rationality is seen to be made up of an assortment of cognitive styles and dispositions. It is not difficult to find clinical examples for each of the specific attributes, some of which are shown in the right column. These describe the operating characteristics and skill set that many would hope to see in a rational physician.

In clinical decision making, some factors will help to improve rationality (facilitators) whereas others will degrade it (inhibitors). A variety of facilitators that serve to optimize rationality are reviewed in Table 8.2. Importantly, rationality may be facilitated through access to specific domains of knowledge (mindware) that have been acquired by the decision maker through specialized training [19].

Mindware

Mindware is a key concept originally used by David Perkins to describe rules, procedures, and other forms of knowledge that are stored in memory and can be retrieved for decision making and problem solving [20]. Nisbett has explored the concept in depth recently, covering a wide range of essential mindware tools for optimal decision making [19], many of which may be construed as facilitators of rational thinking.

Facilitators of rationality are explicitly taught in courses of medical decision making in clinical epidemiology and biostatistics (Chapter 14: Medical Education and the Diagnostic Process), and many of the examples of physician failures in clinical decision making described in some detail by Gigerenzer [21] are due to inadequate knowledge or failed training in these, or failed retention.

TABLE 8.1

Examples of Preferred Attributes of the Rational Decision Maker

Attribute	Clinical Behavior
Resistance to miserly information processing.	Make the effort, and allow adequate time, to obtain a full history of illness from the patient or from a collateral source. Avoid search satisficing, premature closure, status quo bias. Resist base rate neglect. Judge risks and benefits independently.
Resistance to myside thinking. Accurate self-evaluation.	Avoid overconfidence. Be fair to the patient. Hold a broad perspective on diagnostic possibilities. Recognize own limitations and biases. Recognize alternate possibilities and viewpoints.
Absence of irrelevant context effects.	Be aware of influence of context. Ignore distracting patient characteristics, avoid stereotyping: for example, use of terminology such as "frequent flyer." Be aware of framing and anchoring effects.
Belief flexibility. Active open-minded thinking (AOMT).	Construct comprehensive differential diagnosis. Be willing to spend more time on thinking through diagnostic process and changing if necessary. Revising an initial diagnosis is not a sign of weakness.
Value placed on reason and truth.	Believe that following objective and logical reasoning is key to optimizing diagnosis. Seek unbiased evidence. Value the process of critical thinking in the diagnostic process. Understand principles of rationality.
Tendency to seek information, enjoy thought, and fully process information.	Accept and engage the intellectual challenge of diagnostic reasoning. Actively involve the patient. Avoid the cursory exam. Elicit a comprehensive history from the patient and perform an appropriate exam. Engage team members.
Objective reasoning styles.	Be aware of and avoid bias in reasoning. Distinguish facts from opinions and evidence from narrative. Know the logical fallacies.
Sensitivity to contradiction; tendency to seek consistency in belief and argument.	Don't be dismissive of patient's and/or colleagues' input. Ensure congruence of personal beliefs with objective reasoning about patient's illness. Actively listen and make an effort to see both sides of an argument.
Sense of self-efficacy.	Believe in one's own ability and worth in clinical reasoning. Be willing and have the courage to challenge authority gradients. Value your integrity.
Prudently discount the future.	Rule out worst-case scenario to ensure it is not missed. Be prepared to imagine what else the diagnosis might be if not the one you have selected. Provide patient with options if not getting better; that is, consider the possibility of the diagnosis being incorrect.
Self-control skills.	Recognize and deal constructively with visceral bias against certain patients, colleagues, and team members. Relinquish immediate gain from rushing patients and minimizing their complaints versus the longer-term satisfaction of doing the job thoroughly and conscientiously to maximize patient safety.
Fine-grained, controlled emotional regulation.	Be aware of own emotional state and reasons for it. Recognize wide range of affective responsivity required in dealing with patients. Observe principles of mindfulness.
Emotional regulation related to reward.	Think about and identify one's own feelings in a clinical interaction and prioritize the patient's interests over one's own. Take satisfaction in untangling complex diagnostic challenges for their own sake. Avoid cherry picking "easy" patients.

Source: Adapted from Stanovich, K.E. et al., *Cambridge Handbook of Intelligence*, Cambridge University Press, Cambridge, UK, 2012.

TABLE 8.2

Clinical Examples of Facilitators of Rational Thought

Facilitators	Clinical Examples
Probabilistic reasoning	Know clinical epidemiology and biostatistics: understand Bayesian reasoning, importance of sample size and bias, awareness of base rate, sensitivity and specificity, test characteristics, likelihood ratios, number needed to treat, and so on.
Qualitative decision theory insights	Understand the issue of uncertainty in medicine, that quantitative decisions are not always possible, that full rationality may be limited by available information; adopt general decision rules that at least allow the exclusion of must not miss diagnoses and allow open-ended diagnoses; for example, chest pain not-yet-diagnosed (NYD). Take context into account.
Knowledge of scientific reasoning	Basic principles of experimental design, selection of sample and size, need for control groups, blinding, causal variable isolation; understand placebo effects, covariation, belief bias effects, hypothesis generation and testing, importance of falsifiability of a hypothesis; appreciate converging evidence, limitations of personal observation and of single case histories, perception of risk, and so on.
Rules of logical consistency and validity	Be aware of the main logical fallacies; for example, distinguish correlation and causation, avoid *ad hominem* reasoning (some addicted patients have genuine complaints and are not drug seeking), and others.
Economic thinking	Avoid sunk costs in diagnosis, understand gains and losses in treatments, understand spontaneous regression of disease, choose wise strategies in test ordering, and so on.

Source: Adapted from Stanovich, K.E. et al., *Cambridge Handbook of Intelligence*, Cambridge University Press, Cambridge, UK, 2012.

Rationality can be degraded or inhibited. Two main sources of inhibition are *processing* problems and *content* problems (Figure 8.2). All information accessed by the decision maker needs to be processed appropriately. If the decision maker behaves in a miserly fashion (which may be influenced by individual or ambient factors), then rationality may be compromised. Treating information superficially, assuming what you see is all there is (WYSIATI) [22], or not applying sufficient cognitive effort may lead to reduced rational performance. If, instead, cognitive effort is appropriate to the task, but the decision maker is functioning with various *mindware gaps* (for example, deficits in problem-specific knowledge, deficiencies in scientific thinking, lacking knowledge about probability theory, fallacious reasoning, and other factors), then rationality will again be compromised [23]. Another possibility is that the decision maker does have mindware available to solve the problem at hand, but it is contaminated and thus ill-suited to the task.

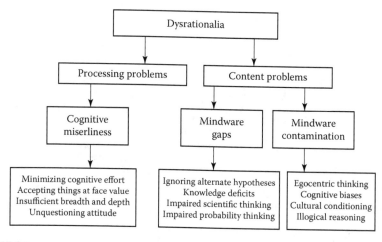

FIGURE 8.2

Factors that compromise rationality. (Adapted from Stanovich, K.E., *Rationality and the Reflective Mind*, Oxford University Press, Toronto, 2011 [23].)

When it is used, it may result in rationality failure. Cognitive biases are a major and inevitable source of contamination. Several of these (Table 8.3) have been reviewed in Chapter 7 (Cognitive and Affective Biases, and Logical Failures) and may lead to irrational decision making. As noted, the extent of an individual's rationality can be measured in terms of the number and extent of cognitive biases to which they are vulnerable [10,12].

Many such inhibitors characterize the practices of alternative medicine (Chapter 5: Complementary and Alternative Medicine), which, together with a virtual absence of facilitators of rationality, explains much of its irrationality. Yet, as the clinical examples in Table 8.3 illustrate, practitioners of conventional medicine may also be vulnerable to the influence of some of these inhibitors.

Basically, the world works in a fairly orderly, scientific way, and a degree of sound rationality should be achievable. The truth is there for the taking, but we often fail because of mindware inhibitors. At the close of *Mindware: Tools for Smart Thinking* by the psychologist Richard Nisbett [20], some of the major mindware inhibitors of rational thinking are reviewed. Among them are:

- Our beliefs about important ways in which the world works and the machinery of our everyday existence are "sorely mistaken." Further, the ways in which we have acquired these beliefs, explicitly or otherwise, is often "fundamentally flawed."

- Our feelings, beliefs, and behavior are affected by a wide variety of stimuli without our knowing or awareness. Even for something as

TABLE 8.3

Clinical Examples of Inhibitors of Rational Thought

Inhibitors	Clinical Examples
Superstitious thinking and belief in the paranormal	Belief in the influence of a full moon on the number of psychiatric patients who present to emergency departments; workplace beliefs about luck; "cloud" theory—belief that some doctors are black clouds (bad things typically happen with them) or white clouds (good things happen); disapproval of anyone in an ED using the "Q" word (commenting on how "quiet" the department is); ritualistic playing of particular pieces of music in the operating room; beliefs that certain items of clothing are lucky.
Belief in the superiority of intuition	Trusting one's gut feelings in diagnosing patients before sufficient evidence has been systematically gathered. Placing too much faith in initial impressions of patients.
Overreliance on folk wisdom and folk psychology	Belief in doing something in a particular way because that's how it was originally taught, despite a lack of evidence; for example, patching an eye for a corneal abrasion. Taking credit for successful diagnoses and ignoring failed ones. Ignorance of, or disbelief in, the impact of a wide range of biases on one's own decision making.
Belief in special expertise	Belief there must be some therapeutic advantage in alternate medical treatments (acupuncture, homeopathy) other than placebo effects, even though the evidence may show no such efficacy.
Incorrigibility of introspection (overoptimistic theories of one's own introspective powers)	Placing undue faith in the power of one's beliefs to the point that they cannot be changed; despite recognizing bias in others, refusing to believe in one's own vulnerability to bias (belief bias); overconfidence in one's intuitive diagnostic abilities.
Dysfunctional personal beliefs	Believing that drug dependency is not a disease but a personal failure (fundamental attribution error). Holding irrational beliefs about some diseases, especially new ones (many physicians were skeptical at the outset about the extent of post-traumatic stress disorder, irritable bowel syndrome, bacterial etiology of stomach ulcers, and others).
A notion of self that encourages egocentric processing	Overconfidence in one's diagnosing abilities; overconfidence in skills; some specialty surgeons come to believe in their all-powerful ability to heal, especially for groundbreaking surgery such as heart transplantation. Such beliefs may lead to egocentric processing.

Source: Adapted from Stanovich, K.E. et al., *Cambridge Handbook of Intelligence*, Cambridge University Press, Cambridge, UK, 2012.

simple as shopping at a store, we are mostly oblivious to the techniques that are being used to manipulate us.

- Without specialized training, most of us are very poor at introspection—knowing what is going on in our minds and being able to express it to ourselves and others. We have poor access to our cognitive processes and their influence on our behavior.

- We are unduly influenced by narrative and anecdote. Many of us do not understand the importance of gathering reliable data and of the effect of sample size.
- We underestimate the role of context even though we may be aware of it.
- We are often oblivious to the power of social influences on our decision making.
- We naïvely believe that we understand what is going on in the real world; such naïve realism is widespread.
- We unconsciously use a variety of heuristics to accomplish the majority of everyday things we do.
- We are not very good at interpreting relationships between events.
- We are flawed as intuitive scientists. We are overly confident about how well we gather evidence to understand how the world works and about the competence of our reasoning skills to deal with it.

Nisbett is an essential meliorist. What he says about the facilitators of rationality is very encouraging. If some basic rules are learned and applied, including awareness of some major biases and context effects, we will be able to perceive the real world more accurately and be more rational; he has demonstrated this in his own research. He differs from others who hold the contrasting view that biases cannot be undone, mindware gaps cannot be repaired, and contaminated mindware cannot be cleansed (see Chapter 15).

Overall, the framework provided by the conceptual structure of rational thought [1,10] together with Nisbett's approach [19] can be used to provide useful tools for evaluating and correcting physicians' rationality in diagnostic decision making. Optimal performance appears to depend primarily on dealing with inhibitors of rationality in the physician's thinking dispositions, but facilitators are a key component too. Importantly, one of the key issues in a physician's rational make-up is how he or she handles intuition (see Chapter 7: Cognitive and Affective Biases, and Logical Failures), and we need to take a closer look at what is actually meant by the term.

Intuition

Intuition means different things to different people. It has a number of everyday uses along a spectrum from soft to more scientifically based. At the soft end are notions expressed in this description: "Like creativity, intuitive inspiration often happens when someone virtually 'fuses' in an activity, when one is highly focused on the respective activity in a state of joy and fulfillment. Intuition can be trained and in its highest level leads into a conscious contact with non-incarnated beings, a process usually called channeling" [24]. Also at the soft end are those intuitions that people trust when they buy lottery tickets, follow their astrology forecast, or engage in Tarot card readings or other examples of magical thinking. Intuition has also been described as a sixth or inner sense, instinct, spiritual guide, or other vague term. In his excellent book on intuition, Myers lists a dozen of intuition's deadly sins [25] that cluster at this end of the spectrum.

Toward the other end of the spectrum, intuition may be used in a default sense whereby if something is complex or difficult to understand, the person simply "goes with their gut," and deep convictions may often be expressed as "gut" feelings. Another meaning implies common sense—something that feels like the right thing to do may be imbued with an "intuitive" logic. This may be a result of past learning, implicit or explicit, that gives the person a feeling of the right thing to do. Another has a sense of specific design; that is, something that ergonomically fits well with its expected use. A car's engineering might be described as "intuitive" when features of the car's function are explicitly designed to provide the best response to the road or changing conditions. Another meaning is again associated with design, but in the sense of user friendliness—in this case, the design may or may not have been intended, but the end result is that it works very well; that is, it is a compliment to good design. An even more sophisticated use of "intuitive" arises in discussions of complex topics, illustrated in this example from the neuroscientist Firestein: "I made a comparison of neuroscience to quantum physics ... about how it is so unintuitive to our brains. To take that a step further, the difference between modern physics and brain science is that the unintuitive thoughts one has to think in physics can be done with the language of mathematics" [26], a far cry, perhaps, from channeling with non-incarnated beings.

Gigerenzer defines intuition as a gut feeling, "a judgment (1) that appears quickly in consciousness, (2) whose underlying reasons we are not fully aware of, yet (3) is strong enough to act upon. A gut feeling is neither caprice nor a sixth sense, nor is it clairvoyance or God's voice. It is a form of unconscious intelligence" [21]. For Myers, intuition is simply defined as "our capacity for direct knowledge, for immediate insight without observation or reason," and for the harder end of the spectrum, he lists evidence of intuition's powers [25].

In the language of cognitive science, intuition is referred to more explicitly as *Type 1 processing* (Chapter 3: Cognitive Approaches to the Diagnostic Process). Originally conceived of as a single channel, it has since been broken down into four channels [23]. Thus, intuitive processes may (1) have hardwired origins; (2) be associated with emotions; (3) result from explicit learning and practice; and (4) be due to implicit learning (Figure 8.3 [27]). An

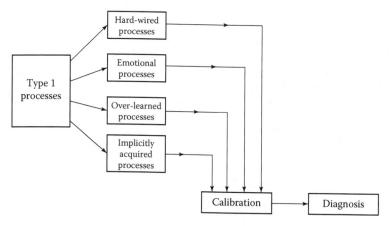

FIGURE 8.3
Differentiation of Type 1 processing into four subgroups. (Reproduced from Croskerry P. et al., *BMJ Qual. Saf.*, 22, ii58–ii64, 2013. With permission from BMJ Publishing Group Ltd. Based on Stanovich, K.E., *Rationality and the Reflective Mind*, Oxford University Press, New York, 2011 [27].)

understanding of what is meant by intuition would be helpful in resolving some of the misunderstandings that have arisen in discussions of intuitive and analytical processing, and perhaps provide a clearer understanding of its role in rational thinking and decision making.

Conclusion

Decision making is the engine that drives all behavior in the animal world and ultimately determines survival. Humans have achieved the highest pinnacle of decision making, having evolved a sophisticated analytic system that can work in parallel with a robust intuitive system. Medicine is a domain that provides a unique opportunity to study efficacy in decision making. Cognitive science has provided the current understanding of the complex processes that govern its decision making and the standard against which medicine's work in decision making will be measured. It remains, however, a work in progress, and we need to know more about the components of rational thought, specifically about the nature of rationality and its facilitators and inhibitors in medicine. These are beginning to take shape. Nevertheless, there is a continuing imperative for medical educators to embrace recent developments in cognitive science and recognize that rationality is the ultimate arbiter of sound clinical reasoning and, therefore, of the diagnostic process. It seems likely that we will find that the truly optimal diagnostician is a rational one.

SUMMARY POINTS

- Converging evidence from a variety of sources suggests that diagnostic failure is due more to how physicians think than to deficits in their knowledge.
- Rationality is the dominant characteristic of cognition and is largely responsible for the calibration of diagnosticians' thinking.
- Rationality comes in two forms: instrumental and epistemic, which are both essential.
- Instrumental rationality gets us what we most want given the available resources; that is, it tells us what to do.
- Epistemic rationality describes how well our beliefs map into the real world; that is, what is true.
- Rationality is made up of a variety of cognitive skills that support well-calibrated decision making.
- Rationality may be undermined by processing problems where there are problems in managing information, and by content problems, which may be due to mindware gaps (essential pieces of necessary knowledge are missing) or contaminated mindware (where the mindware is modified by various biases).
- Rationality may either support good decision making through facilitators, which are explicitly taught in courses of medical decision making in clinical epidemiology and biostatistics, or compromise it through inhibitors, which typically derive from cognitive biases and logical failures.
- Intuition has a variety of meanings to different people. In any discussion of the topic, it would be useful to define how it is being used.

References

1. Stanovich KE, West RF, Toplak ME. *The Rationality Quotient: Toward a Test of Rational Thinking*. Cambridge, MA: MIT Press; 2016.
2. Singh H, Giardina TD, Meyer AND, Forjuoh SN, Reis MD, Thomas EJ. Types and origins of diagnostic errors in primary care settings. *JAMA Intern Med.* 2013 Mar 25;173(6):418–25.
3. Schiff GD, Hasan O, Kim S, Abrams R, Cosby K, Lambert BL, Elstein AS, Hasler S, Kabongo ML, Krosnjar N, Odwazny R, Wisniewski MF, McNutt RA. Diagnostic error in medicine: Analysis of 583 physician-reported errors. *Arch Intern Med.* 2009 Nov 9; 169(20):1881–7.
4. Okafor N, Payne VL, Chathampally Y, Miller S, Doshi P, Singh H. Using voluntary reports from physicians to learn from diagnostic errors in emergency medicine. *Emerg Med J.* 2016; 33(4):245–52.

5. Zwaan L, de Bruijne M, Wagner C, Thijs A, Smits M, van der Wal G, Timmermans DR. Patient record review of the incidence, consequences, and causes of diagnostic adverse events. *Arch Intern Med*. 2010 Jun 28;170(12):1015–21.

6. Winters B, Custer J, Galvagno SM Jr, Colantuoni E, Kapoor SG, Lee H, Goode V, Robinson K, Nakhasi A, Pronovost P, Newman-Toker D. Diagnostic errors in the intensive care unit: A systematic review of autopsy studies. *BMJ Qual Saf*. 2012 Nov; 21(11):894–902.

7. Stanovich KE, West RF. What intelligence tests miss. *The Psychologist*. 2014;27(2):80–3.

8. Chatillon G. Acceptance of paranormal among two special groups. *Skeptical Inquirer*. 1989; 13(2):216–67.

9. Stanovich KE. Dysrationalia: A new specific learning disability. *J Learning Disabilities*. 1993 Oct; 26(8):501–15.

10. Stanovich KE, West RF, Toplak ME, editors. Intelligence and rationality. In: Sternberg R, Kaufman SB, editors. *Cambridge Handbook of Intelligence*. 3rd ed. Cambridge, UK: Cambridge University Press; 2012. pp. 784–826.

11. https://en.wikipedia.org/wiki/Rationality. Accessed July 17, 2016.

12. Bornstein BH, Emler AC. Rationality in medical decision making: A review of the literature on doctors' decision-making biases. *J Eval Clin Pract*. 2001 May;7(2):97–107.

13. Rao G. *Rational Medical Decision Making: A Case-Based Approach*. New York: McGraw Hill Companies; 2007.

14. Sime D, Rennie D. *The Rational Clinical Examination: Evidence-Based Clinical Diagnosis*. New York: McGraw-Hill; 2009.

15. Sackett DL. A primer on the precision and accuracy of the clinical examination. In: Simel D, Rennie D, editors. *The Rational Clinical Examination: Evidence-Based Clinical Diagnosis*. New York: McGraw-Hill; 2009. pp. 1–8.

16. Schmidt HG, van Gog T, Ce Schuit S, Van den Berge K, LA Van Daele P, Bueving H, Van der Zee T, W Van den Broek W, Lcm Van Saase J, Mamede S. Do patients' disruptive behaviours influence the accuracy of a doctor's diagnosis? A randomised experiment. *BMJ Qual Saf*. 2017 Jan;26(1):19–23.

17. Kline JA, Neumann D, Hall CL, Capito J. Role of physician perception of patient smile on pretest probability assessment for acute pulmonary embolism. *Emerg Med J*. 2017; 34: 82–88.

18. Stanovich KE. *2010: Decision Making and Rationality in the Modern World*. New York: Oxford University Press; 2010.

19. Nisbett RE. *Mindware: Tools for Smart Thinking*. New York: Farrar, Straus and Giroux; 2015.

20. Stanovich K. *What Intelligence Tests Miss: The Psychology of Rational Thought*. New Haven, CT: Yale University Press; 2009.

21. Gigerenzer G. *Risk Savvy: How to Make Good Decisions*. London: Penguin Books; 2014.

22. Kahneman D. *Thinking Fast and Slow*. New York: Farrar, Straus and Giroux; 2011.

23. Stanovich KE. *Rationality and the Reflective Mind*. Toronto: Oxford University Press; 2011. p. 99.

24. http://timeforchange.org/definition-of-intuition-intuitive. Accessed August 22, 2016.

25. Myers DG. *Intuition: Its Powers and Perils*. New Haven, CT: Yale University Press; 2002.
26. Firestein S. *Ignorance. How It Drives Science*. New York: Oxford University Press; 2012.
27. Croskerry P, Singhal G, Mamede S. Cognitive debiasing 1: Origins of bias and theory of debiasing. *BMJ Qual Saf*. 2013 Oct; 22 Suppl 2: ii58–ii64.

9

Individual Variability in Clinical Decision Making and Diagnosis

Pat Croskerry

CONTENTS

Introduction

The study of individual decision making (differential psychology) falls in the domain of the cognitive sciences. Psychology is the study of a variety of mental processes such as perception, memory, cognition, and reasoning and the relationship of these processes with behavior. They all impact decision making and how we live from moment to moment. From the time we wake up in the morning till we sleep at night, we are making decisions. The complexity of mental processes will vary from fairly mindless automatic acts, such as taking a shower or getting dressed, all the way through to the focused and deliberative efforts of the annual tax return. Decision making is the very essence of our existence. If we were all of the same genetic and experiential make-up, our behavior would be as predictable as the daily routine of a worker ant, but our genetic diversity, our variable learning environments, and a long developmental period combine to make human behavior the most complex of any species. The basic thrust of this chapter is that all physicians are individuals and as such are all different and, therefore, may be expected to make different kinds of decisions. In psychology, the observation has been made that "the lack of research being conducted from an individual differences perspective limits the applied use of relevant measures for predictive, selection/profiling purposes" [1]. A similar constraint applies to the assessment of individual differences in medical diagnostic decision making. This remains an underappreciated and underresearched area. Several aspects of this issue have been discussed previously in some detail [2].

As far as the individual decision maker goes, much depends on the nature of the task. If it was rocket science, for example, there would be minimal flexibility in decision making—the hard rules of science and logic would have to be followed, and there would be little room for idiosyncrasy or individual interpretation in putting a spacecraft into orbit. Contrast that with the work of an instructor in art school: certainly, there are rules to follow and various organizational requirements of the job, but the content might vary considerably from one class to the next. The difficulty for medical decision makers is that their work is both an art and a science. Over the last century, medicine has gained more credibility and efficacy as a science, especially through the advent of evidence-based medicine, but the art of physicians interacting with patients remains as valuable as ever. It involves concern, empathizing, caring, helping, ethical considerations, compassion, communication, humanitarianism, beneficence, assurance, listening, and advocacy. If physicians were simply the robotic conveyors of scientific information, there might be little variation in decision making. Various algorithms could be constructed to deal with various situations, but we know that helping and healing require both science and art. It is in the art portion of the physician make-up that much of the variation in decision making originates. Human behavior is infinitely variable. We are all very different from one another, yet the default approach taken in medical education is that medical trainees, as a group, are comparatively homogeneous and whatever is inputted to them will be absorbed and reproduced fairly faithfully by them at a later date. Typically, no significant influence is ascribed to gender, personality, aging, intellect, critical thinking, rationality, or other core aspects of human behavior that psychologists, on the other hand, view as interesting, relevant, and worthy of study. As Berger notes, "There is a strong presumption in the medical literature that clinicians are neutral operators governed by objective science and are unaffected by personal variables. Yet, there is a body of research that finds physicians' practice patterns are influenced by their own

TABLE 9.1

Characteristics That May Impact Decision Making

Individual Characteristic
• Gender
• Religion
• Cognitive ability
• Aging
• Experience, competence, and expertise
• Personality, states, and traits
• Cognitive and decision styles

demographic characteristics, and patient care is affected by the demographic concordance or discordance of the physician–patient dyad" [3]. These demographic variables might be especially important in clinical decision making and in how physicians make diagnoses, yet this area has attracted very little attention. For the most part, physicians are seen as consistent, objective, neutral, predictable, and invariable in their decision making. But we do not have to look far in our workplace to see that this is not so. This chapter explores some of the sources of individual variability in clinical decision making that might affect diagnostic calibration (Table 9.1).

Gender

Historically, when the overwhelming majority of physicians were male, gender was not an issue in clinical decision making. "Paternalism" prevailed—the physician was parental in that he decided what was wrong with the patient and what would be the best treatment, and paternal in that the parental role was almost exclusively male. Having a paternalistic approach meant assuming a therapeutic privilege to decide what was best for the patient and that, in some cases, it might be in the patient's interest not to know certain things—there was a focus on care rather than patient autonomy. In a study in 1961, for example, 88% of doctors would not have routinely informed a patient of a diagnosis of cancer [4]. This degree of paternalism is now seen as inappropriate. Nevertheless, in some situations in medicine, a certain amount of paternalism may be appropriate, especially with regard to patient competency. "Asymmetric paternalism" has been advocated [5] following the rule of thumb that the degree of paternalistic intervention required should be inversely proportional to the amount of autonomy present. Some will need more nudging than others [6].

The face of medicine changed when Elizabeth Blackwell came along. An immigrant to the United States from Bristol in England, she graduated from

Geneva Medical College in New York in 1849 as the first female doctor in the United States. Parenthetically, the first female doctor to graduate in medicine in the United Kingdom was Margaret Bulkley in 1809, but she accomplished this by masquerading as a male, James Barry, at the University of Edinburgh [7]. Increasing numbers of women are now entering medicine. In the last decade, in some medical schools in Canada, female enrollment outnumbered male by 3:1. Not surprisingly, studies are beginning to emerge comparing the practices of female physicians with those of their male counterparts.

Perhaps the first question should be: Is there any evidence that female physicians behave differently in their clinical practice toward their patients? The answer is a definite yes. Female physicians have been found to have more apprehension and less self-assurance and worry more than their male counterparts [8], yet also reported more satisfaction with their relationships with patients [9]. Female physicians are also found to place more emphasis on their patients' needs and opinions, and are more effective in encouraging cooperation and questioning from their patients than their male counterparts [10]. Further, they are more likely than male physicians to discuss psychosocial issues and more thoroughly investigate their patient's condition [11,12].

The second question, then, should be: If such gender differences in physician behavior exist, do they impact decision making? This would be expected, and has been confirmed in several studies. Compared with male physicians, female physicians made more referrals [13,14], experienced higher levels of anxiety with medical uncertainty [15], were more compliant with guideline recommendations and prescribed more effectively in patients with chronic heart failure [16], discussed hormone replacement therapy less with female patients [17], more frequently used lethal drugs in end-of-life decisions (vs. intensified use of opioids, or simply withdrawing or withholding treatment) and were less likely to discuss the issue with competent patients [18], influenced older female patients' choice of treatment for breast cancer [19], were less likely to agree to a patient's request for a cesarean section [20], had less implicit racial bias toward patients compared with male physicians [21], delivered more preventive services than males [22–25], and were more likely to order diagnostic imaging tests [26].

Thus, there is strong evidence of gender differences in decision making, but does this translate into differing diagnostic performance? In an experimental study, female physicians asked more questions about history and were more likely to make a psychiatric diagnosis, and be more certain of the diagnosis, than male physicians [27]. In another study using vignettes, female physicians were more likely to make a psychiatric referral than their male counterparts—that is, they were more likely to consider a psychiatric diagnosis [28]—while a survey of Australian general practitioners found that females reported more difficulty than male physicians in diagnosing depression in male patients [29]. A Swiss clinical study of coronary heart disease (CAD) found gender differences in diagnostic accuracy among cardiologists.

Based on history taking alone, overall diagnostic accuracy was comparable between the sexes; however, female cardiologists' diagnostic accuracy was more than 20% better for male than for female patients (85% vs. 66%) [30].

A gender difference in misdiagnosis is also suggested by data showing that male physicians are twice as likely to get sued during the course of their careers than female physicians, but factors other than medical error may be involved [31,32]. Studies in the United States [33,34] and Australia [35], for example, found differences in practice patterns, with female physicians seeing differing types of problems compared with males. In a U.S. study of patients visiting an emergency department, female patients were more satisfied with the care received by female physicians, rating them as more caring and willing to spend more time with them [36]. It has been observed, too, that communication is better between female physicians and their patients, and perhaps they are less likely to be sued for that reason alone. However, given the observation by Schiff et al. [37] implicating failure to order laboratory tests as a major contributor to diagnostic error, coupled with female physicians being more likely to order diagnostic tests [26], it appears that physician gender could be an important factor in diagnostic error. There is direct evidence of this. Patients in family medicine and internal medicine were less likely to get screening pap tests, breast exams, and mammograms from male than from female physicians [22,38] and therefore, would be less likely to be diagnosed with underlying disease. Further, given the tendency of males to be generally less risk-averse than females, it might be expected that male physicians would be inclined to take more risk in uncertain situations, but no studies have reported on this.

Studies of the determinants of diagnostic failure have begun in earnest only over the last few years, and, as yet, there are few direct clinical studies of the impact of gender on diagnostic failure. However, the many studies showing gender differences in clinical decision making, the experimental studies showing differing diagnostic preferences, and the clinical studies reviewed here strongly suggest that physician gender is a relevant variable in diagnostic decision making.

Religion

Western medicine is generally held to be objective, scientific, and secular, yet it would not be surprising if deeply held religious convictions by a physician might impact his or her clinical practice. As Berger notes, "The role of the physician and, more specifically, the influence of physicians' demographic characteristics on clinical care—has been relatively overlooked" [3]. Physicians are not demographically homogeneous and cannot be assumed to be "neutral operators" [2]. In a Medscape 2016 survey, no significant differences in bias toward patients were reported by physicians holding spiritual or religious beliefs compared with those who did not [39]. However, personal beliefs and attitudes of psychologists and psychiatrists have been shown to bias their decision making [40]. In a 2005 survey, 55% of U.S. physicians believed that their religious beliefs influenced their practice [41]. One can readily see how religious beliefs might influence physicians' decisions about certain specific issues such as circumcision, abortion, palliative care, and end-of-life issues, but there is evidence, too, that truth telling, beneficence, paternalism, and full disclosure of diagnosis and prognosis may be influenced by religion. In Islamic ethics, the belief is held that death only occurs when God permits it [42]. Thus, communication between physician and patient may be less open and objective than in a secular relationship, or at least in one that is free of religious influence. Thus far, a direct effect of physician demographics on diagnostic outcomes does not appear to have been studied, but given the very clear influence of physician demographics on clinical practice, some impact on diagnostic outcomes would be expected.

Cognitive Ability: Intellect, Critical Thinking, and Rationality

In a thoughtful essay by Laqueur, parallels are drawn between the challenges of studying political intelligence and those of clinical judgment, suggesting that intelligence analysts would do well to contemplate "the principles of medical diagnosis" [43]. Certainly, the degree of complexity of the diagnostic process ranges from very low to extremely high, and whereas pathognomonic signs of straightforward conditions might not be expected to require extraordinary mental effort, those that are obscure, undifferentiated, and with a high degree of uncertainty might well do so. Making diagnoses generally is no trivial intellectual exercise. Although the correlates and consequences of cognitive abilities have attracted considerable political sensitivity [44], it nevertheless seems reasonable to ask whether accomplished diagnosticians have higher levels of cognitive ability than those who are more prone to diagnostic failures. There is a wide range of cognitive abilities, but most prominent among them are measures of fluid and crystallized intelligence, which define two broad ability domains related to levels of knowledge, reasoning ability, judgment, and decision making.

How doctors think might have a lot to do with their ability to make an accurate diagnosis, and one might expect that a good thinker would require at least normal intelligence. Usually, by the time someone has made it to medical school, they will have demonstrated their intellectual ability in multiple ways up to and including the Medical College Admission Test (MCAT). Although many factors are involved in academic success, it is likely that general intelligence (g) plays an important part. Gottfredson described g as "a highly general information processing capacity that facilitates reasoning, problem solving, decision making, and other higher order thinking skills" [45] and would be essential in dealing with the complexity of many diagnostic decisions.

If intellectual ability proved to be a limiting factor on the efficacy of diagnosing ability, it should at least be reassuring that a major study found that the group with the highest range of intelligence quotients (IQs) were physicians: with a general population average of 100, physician IQs were in the range from 106 (10th percentile) to 133 (90th percentile) [46].

Further, insofar as academic achievement is associated with intellectual ability, there may be some additional differentiation through medical training. A study involving 883 graduates over a 10-year period found that academic achievement (measured by grade point average [GPA] and scores on steps 1 and 2 of the United States Medical Licensing Examination) was correlated with career choice [47]. Lower levels of academic achievement were significantly associated with the general residencies (family medicine, general practice, obstetrics-gynecology, general pediatrics, and general psychiatry), whereas the higher levels were associated with the specialized residencies (diagnostic radiology, surgery, anesthesiology, medicine pediatrics, ophthalmology, pathology, emergency medicine, and other surgical subspecialties) [47]. While there are probably many other factors at play here, it does at least raise the question of whether or not diagnoses are more challenging and therefore require greater intellectual effort in specialized medicine. Given that vulnerability to cognitive bias has been shown to be related to intellectual ability, and dispositions to engage in analytic processing (System 2) have been found to be negatively related to biases [48], it is interesting that diagnostic failure is reported to be highest in the general disciplines where diagnostic problems are the least differentiated (family practice, internal medicine, and emergency medicine) [49], and these same specialties are among those with the highest self-reported biases toward patients: emergency medicine at 62%, family medicine at 47%, and internal medicine at 40% (Figure 9.1) [50].

A second important aspect of cognition is critical thinking. In recent years, a variety of domains have begun to acknowledge and promote the concept of critical thinking and its influence on problem solving, reasoning, and decision making. It is a significant contributor to overall rationality and is discussed in three other chapters in this book: Chapter 2 (Medical Decision Making), Chapter 8 (The Rational Diagnostician), and Chapter 14 (Medical Education and the Diagnostic Process).

Rationality is of all-encompassing importance for clinical decision making. While the folk psychology of intelligence sees the performance on an IQ test as a reliable measure of intelligence characteristics such as reasoning and problem solving, others have challenged this view. We can all think of numerous examples of intelligent people exercising poor judgment and decision making. It appears that IQ tests may be missing some important areas of cognitive function. They do not measure certain characteristics such as interpersonal

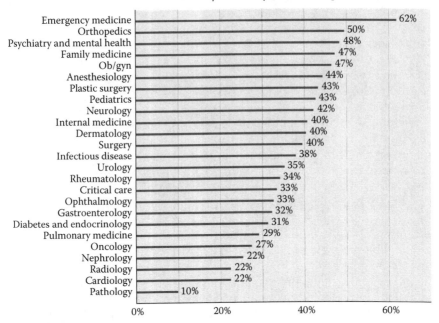

FIGURE 9.1
Self-reported biases against patients by specialty. (Figure reprinted with permission from Medscape, 2016, available at: www.medscape.com/features/slideshow/lifestyle/2016/public/overview [50].)

skills, empathy, emotional intelligence, and other socioemotional abilities; nor do they measure a very important aspect of cognitive function, rationality [51]. Stanovich has described two major aspects of Type 2 processing: the reflective mind and the algorithmic mind [52]. Importantly, most individual variability in rational thought arises from variation in the reflective mind, which characterizes fluid intelligence (Gf: intelligence as process)—the ability to reason across a variety of domains. The algorithmic mind, in contrast, is what IQ tests mostly measure, referred to as crystallized intelligence (Gc: intelligence as knowledge)—cognitive skills acquired through learning experiences (general knowledge, vocabulary skills, verbal comprehension). The thinking dispositions that underlie rationality and action, argue Stanovich et al. [51], are

> the tendency to collect information before making up one's mind, the tendency to seek various points of view before coming to a conclusion, the disposition to think extensively about a problem before responding, the tendency to calibrate the degree of strength of one's opinion to the degree of evidence available, the tendency to think about future consequences before taking action, the tendency to explicitly weight pluses

and minuses of situations before making a decision, and the tendency to seek nuance and avoid absolutism.

If we are looking for optimal diagnostic problem solving, reasoning, and decision making, we surely would want someone who not only has the requisite knowledge and information, and critical thinking skills, but also possesses the rational characteristics described above. Thus, a certain level of Gc is necessary but not sufficient, whereas a good helping of Gf is essential for the well-calibrated diagnostician. We would want to know, too, that the decision maker is a competent critical thinker. Overall, we need knowledgeable physicians who are critical thinkers and rational. As things stand, a RQ (rationality quotient) test is not yet available, but there is a burgeoning interest and growth of knowledge about the topic. Rationality is superordinate to critical thinking as well as to intelligence [51] and may ultimately prove to be a key factor in optimal diagnostic decision making (see Chapter 8: The Rational Diagnostician).

Aging

Generally, we tend to think of older physicians as wiser, at least up to a point. As physicians get older, they certainly are more experienced, and we assume that with experience comes wisdom. However, it is sometimes difficult to differentiate the impact of aging and experience on diagnostic decision making. Also, the two may be confounded, such that cognitive deficits associated with aging are masked or compensated by the benefits of experience as well as the acquisition of learning, reasoning, and critical thinking skills.

Given that the clear majority of diagnostic errors are due to various cognitive failures [53–56], we would expect that anything associated with cognitive impairment would, in turn, impair diagnostic performance. Some neurodegenerative diseases and syndromes that are age-related, such as dementia, Alzheimer's disease, frontotemporal dementia, vascular dementia, and transient global amnesia [57], would all be associated with cognitive deficits to varying degrees and differentially affect particular areas of the brain and their associated cognitive functions. It would be expected that these conditions would quickly become evident, but early changes may be subtle and attributed to the effects of fatigue, stress, sleep loss, and other variables. It is said that "When true impairment in clinical skills is apparent, the illness is usually severe and longstanding" [58]. Once disease is diagnosed, neuropsychological assessment is useful in localizing brain lesions and assessing functional abilities, and neuropsychiatric assessment can focus on the relationship between the disorder and mental symptoms. Diagnosis may be difficult, with some cases requiring positron emission tomography scans before the anatomic lesion is identified. So, the first point is that if a physician has an age-related neurodegenerative disease, deficits in cognition might impair decision making and lead to diagnostic error before the condition is diagnosed.

If we assume that physician expertise is probably asymptotic at about 10 years following completion of training, as has been found in other domains [59], then it may be possible for aging effects to be separated out following that period. If medical graduation occurs at about age 25, and completion of specialty training by about 30, then "expertise" should be reached at about 40. Retirement from medical practice tends not to follow the abrupt cessation that occurs traditionally in other fields at age 65. Doctors appear to retire later, but mitigate their workload or take on other nonclinical responsibilities in their 60s, and hold off retirement proper till around 70 or later. Thus, the critical interval we are looking at is approximately 40–65 years.

In healthy, educated adults, there is a decline in cognitive function with aging from the mid-20s onward, accelerating after the mid-60s (Figure 9.2) [60]. Memory, overall speed of thinking, spatial visualization, and reasoning all show almost monotonic negative age trends that begin in early adulthood. From the age of 61 to 96, the rate of decline accelerates [60].

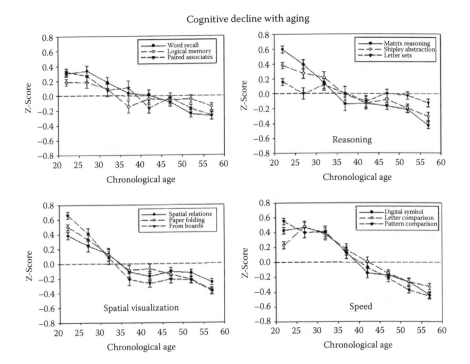

FIGURE 9.2
The decline of various brain functions with aging. (Reprinted from *Neurobiology of Aging*, 30(4), Salthouse, T.A., When does age-related cognitive decline begin? Copyright 2009, with permission from Elsevier [60].)

It is important to note that other characteristics of cognitive function, such as accumulated knowledge (vocabulary, general information), actually increase till at least the age of 60 [60].

There are also declines in general intelligence with aging. As noted above, general intelligence is made up of *fluid* intelligence and *crystallized* intelligence. Fluid intelligence peaks at an earlier age and declines faster than crystallized intelligence (Figure 9.3) [61]. These patterns follow what one might expect with normal development and aging; that is, it would appear preferable to have fluid intelligence at its peak in the formative years of learning and reasoning, and have the skills associated with crystallized intelligence peak later and be more sustained through later years. Finally, multitasking of cognitive and motor functions, an important requirement of clinical practice in several settings, declines through middle age [62]. So-called "senior moments" appear to be due to the brain's diminished capacity to ignore distractions with age. Older people have more difficulty disengaging from a distraction as well as getting back to what they were doing before the distraction occurred [63].

To summarize, during the interval of concern (40–65 years), there is a decline in most areas of cognitive function. The important question is whether or not this translates into compromises in clinical reasoning that may lead to diagnostic failure. To date, there are no direct studies that have reported any escalation of diagnostic failure through this period, and it would seem unlikely for several reasons: first, accumulated knowledge and other skills increase till the 60s, and one would also expect that a variety of short-cuts and efficiencies would have been built into clinical practice through the years. As far as diagnostic reasoning goes, specific deficits such as memory and speed are unlikely to adversely affect performance, as the presenting features of specific diseases have been repeatedly experienced over the years, and what needs to be remembered will have become well

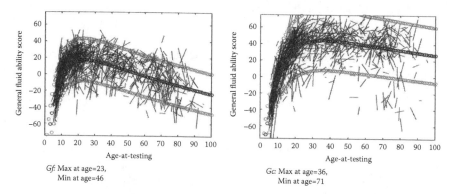

Gf: Max at age=23, Min at age=46

Gc: Max at age=36, Min at age=71

FIGURE 9.3

Changes in Gf (fluid intelligence) and Gc (crystalline intelligence) with aging. (Reprinted from McArdle, J.J. et al., *Developmental Psychology*, 38(1), 133, 2002, American Psychological Association, reprinted with permission [61].)

consolidated. It would be expected, too, that sophisticated pattern recognition would be well established for the majority of conditions. Further, the increased availability and ready access to computerized information (checklists, clinical guidelines, and decision support) significantly reduce reliance on memory and may even lead to overall improvements in performance. Although fluid intelligence will be on the wane through this period, crystallized intelligence, which has significant value through this period, peaks later and is declining only slowly. Much will depend on the particular medical environment. In the pattern-recognition specialties such as radiology, dermatology, and clinical pathology, the impact of aging on diagnostic reasoning is probably minimal, whereas multitasking, maintaining speed, and functioning well in a setting such as emergency medicine would be more challenging and diagnostic failure more likely.

Experience, Competence, and Expertise

We do not know many young experts. We usually associate expertise with older people, and as noted earlier, age, experience, and expertise all go together. It is clear that expertise differs from one domain to another. Generally, those domains with a strong science basis allow development of expertise that is reliable and measurable; for example, the parameters of expertise for a nuclear physicist can be very clearly described. For other areas that are less scientific and less rational, for example, economic forecasting and political science, there is considerably less certainty, and expert performance may be very poor, even worthless [64]. Medicine is a combination of both extremes. A considerable part of it is now scientific, and expertise in those areas is more predictable. A significant part of medical practice, however, is nonscientific and construed as an art. Expertise in this aspect is presently more difficult to describe and measure. Most of us are aware of diagnosticians with strong reputations, and it is usually clear that their expertise is not accounted for solely by their knowledge of scientific aspects of medicine.

The range of activities within any domain is also important. Higher levels of expertise can be achieved where the domain is focused and restricted. Chicken-sexing, arguably one of the most specialized careers, is a complex visual pattern-recognition skill that requires 2–3 years to achieve expertise, the expert chicken-sexer expecting to achieve 100% accuracy [65]. Dermatologists, in contrast, require double that time after completing medical training and a few more years after that before approaching anywhere near the success rate of the chicken-sexers. Whereas chicken-sexers need to recognize a limited number of configurations of the anal vent, dermatologists have to learn to recognize a much wider range of patterns. Specialists who work in specialized hospitals, which restrict themselves to a very limited range of diseases, achieve higher rates of success with fewer adverse events. The Shouldice Hospital in Toronto, Canada, for example, does nothing but hernia repairs—over 7000 a year—and does them very well. Whereas a hernia operation typically takes about 90 min in a regular hospital and fails in about 10%–15% cases, at Shouldice it can be done in half the time with a failure rate of 1%. It is also done at about half the cost [66].

Acquisition of expertise is biphasic. The initial shorter basic training phase involves learning basic schemas, whereas the secondary protracted phase is primarily experiential, requiring continuing learning in the gradual refinement of schemas. It is preferable that the experiential phase starts early, with students seeing patients early on in the first phase. Figure 9.4 shows the relationship between time spent in training and practice. The abscissa is on a logarithmic scale. Before entering medical school, candidates are mostly at the *consciously incompetent* stage; that is, they are aware of what they don't know. After 3–4 years, they have traveled up the steep part of the learning curve, meriting the description *consciously competent* [67,68].

As experience and further learning occur beyond medical school, they will pass a further phase of *unconscious competence*, in which the performance of a psychomotor task may be completely relegated to System 1 processing, and perhaps then to a final stage of *reflective competence*, in which System 2 processes can be used to reflect on or monitor the unconscious output from System 1.

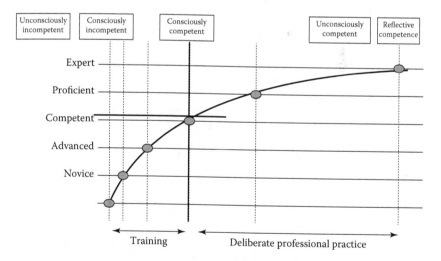

FIGURE 9.4
The learning curve and stages of competency. The first four stages are attributed to Noel Burch, an employee at Gordan Training International in the 1970s (from Robinson, W.L., *The Personnel Journal*, 53(7), 538–539, 1974 [67]), and the fifth stage to Baume, 2004 (from Adams, L., *Learning a New Skill Is Easier Said than Done* [Internet], Gordon Training International, Solana Beach, CA, c2016 [68].)

Development of Clinical Expertise

The progressive elaboration on patterns of disease is reminiscent of schema theory, developed by the early psychologists. An initial schema is established, much like a cognitive scaffold. This is gradually elaborated on and progressively refined with use and experience [69].

The process is described by Schmidt and Rikers [70] as beginning with a scaffold foundation of biology, anatomy, and pathophysiology, on which causal networks are elaborated that explain the causes and consequences of disease. Through repetitive application of this burgeoning knowledge and exposure to patient clinical problems, the networks become encapsulated into diagnostic labels that are used to explain signs and symptoms. Other labels for *encapsulated knowledge* are syndrome, stigmata, disease prototypes, pathognomonic features, toxidrome, and others. The labels become a type of shorthand to summarize the detailed pathophysiology that underlies the disease in question and speed up communication.

The next stage is a reorganization of encapsulated knowledge into *illness scripts* [71]. These contain important, clinically relevant information about the enabling conditions for disease. During the diagnostic process, an illness script may be matched to the patient presentation and then instantiated in the course of script verification; that is, the particular patient's illness is accepted (or rejected) as a valid example (instance) of the illness script that describes a particular disease entity.

Ericsson et al. estimated that expertise requires 10,000 hours of deliberate practice [72]. Much will depend on the individual—their motivation, intelligence, rationality, personality, and other factors discussed in this chapter, as well as the characteristics of the environment in which they work—the quality of support, feedback, resources, team member performance, and others. As physicians see more and more patients in their clinical setting, the number and richness of enabling conditions associated with a particular disease increase [73–76], and illness script patterns become increasingly refined and sophisticated. Whereas a physician's initial encapsulated knowledge and diagnostic labels are based on prototypical manifestations of illness, with experience they increasingly recognize atypical manifestations and subtler enabling conditions for particular diseases. Not only do they deliberately maintain and expand their explicit knowledge of the diseases in their area of specialty, but they also acquire tacit knowledge of them. Simply being present in the clinical context in which patients repeatedly present themselves allows the passive accumulation of additional knowledge (implicit learning) about patients and their diseases. In the process, they move more and more toward the unconsciously competent state and, accordingly, spend increasing time in System 1 (Figure 9.5). This allows them to become faster, use fewer resources, and become increasingly accurate in their diagnoses. As noted, reflective competence is achieved when the decision maker has the

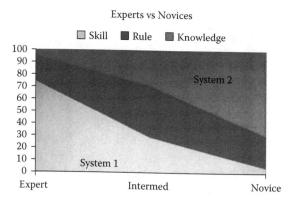

FIGURE 9.5

The respective times spent by novices and experts in Systems 1 and 2. (Courtesy of Catherine Lucey. Adapted by Catherine Lucey from Reason J. Overview of unsafe acts. Presented at the Second Halifax Symposium on Healthcare Error, 2002, Dalhousie University, Halifax, Nova Scotia, Canada.)

capacity to attend to and reflect on their own unconscious competence, as well as that of others. This is necessary to monitor, and modify if necessary, the unconscious output from System 1. This faculty is the essence of the rational thinker and fundamental to cognitive bias mitigation (see Chapter 15: Cognitive Bias Mitigation: Becoming Better Diagnosticians).

We can see that competence and expertise are important characteristics of the individual decision maker. If it were the case that one simply had to complete the two phases of expertise acquisition to become an expert, then all would be well. But we know that experience alone is not sufficient to attain expertise. There are some experienced nonexperts who do not make optimal diagnostic decisions. Further, there are other phenomena associated with experience that may be counter-productive, such as overconfidence, over-familiarity, burnout, complacency, altered states of emotional well-being, reduced motivation, and many others. A number of these issues have been reviewed in the context of clinical decision making [77].

Personality: Traits and States

There are significant individual differences among people in their behavior patterns, cognition, and emotion. Each of us in our own workplace can think of colleagues who are cheerful, optimistic, tireless, careful, thorough, and agreeable, as well as others who are pessimistic, hasty, inward looking, superficial, and disagreeable, and others who are shades in between. It would be surprising if clinical decision making were not influenced by such individual characteristics. In fact, there is substantial evidence that personality does influence general performance and that it also influences the ways in which we make decisions [78]. As Bensi et al. note, "Personality can play a crucial role in how people reason and decide. Identifying individual differences related to how we actively gather information and use evidence could lead to a better comprehension and predictability of human reasoning" [79]. Overall, leadership, job performance, and career success are all associated with personality characteristics [2,80].

Generally, personality is stable over time and can be broken down into five major *traits*: conscientiousness (C), openness to experience (O), extraversion (E), agreeableness (A), and neuroticism (N) (or emotionality). *States*, such as fear, happiness, joy, excitement, disappointment, and many others, are moment-to-moment passing conditions that may last for a few hours or a day and may change quickly, whereas traits (in addition to the big five), such as empathy, mindfulness, metacognition, confidence, and reliability, are more enduring.

Of the big five, extraversion is associated with making decisions impulsively [81] and therefore, one would think, less competently. Neuroticism (emotional stability) is also associated with appropriate decision making in paramedics [82]. Conscientiousness appears particularly relevant to diagnostic decision making. Conscientious individuals are careful and thorough, and, importantly, they are deliberate; that is, they think before acting and are less likely to show spontaneous, impulsive, reflexive behaviors. Conscientiousness, followed by extraversion, is a significant predictor of job performance in most domains [83]. Given the choice, many of us would prefer to have conscientious decision makers making decisions about our diagnosis, as they are more careful and thorough. Conscientiousness in medical students is a predictor of exam success in preclinical training [84]. It is also a significant predictor of a preference for rational thinking and an inverse predictor of intuitive thinking.

Certain medical specialties are associated with characteristic personality differences [8,85–87]. Anesthetists generally were found to be shy, withdrawn, inhibited, and cold, whereas surgeons were confident, domineering, and aggressive [88]. It seems that surgeons are less likely than anesthetists to accept that stress, fatigue, or personal problems have an impact on decision making and performance [89]. This is interesting in light of the finding from a study of Israeli judges, which found that favorability of decisions for parole appeared to be related to decision fatigue [90]. Seventy percent of prisoners whose cases were heard early in the day received a favorable decision compared with only 10% of those whose appeals were heard late in the afternoon. This has also been referred to as "ego depletion" [91] and has been related to more activity in the nucleus accumbens (part of Type 1 processing) and less in the amygdala, which helps in impulse control. This and other studies suggest that for each individual, there is a finite store of energy that fuels willpower or mental self-control [92]. As it becomes exhausted, we tend to fall back into well-established habits and preserve the status quo; in the process, we become less inclined toward Type 2 decision making. Glucose levels appear to be involved as part of the substrate for ego depletion [93]. That similar phenomena affect medical decision making is suggested by two studies. A study of 477 colonoscopies found that more polyps were detected in patients receiving colonoscopies early in the morning compared with later in the day, and that the likelihood of detection and diagnosis of adenomas reduced as the day progressed [94]. In another study, the tendency for primary care clinicians to prescribe marginal or unnecessary antibiotics for acute respiratory infections significantly increased through the clinic session [95]. In a meta-analytic study effort, perceived difficulty, negative affect, subjective fatigue, and blood glucose levels were all found to be significantly associated with ego depletion [96]. Thus, it might be a relatively simple exercise to demonstrate such circadian diagnostic failure increases with decision fatigue/ego depletion/self-control failure in other domains of clinical medicine.

Other examples of personality variables biasing decision making come from studies showing that raters' scoring tendencies of criminal offenders were related to their own personality traits [97]. In another study, physician personality variables were associated with the likelihood of a diagnosis of depression in their patients. Physicians who were rated more dutiful and more vulnerable on a personality assessment (Neuroticism-Extraversion-Openness Personality Inventory [NEO-PI-R]) were more likely to diagnose depression in standardized patients [98]. In an extensive review, Boerebach et al. [99] looked at the impact of clinicians' personality and their interpersonal behaviors on the quality of patient care. Most of the outcome measures of quality involved patient satisfaction, reduction of patient anxiety, patient compliance, change in patient health behaviors, quality of care rated by judges, and others. Although there were very few observations on diagnostic outcomes, it would be surprising, given the above observations, if physician personality were not an important variable. Several personality traits in particular might be expected to influence diagnostic decision making: mindfulness, need for cognition (NFC), reflective coping (RC), metacognitive awareness, and others [100].

Mindfulness requires both awareness and attention in a particular situation. It reflects the capacity of some individuals for *enhanced* attention to and awareness of aspects of the environment, and generally of life's experiences. The trait is generally associated with an open, nonjudgmental, and receptive overall attitude. It is the opposite of automatic, impulsive behavior. Studies of trait mindfulness suggest it reflects aspects of executive function and emotional regulation. These are important cognitive functions that depend on the functional integrity of the prefrontal cortex [101], which is where Type 2 processes such as cognitive bias mitigation (CBM) originate. Training in mindfulness has been shown to reduce sunk cost bias [102], implicit age and race bias [103], and the effects of negativity bias [104]. Interestingly, Sibinga and Wu have proposed a clinical connection between mindfulness and CBM [105]. Thus, mindfulness state may be an important predictor of successful diagnostic decision making; it can be measured using the Mindful Attention Awareness Scale [106].

Need for cognition trait motivates individuals to engage in and enjoy effortful cognitive activities [107]; that is, they are more likely to be what we would call thinkers. NFC in jurors influences their legal decisions [108]. NFC is correlated with metacognition [109], academic motivation [110], and academic achievement [111]. Those low on NFC are more likely to default toward the cognitive miser function of Type 1 processing [112] and less likely to see the need for Type 2 effortful processing. Thus, it might be expected that individuals with NFC trait would be less vulnerable to diagnostic failure. In one study, obstetricians' NFC was associated with superior maternal and/or neonatal outcomes [100].

Reflective coping is a subscale on the Proactive Coping Inventory [113]. Its items are focused on optimal decision making and include brainstorming, thinking analytically about problems at hand, considering alternatives and their perceived effectiveness, resource utilization, and imagining hypothetical plans of action. In a study by Dunphy et al., obstetricians high in RC more effectively identified clinical problems early, before the onset of the second stage of labor, which suggested superior diagnostic skills leading to superior maternal and/or neonatal outcomes [100].

Metacognition is an important trait in the context of decision making. It is the ability to detach yourself from the immediate pull of the stimulus or pattern in front of you to take a broader perspective. It is a trait that reflects an individual's ability to think about their own thinking, beliefs, feelings, and motives and goes by a number of other names (cognitive awareness, mindfulness, reflection, self-regulation, executive cognition, executive control, meta-knowledge, and others). It is a concept somewhat related to mindfulness. We might imagine that metacognitive skills would be among the most desirable in optimal diagnostic decision making. The original concept of metacognition embraced knowledge about cognitive tasks and strategies and referred to self-monitoring and regulation of cognition [114]. It can be measured by the Metacognitive Awareness Inventory [115]. Further, confidence trait appears to augment metacognition [1]. Preliminary studies using the Solomon Questionnaire [116], designed to assess metacognitive knowledge about behaviors and mental processes involved in decision making, have begun looking at different medical specialties [117]. Thus, metacognition is proposed as a framework for physicians to gain insight into their decision making, understanding the limitations of their cognition; the inherent uncertainty of many clinical problems and associated risk; the constant need to be current in knowledge; the need to ration one's cognitive effort and know when heuristics may be safely used; the overall impact of stress, time, and other resource limitations; the intrusion of one's own feelings into decision making; the level of confidence one can safely attach to a particular decision; and other factors [117]. It is akin to a *cognitive manager* role [118]. A metacognition assessment questionnaire was administered to small groups of emergency physicians, internists, and surgeons. The density of decision making was highest in emergency medicine, and emergency physicians made more decisions about themselves—that is, reflecting on their own performance—than the other two groups. Approximately half of all physicians considered feeling regret to be an important feature of good decision making; the more difficult the decision, the more likely were feelings of regret. The emotional experience of regret was considered an important characteristic of a good decision maker; that is, in the process of making a decision, individuals may anticipate feeling regret after the uncertainty is resolved, and therefore build into their choice their desire to eliminate or reduce this possibility. Thus, anticipated regret may lead the decision maker to more thinking and reflection during the stages of the decision-making process [117]. Colbert et al. have detailed a number of explicit strategies to actively enhance the metacognitive skills of medical students to improve their clinical reasoning [118].

Cognitive and Decision Styles

People are noticeably different in the way they make decisions. Some make rapid shoot-from-the-hip decisions with confidence, whereas others may be more hesitant, reflective, and deliberate before committing themselves. Cognitive styles are classified into three groups: knowing, planning, and creating. Knowing is a style that places emphasis on precision, objectivity, and logic; planning emphasizes structure, control, and routines; and creating involves subjectivity, impulsivity, and openness to possibilities [119]. Dewberry et al. describe a variety of decision styles, such as anxiety, avoidant, dependent, brooding, vigilant, intuition, and spontaneous [78]. In terms of the predictive value of cognitive style toward characterizing individual decision making, it does not appear to add anything to the effects of personality and decision style. Some of these styles are measured in tests that examine preferences for rationality or intuition, ways in which individuals prefer to process information to make decisions. However, decision-making styles do predict decision-making competence [78]. Individual differences in decision-making competence can be measured with the Decision Outcomes Inventory [120].

The Rational-Experiential Inventory (REI) is a widely used tool that was developed by Pacini and Epstein [121] to assess preferences for particular decision styles. It can be used to distinguish between people who have a greater tendency to make Type 1 decisions (experiential) and those more inclined to make their decisions using Type 2 processing (rational). The German Preference for Intuition versus Deliberation scale (PID) [122] measures similar parameters as the REI, but there are differences. The REI-E and PID-I appear to be measuring the same decision style, which follows feelings and affect, whereas there is a lower correlation between REI-R and PID-D in the tendency to have thoughtful, analytical cognitions; that is, rationality overlaps less with deliberation than experientiality does with intuition [123]. The REI has been used in several studies in medicine [124–128], but, thus far, none have looked at a potential relationship between REI scores and efficacy of diagnostic decision making. It seems likely that such a relationship might exist.

The individual preferences that are measured in both the REI and the PID are germane to the present discussion in that they are predictive of a range of decision-making behaviors. Higher scores in rationality (analytical thinking), for example, are associated with superior reasoning skills, better syllogistic reasoning, and a reduced vulnerability to cognitive biases, whereas higher experiential (intuitive) scores are associated with poorer reasoning skills and susceptibility to cognitive biases, superstitiousness, and categorical thinking [129,130]. The ability to control attention is strongly dependent on working memory capacity, which imposes a constraint on rational processing. Controlling attention is essential for sustaining motivation and goal-directedness during complex tasks. Importantly, it promotes concentration on relevant stimuli, allowing irrelevant stimuli to be ignored, and it inhibits the characteristic stereotypical and reflexive processes of System 1. The net expectation would be that optimal diagnostic decision making would be expected with individuals who score higher on the rational (analytical) scale.

It should be noted that these measures of decision style are not categorical in the sense that individuals are either rational or intuitive. There are gradients of each characteristic both within and between people. Further, the degree to which one might be rational or intuitive may vary with ambient conditions. This appears to be the case in the antibiotic prescribing study [95], where the prescribers' "rationality" appeared to be compromised by fatigue; that is, as the day wore on, the degree of executive control over decisions (Type 2 processing) appeared to be compromised. Other studies on this effect are probably forthcoming in medicine. Also, individual differences in the degree of competence of intuition and analytical reasoning can be expected. Thus, some analytical thinkers will be more normative and rational than others, while some who prefer to make predominantly intuitive decisions may be particularly good at it and others may not. Good intuitive decision makers may be those who have learned well from life's lessons.

Finally, a test of decision style that shows particular promise and relevance to medical decision making is the actively open-minded thinking (AOMT) test. Originally developed by Baron in 1993 [131], it was designed to identify thinkers who have a tendency to weigh new evidence against currently preferred beliefs, to persist with effort and devote more time to solving a problem rather than giving up, and to dialectically take into consideration the opinions of others in the process of forming one's own opinions. It has been shown to significantly predict failures in rationality due to contaminated mindware [132], which may significantly compromise reasoning (see Chapter 8: The Rational Diagnostician), and is a good predictor of both the short form and the full form of the recently developed Comprehensive Assessment of Rational Thinking (CART) test [132]. Importantly, there is evidence that thinking skills can be improved by teaching adaptive cognitive thinking styles, including AOMT. Baron et al. [133] developed an 8 month course of decision making, with hypothetical examples, practice exercises and feedback, which effectively reduced the susceptibility of students to bias and improved thinking skills. Similar findings have been obtained in other studies [134,135].

In summary, there is good evidence that personality traits and cognitive and decision style are related to decision making, and therefore likely related to the decisions that are made in the process of diagnosis. To date, work on the AOMT and the REI is most promising. It seems likely that other personality characteristics, such as metacognitive awareness, RC, NFC, mindfulness, conscientiousness, and other traits are all involved. Further study of these important features of decision making is warranted.

Conclusions

The rich variety of studies reviewed here on individual differences in decision making suggest that this area is particularly relevant to the diagnostic process. Although it appears to have received wide consideration in the literature, it has not hitherto factored in the general discussion of diagnostic performance and receives very little mention in the recent report from the National Academies of Sciences, Engineering and Medicine on *Improving Diagnosis in Healthcare* [136]. There is sufficient evidence from these studies, however, to begin to embrace the topic in the context of diagnostic failure. It is of more than academic relevance. Further research will lead to a greater understanding of how these variables (gender, aging, personality traits and states, and religion and other belief systems) and cognitive processes (rationality, intellect, mindfulness, reflection, metacognition, cognitive and decision styles, and others) impact diagnostic decision making. This may lead, in turn, to greater attention to these factors in medical education, and in particular, to our understanding of clinical reasoning and decision making.

SUMMARY POINTS

- Individual variability in decision making refers to characteristics of the individual decision maker that may influence the quality of decision making.

- In mainstream discussions of diagnostic decision making, individual variability has not attracted much interest, even though a number of studies support its relevance.

- Both physician gender and religion, and other demographic variables, would be expected to exert some overall influence on diagnostic decision making of the physician.

- Rationality, critical thinking, and intellect all exert influence on diagnostic decision making.

- Rationality likely has the most powerful impact of all cognitive abilities on diagnostic decision making.

- Although a number of important cognitive faculties decline with aging, the development of expertise with experience probably offsets any significant adverse effect of aging on the calibration of diagnostic reasoning in physicians.

- The development of clinical expertise with experience begins with a scaffold foundation of biology, anatomy, and pathophysiology, on which causal networks are elaborated that explain the causes and consequences of disease. The networks become encapsulated into diagnostic labels that are used to explain the signs and symptoms of disease. This encapsulation ultimately results in illness scripts that undergo progressive refinement with experience.

- Traits and states of the physician's personality may influence decision making; mindfulness, need for cognition, reflective coping, and metacognition may be especially relevant.

- Certain medical specialties are associated with characteristic personality differences that may, in turn, influence clinical decision making.

- Individuals differ in their cognitive and decision styles. The Rational-Experiential Inventory is a widely used tool that can be used to assess preferences for particular decision styles.

- The actively open-minded thinking test appears particularly relevant to the study of medical decision making.

References

1. Jackson SA, Kleitman S. Individual differences in decision-making and confidence: Capturing decision tendencies in a fictitious medical test. *Metacogn Learn*. 2014 Apr;9(1):25–49.

2. Croskerry P, Musson D. Individual factors in patient safety. In: Croskerry P, Cosby KS, Schenkel S, Wears R, editors. *Patient Safety in Emergency Medicine.* Philadelphia, PA: Lippincott Williams & Wilkins; 2008. pp. 269–76.
3. Berger JT. The influence of physicians' demographic characteristics and their patients' demographic characteristics on physician practice: Implications for education and research. *Acad Med.* 2008 Jan;83(1):100–5.
4. Oken D. What to tell cancer patients: A study of medical attitudes. *JAMA.* 1961;175:120–8.
5. Camerer C, Issacharoff S, Loewenstein G, O'Donoghue T, Rabin M. Regulation for conservatives: Behavioral economics and the case for "asymmetric paternalism". *Univ PA Law Rev.* 2003 Jan;151(3):1211–54.
6. Thaler RH, Sunstein CR. *Nudge: Improving Decisions about Health, Wealth, and Happiness.* New York: Penguin Books; 2008.
7. du Preez HM. Dr James Barry (1789–1865): The Edinburgh years. *J R Coll Physicians Edinb.* 2012;42(3):258–65.
8. Borges NJ, Osmon WR. Personality and medical specialty choice: Technique orientation versus people orientation. *J Vocat Behav.* 2001 Feb;58(1):22–35.
9. McMurray JE, Linzer M, Konrad TR, Douglas J, Shugerman R, Nelson K. The work lives of women physicians results from the physician work life study. The SGIM Career Satisfaction Study Group. *J Gen Intern Med.* 2000 Jun;15(6):372–80.
10. Roter DL, Hall JA, Aoki Y. Physician gender effects in medical communication: A meta-analytic review. *JAMA.* 2002 Aug 14;288(6):756–64.
11. Kaplan SH, Gandek B, Greenfield S, Rogers W, Ware JE. Patient and visit characteristics related to physicians' participatory decision-making style. Results from the Medical Outcomes Study. *Med Care.* 1995 Dec;33(12):1176–87.
12. Bertakis KD, Franks P, Azari R. Effects of physician gender on patient satisfaction. *J Am Med Womens Assoc.* 2003 Spring;58(2):69–75.
13. Franks P, Williams GC, Zwanziger J, Mooney C, Sorbero M. Why do physicians vary so widely in their referral rates? *J Gen Intern Med.* 2000 Mar;15(3):163–8.
14. Ingram JC, Calnan MW, Greenwood RJ, Kemple T, Payne S, Rossdale M. Risk taking in general practice: GP out-of-hours referrals to hospitals. *Br J Gen Pract.* 2009;59(558):e16–24.
15. Bachman KH, Freeborn DK. HMO physicians' use of referrals. *Soc Sci Med.* 1999 Feb;48(4):547–57.
16. Baumhäkel M, Müller U, Böhm M. Influence of gender of physicians and patients on guideline-recommended treatment of chronic heart failure in a cross-sectional study. *Eur J Heart Fail.* 2009 Mar;11(3):299–303.
17. Huston S, Sleath B, Rubin RH. Physician gender and hormone replacement therapy discussion. *J Womens Health Gend Based Med.* 2001 Apr;10(3):279–87.
18. Mortier F, Bilsen J, Vander Stichele RH, Bernheim J, Deliens L. Attitudes, sociodemographic characteristics, and actual end-of-life decisions of physicians in Flanders, Belgium. *Med Decis Making.* 2003 Nov–Dec;23(6):502–10.
19. Cyran EM, Crane LA, Palmer L. Physician sex and other factors associated with type of breast cancer surgery in older women. *Arch Surg.* 2001 Feb;136(2):185–91.
20. Ghetti C, Chan BK, Guise JM. Physicians' responses to patient-requested cesarean delivery. *Birth.* 2004 Dec;31(4):280–4.
21. Sabin J, Nosek BA, Greenwald A, Rivara FP. Physicians' implicit and explicit attitudes about race by MD race, ethnicity, and gender. *J Health Care Poor Underserved.* 2009 Aug;20(3):896–913.

22. Lurie N, Slater J, McGovern P, Ekstrum J, Quam L, Margolis K. Preventive care for women: Does the sex of the physician matter? *N Engl J Med*. 1993 Aug 12;329(7):478–82.

23. Cassard SD, Weisman CS, Plichta SB, Johnson TL. Physician gender and women's preventive services. *J Womens Health*. 1997 Apr;6(2):199–207.

24. Kreuter MW, Strecher VJ, Harris R, Kobrin SC, Skinner CS. Are patients of women physicians screened more aggressively? A prospective study of physician gender and screening. *J Gen Intern Med*. 1995 Mar;10(3):119–25.

25. Flocke SA, Gilchrist V. Physician and patient gender concordance and the delivery of comprehensive clinical preventive services. *Med Care*. 2005 May;43(5):486–92.

26. Freeborn DK, Levinson W, Mullooly JP. Medical malpractice and its consequences: Does physician gender play a role? *J Gender Cult Health*. 1999 Sep;4(3):201–14.

27. Lutfey KE, Eva KW, Gerstenberger E, Link CL, McKinlay JB. Physician cognitive processing as a source of diagnostic and treatment disparities in coronary heart disease: Results of a factorial priming experiment. *J Health Soc Behav*. 2010 Mar;51(1):16–29.

28. Ross S, Moffat K, McConnachie A, Gordon J, Wilson P. Sex and attitude: A randomized vignette study of the management of depression by general practitioners. *Br J Gen Pract*. 1999 Jan;49(438):17–21.

29. Lyons Z, Janca A. Diagnosis of male depression: Does general practitioner gender play a part? *Aust Fam Physician*. 2009 Sep;38(9):743–6.

30. Bürgi Wegmann B, Sütsch G, Rickli H, Seifert B, Muntwyler J, Lüscher TF, Kiowski W, Attenhofer Jost CH. Gender and noninvasive diagnosis of coronary artery disease in women and men. *J Womens Health (Larchmt)*. 2003 Jan–Feb;12(1):51–9.

31. Rosen MP, Davis RB, Lesky LG. Utilization of outpatient diagnostic imaging: Does the physician's gender play a role? *J Gen Intern Med*. 1997 Jul;12(7):407–11.

32. Krupa C. Medical liability: By late career, 61% of doctors have been sued. *American Medical News* [Internet]. 2010 August 16. Available from: http://www.amednews. com/article/20100816/profession/308169946/2/. Accessed September 11, 2016.

33. Ellsbury K, Schneeweiss R, Montano DE, Gordon KC, Kuykendall D. Gender differences in practice characteristics of graduates of family medicine residencies. *J Med Educ*. 1987 Nov;62(11):895–903.

34. Maheux B, Dufort F, Lambert J, Berthiaume M. Do female general practitioners have a distinctive type of medical practice? *CMAJ*. 1988 Oct 15;139(8):737–40.

35. Britt H, Bhasale A, Miles DA, Meza A, Sayer GP, Angelis M. The sex of the general practitioner: A comparison of characteristics, patients, and medical conditions managed. *Med Care*. 1996 May;34(5):403–15.

36. Derose KP, Hays RD, McCaffrey DF, Baker DW. Does physician gender affect satisfaction of men and women visiting the emergency department? *J Gen Intern Med*. 2001 Apr;16(4):218–26.

37. Schiff GD, Kim S, Abrams R, Cosby K, Lambert B, Elstein AS, Hasler S, Krosnjar N, Odwazny R, Wisniewski MF, McNutt RA. Diagnosing diagnosis errors: Lessons from a multi-institutional collaborative project. In: Henriksen K, Battles JB, Marks ES, Lewin DI, editors. *Advances in Patient Safety: From Research to Implementation. Volume 2: Concepts and Methodology*. Rockville, MD: Agency for Healthcare Research and Quality (US); 2005. pp. 255–78.

38. Woodward CA, Hutchison BG, Abelson J, Norman G. Do female primary care physicians practise preventive care differently from their male colleagues? *Can Fam Physician*. 1996 Dec;42:2370–9.

39. Peckham C. *Medscape Lifestyle Report 2016: Bias and Burnout* [Internet]. New York: Medscape Inc.; 2016 January 13. p. 14. Available from: www.medscape. com/features/slideshow/lifestyle/2016/public/overview#page=14. Accessed September 11, 2016.

40. Deitchman MA, Kennedy WA, Beckham JC. Self-selection factors in the participation of mental health professionals in competency for execution evaluations. *Law Hum Behav*. 1991 Jun;15(3):287–303.

41. Curlin FA, Lantos JD, Roach CJ, Sellergren SA, Chin MH. Religious characteristics of U.S. physicians: A national survey. *J Gen Intern Med*. 2005 Jul;20(7):629–34.

42. de Pentheny O'Kelly C, Urch C, Brown EA. The impact of culture and religion on truth telling at the end of life. *Nephrol Dial Transplant*. 2011 Dec;26(12):3838–42.

43. Laqueur W. The question of judgment: Intelligence and medicine. *J Contemp Hist*. 1983;18(4):533–48.

44. Herrnstein RJ, Murray C. *The Bell Curve: Intelligence and Class Structure in American Life*. New York: Free Press; 1994.

45. Gottfredson LS. Why g matters: The complexity of everyday life. *Intelligence*. 1997 Jan–Feb;24(1):79–132.

46. Hauser RM. *Meritocracy, Cognitive Ability, and the Sources of Occupational Success*. Center for Demography and Ecology Working Paper No. 98-07. Madison, WI: University of Wisconsin-Madison; 2002. Available from www.ssc.wisc.edu/ cde/cdewp/98-07.pdf. Accessed March 4, 2017.

47. Rubeck RF, Witzke DB, Jarecky RK, Nelson B. The relationship between medical students' academic achievement and patterns of initial postgraduate placement. *Acad Med*. 1998 Jul;73(7):794–6.

48. Klaczynski PA, Robinson B. Personal theories, intellectual ability, and epistemological beliefs: Adult age differences in everyday reasoning biases. *Psychol Aging*. 2000 Sep;15(3):400–16.

49. Berner ES, Graber ML. Overconfidence as a cause of diagnostic error in medicine. *Am J Med*. 2008 May;121(5 Suppl):S2–23.

50. Peckham C. *Medscape Lifestyle Report 2016: Bias and Burnout* [Internet]. New York: Medscape. Inc.; 2016 Jan 13 [cited 2016 Sep 11]. p. 6. Available from: http://www. medscape.com/features/slideshow/lifestyle/2016/public/overview#page=6.

51. Stanovich KE, West RF, Toplak ME. Intelligence and rationality. In: Sternberg R, Kaufman SB, editors. *Cambridge Handbook of Intelligence*. 3rd ed. Cambridge, UK: Cambridge University Press; 2012. pp. 784–826.

52. Stanovich KE. *Rationality and the Reflective Mind*. New York: Oxford University Press; 2011. p. 19.

53. Graber M, Gordon R, Franklin N. Reducing diagnostic errors in medicine: What's the goal? *Acad Med*. 2002 Oct;77(10):981–92.

54. Wilson RM, Harrison BT, Gibberd RW, Hamilton JD. An analysis of the causes of adverse events from the Quality in Australian Health Care Study. *Med J Aust*. 1999 May 3;170(9):411–15.

55. Zwaan L, de Bruijne M, Wagner C, Thijs A, Smits M, van der Wal G, Timmermans DR. Patient record review of the incidence, consequences, and causes of diagnostic adverse events. *Arch Intern Med*. 2010 Jun 28;170(12):1015–21.

56. Singh H, Giardina TD, Meyer AND, Forjuoh SN, Reis MD, Thomas EJ. Types and origins of diagnostic errors in primary care settings. *JAMA Intern Med.* 2013;173:418–25.

57. Pitkanen M, Hurn J, Kopelman MD. Doctors' health and fitness to practise: Performance problems in doctors and cognitive impairments. *Occup Med (Lond).* 2008 Aug;58(5):328–33.

58. Boisaubin EV, Levine RE. Identifying and assisting the impaired physician. *Am J Med Sci.* 2001 Jul;322(1):31–6.

59. Ericsson KA, editor. *The Road to Expert Performance: Empirical Evidence from the Arts and Sciences, Sports, and Games.* Mahwah, NJ: Erlbaum; 1996.

60. Salthouse TA. When does age-related cognitive decline begin? *Neurobiol Aging.* 2009 Apr;30(4):507–14.

61. McArdle JJ, Ferrer-Caja E, Hamagami F, Woodcock RW. Comparative longitudinal structural analyses of the growth and decline of multiple intellectual abilities over the life span. *Dev Psychol.* 2002 Jan;38(1):115–42.

62. Li KZ, Lindenberger U, Freund AM, Baltes PB. Walking while memorizing: Age-related differences in compensatory behavior. *Psychol Sci.* 2001 May;12(3):230–7.

63. Clapp WC, Rubens MT, Sabharwal J, Gazzaley A. Deficit in switching between functional brain networks underlies the impact of multitasking on working memory in older adults. *Proc Natl Acad Sci USA.* 2011 Apr 26;108(17):7212–17.

64. Tetlock PE. *Expert Political Judgment: How Good Is It? How Can We Know?* Princeton, NJ: Princeton University Press; 2006.

65. Osava M. Sexing chickens: Say again?—Still a Japanese fraternity. *Online Asia Times* [Internet]. 2001 January 27. Available from: www.atimes.com/japan-econ/CA27Dh02.html. Accessed September 11, 2016.

66. Gawande A. Medical dispatch: No mistake. *New Yorker.* 1998 March 30: 74–81. Available at: www.newyorker.com/magazine/1998/03/30/no-mistake. Accessed March 4, 2017.

67. Robinson WL. Conscious competency: The mark of a competent instructor. *The Personnel Journal.* 1974 July:53(7):538–9.

68. Adams L. *Learning a New Skill Is Easier Said than Done* [Internet]. Solana Beach, CA: Gordon Training International; c2016. Available from: www.gordontraining.com/free-workplace-articles/learning-a-new-skill-is-easier-said-than-done/. Accessed September 11, 2016.

69. Bartlett FC. *Remembering: A Study in Experimental and Social Psychology.* Cambridge, UK: Cambridge University Press; 1932.

70. Schmidt HG, Rikers RM. How expertise develops in medicine: Knowledge encapsulation and illness script formation. *Med Educ.* 2007 Dec:41(12):1133–9.

71. Feltovich PJ, Barrows HS. Issues of generality in medical problem solving. In: Schmidt HG, de Volder ML, editors. *Tutorials in Problem-Based Learning: New Directions in Training for the Health Professions.* Assen, Netherlands: Van Gorcum; 1984. pp. 128–42.

72. Ericsson KA, Krampe RT, Tesch-Romer C. The role of deliberate practice in the acquisition of expert performance. *Psychol Rev.* 1993;100(3):363–406, 393–4.

73. Custers E, Boshuizen PA, Schmidt HG. The role of illness scripts in the development of medical diagnostic expertise: Results from an interview study. *Cognit Instr.* 1998;16(4):367–98.

74. Custers EJ, Boshuizen HP, Schmidt HG. The influence of medical expertise, case typicality, and illness script component on case processing and disease probability estimates. *Mem Cognit.* 1996 May;24(3):384–99.

75. van Schaik P, Flynn D, van Wersch A, Douglass A, Cann P. Influence of illness script components and medical practice on medical decision making. *J Exp Psychol Appl.* 2005 Sep;11(3):187–99.

76. Croskerry P. The theory and practice of clinical decision making. *Can J Anesth* 2005; 52(6): R1–R8.

77. Moulton CA, Regehr G, Mylopoulos M, MacRae HM. Slowing down when you should: A new model of expert judgment. *Acad Med.* 2007 Oct;82(10 Suppl):S109–16.

78. Dewberry C, Juanchich M, Narendran S. Decision-making competence in everyday life: The roles of general cognitive styles, decision-making styles and personality. *Pers Individ Dif.* 2013 Oct;55(7):783–8.

79. Bensi L, Giusberti F, Nori R, Gambetti E. Individual differences and reasoning: A study on personality traits. *Br J Psychol.* 2010 Aug;101(Pt 3):545–62.

80. Seibert SE, Kraimer ML. The five-factor model of personality and career success. *J Vocat Behav.* 2001 Feb;58(1):1–21.

81. Campbell JB, Heller JF. Correlations of extraversion, impulsivity and sociability with sensation seeking and MBTI-introversion. *Pers Individ Dif.* 1987;8(1):133–6.

82. Pilarik L, Sarmany-Schuller I. Personality predictors of decision-making of medical rescuers. *Studia Psychologica.* 2011;53:175–84.

83. Barrick MR, Mount MK. The big five personality dimensions and job performance: A meta analytic review. *Pers Psychol.* 1991;44:1–26.

84. Lievens F, Coetsier P, De Fruyt F, De Maeseneer J. Medical students' personality characteristics and academic performance: A five-factor model perspective. *Med Educ.* 2002 Nov;36(11):1050–6.

85. Coombs RH. *Mastering Medicine: Professional Socialization in Medical School.* New York: Free Press; 1978.

86. Mowbray RM, Davies B. Personality factors in choice of medical specialty. *Br J Med Educ.* 1971 Jun;5(2):110–17.

87. Schwarzt RW, Barclay JR, Harrell PL, Murphy AE, Jarecky RK, Donnelly MB. Defining the surgical personality: A preliminary study. *Surgery.* 1994 Jan;115(1):62–8.

88. Gaba DM, Howard SK, Jump B. Production pressure in the work environment: California anesthesiologists' attitudes and experiences. *Anesthesiology.* 1994 Aug;81(2):488–500.

89. Sexton JB, Thomas EJ, Helmreich RL. Error, stress, and teamwork in medicine and aviation: Cross sectional surveys. *BMJ.* 2000 Mar 18;320(7237):745–9.

90. Danziger S, Levav J, Avnaim-Pesso L. Extraneous factors in judicial decisions. *Proc Natl Acad Sci USA.* 2011 Aug 26;108(17):6889–92.

91. Baumeister RF. Ego depletion and self-control failure: An energy model of the self's executive function. *Self and Identity.* 2002;1:129–36.

92. www.nytimes.com/2011/08/21/magazine/do-you-suffer-from-decision-fatigue.html?_r=0. Accessed March 4, 2017.

93. Gailliot MT, Baumeister RF, DeWall CN, Maner JK, Plant EA, Tice DM, Brewer LE, Schmeichel BJ. Self-control relies on glucose as a limited energy source: Willpower is more than a metaphor. *J Pers Soc Psychol.* 2007 Feb; 92(2): 325–36.

94. Chan MY, Cohen H, Spiegel BM. Fewer polyps detected by colonoscopy as the day progresses at a Veteran's Administration teaching hospital. *Clin Gastroenterol Hepatol.* 2009 Nov;7(11):1217–23.

95. Linder JA, Doctor JN, Friedberg MW, Reyes Nieva H, Birks C, Meeker D, Fox CR. Time of day and the decision to prescribe antibiotics. *JAMA Intern Med.* 2014 Dec;174(12):2029–31.

96. Hagger MS, Wood C, Stiff C, Chatzisarantis NL. Ego depletion and the strength model of self-control: A meta-analysis. *Psychol Bull.* 2010 Jul;136(4):495–525.

97. Miller AK, Rufino KA, Boccaccini MT, Jackson RL, Murrie DC. On individual differences in person perception: Raters' personality traits relate to their psychopathy checklist–revised scoring tendencies. *Assessment.* 2011 Jun;18(2):253–60.

98. Duberstein PR, Chapman BP, Epstein RM, McCollumn KR, Kravitz RL. Physician personality characteristics and inquiry about mood symptoms in primary care. *J Gen Intern Med.* 2008 Nov;23(11):1791–5.

99. Boerebach BC, Scheepers RA, van der Leeuw RM, Heineman MJ, Arah OA, Lombarts KM. The impact of clinicians' personality and their interpersonal behaviors on the quality of patient care: A systematic review. *Int J Qual Health Care.* 2014 Aug;26(4):426–81.

100. Dunphy BC, Cantwell R, Bourke S, Fleming M, Smith B, Joseph KS, Dunphy SL. Cognitive elements in clinical decision-making: Toward a cognitive model for medical education and understanding clinical reasoning. *Adv Health Sci Educ Theory Pract.* 2010 May;15(2):229–50.

101. Lyvers M, Makin C, Toms E, Thorbery FA, Samios C. Trait mindfulness in relation to emotional self-regulation and executive function. *Mindfulness.* 2014 Dec;5(6):619–25.

102. Hafenbrack AC, Kinias Z, Barsade SG. Debiasing the mind through meditation: Mindfulness and the sunk-cost bias. *Psychol Sci.* 2014 Feb;25(2):369–76.

103. Lueke A, Gibson B. Mindfulness meditation reduces implicit age and race bias: The role of reduced automaticity of responding. *Soc Psychol Pers Sci.* 2015;6:284–91.

104. Ho NS, Sun D, Ting KH, Chan CC, Lee TM. Mindfulness trait predicts neurophysiological reactivity associated with negativity bias: An ERP study. *Evid Based Complement Alternat Med.* 2015;2015:212368.

105. Sibinga EM, Wu AW. Clinician mindfulness and patient safety. *JAMA.* 2010 Dec 8; 304(22):2532–33.

106. Brown KW, Ryan RM. The benefits of being present: Mindfulness and its role in psychological well-being. *J Pers Soc Psychol.* 2003 Apr;84(4):822–48.

107. Cacioppo JT, Petty RE. The need for cognition. *J Pers Soc Psychol.* 1982;42(1):116–31.

108. Bornstein BH. The impact of different types of expert scientific testimony on mock jurors' liability verdicts. *Psychol Crime Law.* 2004 Dec;10(4): 429–46.

109. Coutinho SA. The relationship between the need for cognition, metacognition and intellectual task performance. *Educ Res Rev.* 2006 Aug;1(5):162–4.

110. Fagela-Tiango C. College students' need for cognition, academic motivation, performance, and well-being. *The Mindanao Forum.* 2012 Dec;25(2):63–81.

111. Dwyer M. Need for cognition, life satisfaction and academic achievement. *Epistimi.* 2008;3:12–13.

112. Cacioppo JT, Petty RE, Feinstein JA, Jarvis WBJ. Dispositional differences in cognitive motivation: The life and times of individuals varying in need for cognition. *Psychological Bulletin.* 1996 Mar;119(2):197–253.
113. Greenglass E, Schwarzer R. The proactive coping inventory (PCI). In: Schwarzer R, editor. *Advances in Health Psychology Research* (CD-ROM). Berlin: Free University of Berlin. Institut for Arbeits, Organizations-und Gesundheipsychologie; 1998.
114. Flavell JH. Metacognition and cognitive monitoring: A new area of cognitive–developmental inquiry. *Am Psychol.* 1979 Oct;34(10):906–11.
115. Schraw G, Dennison RS. Assessing metacognitive awareness. *Contemp Educ Psychol.* 1994;19:460–75.
116. Colombo B, Iannello P, Antonietti A. Metacognitive knowledge of decision-making: An explorative study. In: Efklides A, Misailidi P, editors. *Trends and Prospects in Metacognition Research.* New York: Springer; 2010. pp. 445–72.
117. Iannello P, Perucca V, Riva S, Antonietti A, Pravettoni G. What do physicians believe about the way decisions are made? A pilot study on metacognitive knowledge in the medical context. *Eur J Psychol.* 2016 Nov 27;11(4):691–706.
118. Colbert CY, Graham L, West C, White BA, Arroliga AC, Myers JD, Ogden PE, Archer J, Mohammad ZT, Clark J. Teaching metacognitive skills: Helping your physician trainees in the quest to "know what they don't know." *Am J Med.* 2015 Mar;128(3):318–24.
119. Cools E, Van den Broeck H. Development and validation of the cognitive style indicator. *J Psychol.* 2007 Jul;141(4):359–87.
120. Bruine de Bruin W, Parker AM, Fischhoff B. Individual differences in adult decision-making competence. *J Pers Soc Psychol.* 2007 May;92(5):938–56.
121. Pacini R, Epstein S. The relation of rational and experiential information processing styles to personality, basic beliefs, and the ratio-bias phenomenon. *J Pers Soc Psychol.* 1999 Jun;76(6):972–87.
122. Betsch C. Präferenz für Intuition und Deliberation. Inventar zur Erfassung von affekt- und kognitionsbasiertem Entscheiden. [Preference for Intuition and Deliberation (PID): An inventory for assessing affect- and cognition-based decision-making]. *Zeitschrift für Differentielle und Diagnostische Psychologie.* 2004;25:179–97.
123. Witteman C, van den Bercken J, Claes L, Godoy A. Assessing rational and intuitive thinking styles. *Eur J Psychol Assess.* 2009;25:39–47.
124. Sladek RM, Bond MJ, Huynh L, Chew D, Phillips PA. Thinking styles and doctors' knowledge and behaviours relating to acute coronary syndrome guidelines. *Implement Sci.* 2008;3:23.
125. Sladek RM, Bond MJ, Phillips PA. Why don't doctors wash their hands? A correlational study of thinking styles and hand hygiene. *Am J Infect Control.* 2008 Aug;36(6):399–406.
126. Sladek RM, Bond MJ, Phillips PA. Age and gender differences in preferences for rational and experiential thinking. *Pers Individ Dif.* 2010 Dec; 49(8):907–11.
127. Calder LA, Forster AJ, Stiell IG, Carr LK, Brehaut JC, Perry JJ, Vaillancourt C, Croskerry P. Experiential and rational decision making: A survey to determine how emergency physicians make clinical decisions. *Emerg Med J.* 2012 Oct;29(10):811–16.
128. Jensen JL, Croskerry P, Travers AH. Paramedic clinical decision making. *International Journal of Paramedic Practice.* 2011;1:63–71.

129. Marks ADG, Hine DW, Blore RL, Phillips WJ. Assessing individual differences in adolescents' preference for rational and experiential cognition. *Pers Individ Dif.* 2008 Jan;44(1):42–52.

130. Fletcher JM, Marks ADG, Hine DW. Working memory capacity and cognitive styles in decision-making. *Pers Individ Dif.* 2011;50:1136–41.

131. Baron J. Why teach thinking? An essay. *Appl Psychol: Int Rev.* 1993;42:191–214.

132. Stanovich KE, West RF, Toplak ME. *The Rationality Quotient: Toward a Test of Rational Thinking.* Cambridge, MA: MIT Press; 2016. pp. 225–6.

133. Baron J, Badgio PC, Gaskins IW. Cognitive style and its improvement: A normative approach. In: Sternberg RJ, editor. *Advances in the Psychology of Human Intelligence.* London: Lawrence Erlbaum Associates; 1986. pp. 173–220.

134. Perkins D, Bushey B, Faraday M. Learning to reason. Final report, Grant No. NIE-G-83-0028, Project No 030717. Harvard Graduate School of Education; 1986.

135. Haran U, Ritov I, Mellers BA. The role of actively open-minded thinking in information acquisition, accuracy, and calibration. *Judgment and Decision Making.* 2013 May;8(3):188–201.

136. National Academies of Science, Engineering, and Medicine. *Improving Diagnosis in Health Care.* Washington, DC: National Academies Press; 2015.

Section IV

Challenges and Controversies in Diagnosis

10

Diagnostic Error

Karen Cosby

CONTENTS

Introduction

Diagnostic work is quite amazing. There are myriad diagnoses, yet relatively few physical expressions of disease; there are nearly infinite possibilities, yet only one right answer. The diagnostician often seems to be part scientist, part shaman. The fact that our diagnoses are accurate most of the time is almost surprising considering the conditions of uncertainty under which we operate and the endless demands of a busy clinical practice. The gambler might well hedge his bets on diagnostic success! We would be impressed with ourselves if it were not for the remaining 10% or so of missed diagnoses—cases that reflect human lives that are diminished by our failures. While there are countless steps and processes in diagnostic evaluations, diagnostic errors are generally attributed to cognitive errors, system flaws, or both. This chapter examines processes common to many diagnostic workups, reviews contributing factors to diagnostic errors, and provides suggestions for clinicians, healthcare institutions, and patients to improve the odds of making the right diagnosis.

Incidence of Diagnostic Error

The Institute of Medicine (IOM) report *To Err Is Human* heralded the birth of the patient safety movement in 2000 and awakened the healthcare community's consciousness to the reality that harm comes to many in the course of their medical care [1]. A quiet rumble of concern was raised early on by a few who felt that diagnostic delays and errors should be addressed as part of patient safety initiatives, but was largely ignored to attend to the "low lying fruit" of treatment-related harm [2–7]. Judgments about diagnostic errors were considered controversial, since they inevitably involve some degree of hindsight bias. Admittedly, some of the concern was rooted in the sense that we ought not to judge another person's cognitive process, lest we too be judged. Doing so was almost a betrayal of the inviolable trust among our peers and profession. Some of the hesitancy in addressing diagnosis error is understandable; diagnostic error is hard to define, difficult to detect, and challenging to study [8–12].

We are only just beginning to grapple with the definition and measurement of diagnostic error, and the early numbers reveal what many have suspected: diagnosis is a highly uncertain and imperfect process [13]. Most of us will experience a diagnostic error in our lifetimes [14]; 12 million Americans experience a diagnostic error each year [14,15]. The annual death toll attributable to diagnostic errors has been estimated to be between 40,000 and 80,000 lives each year in the United States [16]. Estimates from adult autopsies suggest that 71,400 of adults who die in hospitals each year (8.4% of all adult deaths) have a major diagnostic error, half of them significant enough to have likely impacted their outcome [17]. More than 34,000 patients die each year in intensive care units (ICUs) with Class I diagnostic errors—major conditions that might have been treatable and survivable had they been identified [18]. Data from pediatric ICUs reveal diagnostic errors in approximately 20% of autopsied cases [19]. Not all diagnostic errors occur in hospitals; half of missed diagnoses occur in ambulatory settings [14]. Five percent of outpatients experience a diagnostic failure each year, most commonly involving delays in the detection of lung, breast, and colon cancer [14,20]. Diagnostic errors occur across the spectrum of healthcare, including hospital wards, emergency departments (EDs), ICUs, and ambulatory settings; they also occur in all specialties. One recent U.S. survey found that 35% of adults recalled having personally experienced a medical error (in either themselves, a family member, or close friend); half of these were diagnostic errors [21,22].

Diagnosis error is the leading source of paid malpractice claims, accounting for the highest proportion of payments and the largest settlements, and involving the cases most likely to experience the worst patient outcomes (death or major disability). In a summary of 25 years of claims data, diagnostic error accounted for $38.8 billion U.S. dollars in settlements [23]. The biggest revelation is perhaps in finally admitting that healthcare providers, institutions, and processes are imperfect. Despite advanced training, commitment, passion, and high-tech care, we have simply failed to recognize and design for fallibility.

Sources of Error in Diagnosis

On the face of it, diagnostic activity would seem to be a mostly cognitive process that is largely dependent on expert clinical reasoning; we have explored the elements of reasoning and clinical decision making in the first half of this book. Diagnosis also relies on coordinated activity with different phases and processes of care provided by the healthcare system. The recent report from the National Academies of Sciences, Engineering, and Medicine, *Improving Diagnosis in Health Care*, describes diagnosis as a process, as illustrated in Figure 10.1 [24]. Attempts to address diagnostic error can be broken down into improvements made in each of these steps, beginning with patient entry into the healthcare system.

Patient Interview

Most diagnoses begin with a conversation—an interview with the patient. No one has yet established how accurate, reproducible, or valid the patient history is. However, clinicians will attest to how variable and difficult even a basic routine history can be. Patients may have difficulty describing their symptoms or use adjectives that differ from standard textbook descriptions. In academic settings, the information obtained from an inexperienced medical student may differ from that of other team members (sometimes better if they take the time to listen, but often not if they approach the history with closed-ended questions). The questions that are asked, how they are asked, and how much of a rapport the examiner has with the patient, can all

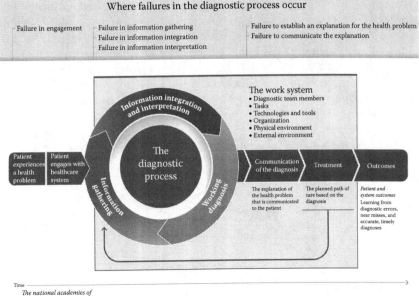

The work system
• Diagnostic team members
• Tasks
• Technologies and tools
• Organization
• Physical environment
• External environment

FIGURE 10.1

Diagnosis as a process, and points in the process where failure can occur. (Reprinted from National Academies of Science, Engineering, and Medicine. *Improving Diagnosis in Healthcare*, National Academy Press, Washington, DC, 2015. With permission from the National Academies Press, Copyright 2016 [24].)

influence the quality of information obtained. Sometimes patients don't offer the most relevant history; a worried or frightened patient may sense that something is wrong but fixate on extraneous sensations, unable to discriminate between those that are relevant and those that may simply be benign or normal variants. In some cases, patients may be unable to contribute to a meaningful interview—too ill, medically illiterate, or simply disengaged.

The history is probably one of the most important yet precarious aspects of diagnosis. The quality of information gained from the clinical history is essential to an accurate diagnosis, but is largely dependent on the expertise of the interviewer. There are two types of information obtained in the clinical history: (1) facts and (2) subjective descriptions. The facts include (in part) past medical events, current medications, timeline of symptoms, and results of previous testing. The subjective description includes the patient's account of their illness as they experience it. Features of the history may have varying degrees of predictive value in establishing a diagnosis. For example, abdominal pain that precedes vomiting is thought to more likely reflect a surgical problem than pain that follows vomiting. An experienced clinician elicits specific details in an attempt to match the patient's symptoms to recognizable patterns or illness scripts. The ability to obtain accurate and meaningful data about the patient is acknowledged as an essential skill by medical educators [25]. However, experts do more than just collect facts; they generate and test hypotheses. They form an overall impression, or gestalt, of the patient and their health. Experts derive contextual information during an interview that influences the quality of the history and the accuracy of hypotheses considered [26]. This is not usually conscious or deliberate, but rather derived from experience and expertise. Even with expertise in gathering and interpreting data, important information from the history is missing in many encounters. In close to a third of ED visits, significant information desired by clinicians is unknown or unavailable, a situation referred to as an *information gap* [27]. Incomplete information may prolong ED lengths of stay [27] and likely compromises diagnostic accuracy [28].

Physical Examination

Once the history is obtained, diagnosis begins with a fundamental clinical assessment of the patient: the physical examination. Historically, the history and physical examination (H&P) has been the most powerful and certainly the most routinely accessible diagnostic tool available to a physician. Over 50 years ago, one practitioner boasted that a diagnosis could be made on the basis of a carefully performed H&P in 88% of cases [29]. Medical science has arguably advanced in the half century since; not only do we recognize more conditions, but we also classify and sub-classify many more conditions. In addition, our armamentarium of diagnostic tools has become more sophisticated. Advanced imaging modalities (e.g., spiral computed tomography scans, magnetic resonance imaging, positron emission scans, echocardiography) have begun to replace the careful and deliberate use of the physical examination for many conditions, especially in the detection and evaluation of heart murmurs and dyspnea, and in the evaluation of abdominal pain and neurological conditions.

Although traditionally valued as a measure of a physician's skill, the physical examination seems to have become less esteemed in recent years [30]. Many lament the "demise of the physical exam" [31], referring to "disuse atrophy" [30] and "hyposkillia" [32] in the physical diagnosis skills of current medical graduates. One junior doctor has even suggested that the clinical exam "is dead" [33]. Concerned clinician–educators argue that the fault may not be in the actual value of the exam, but rather in the proper skill set and methods used for physical diagnosis, and the appropriate application of findings in decision making [34,35]. In recent years, at least one author has demonstrated that the basic H&P can still accomplish diagnosis within the first few hours of a patient's hospital course with about 60% accuracy; the addition of very basic labs and an electrocardiogram can increase the accuracy to about 80% [36], similar to the number cited 50 years ago.

The physical examination begins with the measurement of vital signs that reflect the overall health of the patient (blood pressure, heart rate, respiratory rate, temperature, and pulse oximetry). Triage systems rely almost exclusively on these basic measurements. These time-honored measurements are subject to significant interobserver variability and poor reproducibility [37–39]. Blood pressure measurements by primary care providers in Geneva overestimated the incidence of hypertension in 32% of normal volunteers [39]. In an ED population, independent measures of respiratory rates differed by 35%, and heart rates by 10%–15% [37]. Across different practice settings, a range of providers using different measuring devices at different moments in time may show significant variation. However, clinicians intuitively know this, and are pretty good at discriminating between normal or abnormal, or at least agreeing on which measurements require intervention [37]. Experienced clinicians tend to rely more on clinical context, serial measurements, trends over time, and combinations of findings (e.g., blood pressure *and* heart rate) to assess how meaningful any single value is.

Measurements in biological systems are tricky. Humans are not automatons with fixed responses to stimuli. There is inherent variability from moment to moment that can be impacted by environmental stimuli as well as equipment and technical proficiency. Patients may not understand this and may need reassurance and education in how to participate with in-home monitoring and follow-up to distinguish between measurements that reflect underlying disease and those that may be due to a *white coat effect*, the spurious abnormality known to occur in stressful settings [40].

The diagnostic utility of individual physical findings is not as well-established as for laboratory tests and imaging studies, and competencies in physical diagnosis are certainly not standardized across medical education. The sensitivity of many common findings described in textbooks is shockingly poor, so much so that some are no longer thought to have any value in the detection of disease. Homan's sign, described in the detection of deep venous thrombosis, has a sensitivity of 8%–56%, and a specificity of 39% [41]. The finding of a fluid wave in ascites has a sensitivity of 50%–53% [42]. Many common exam findings have only poor to moderate interobserver agreement, such as the Babinski reflex (kappa 0.17–0.59)[43,44] and crackles on lung exam (kappa 0.3–0.63) [45]. The diagnostic performance of some findings is so poor that it has been argued that they should be abandoned altogether, including the auscultation of bowel sounds in suspected bowel obstructions [46].

While sensitivity may be lacking, some examination findings have excellent specificity (Table 10.1), and their presence can be extremely valuable in establishing diagnoses [44,47–56]. Clinical vignettes collected from a physician survey about diagnostic errors found that the most common reason why physical findings were missed was the failure to look for them in the first place [57]! In examples like those in Table 10.1, the absence of an abnormality on exam may be unreliable in excluding disease, but the presence of a finding is quite good evidence for disease.

TABLE 10.1

Diagnostic Utility for Common Physical Findings

Physical Sign	Diagnosis	Sensitivity (%)	Specificity (%)	Reference
Palpable spleen	Splenomegaly	58	92	38
Pulse pressure over 80	Moderate to severe aortic regurgitation	57	95	39
Breast exam	Breast cancer	54	94	40
S3 heart sound	Ejection fraction under 50%	51	90	41
Murphy's sign	Cholecystitis	50–97	80	42
Phalen's test	Carpal tunnel	40–90	80	43
Initial impression	COPD	25	95	35, 44
Hepatojugular reflex	Congestive heart failure	24–33	95	35, 45
Femoral artery bruit	Peripheral arterial disease	20–29	95	46
Kernig's sign	Meningitis	5	95	47

Source: Adapted from Joshua, A.M., Celermajer, D.S., Stockler, M.R, *Int Med J.*, 35, 178–87, 2005; Hansen, M. et al., *Acta Neurol Scand.*, 90(3), 145–9, 1994.

Studies that evaluate the accuracy of physical findings in isolation may not fairly assess their clinical utility. Exam findings are sought based in part on the patient's narrative account of illness, thus they have context. If the leg hurts, searching for the area of tenderness helps localize the problem (skin, subcutaneous tissue, muscle, bone, or joint) and refine the differential diagnosis. And at the very least, looking at the area of interest shows concern! In addition, the value of any one part of the exam is enhanced by the presence or absence of other physical findings. For example, when percussion dullness is noted over the spleen, the sensitivity of palpation for splenomegaly is improved to 87% [58]. While no single finding may be diagnostic, a combination of findings can be more predictive, and a combination of physical findings interpreted in light of the patient's history can collectively improve diagnosis, as in the Alvarado score for appendicitis [59]. In addition, repeated examinations over time may be useful. Surgeons have long advocated for serial exams in the assessment of abdominal pain. With repeated examinations, a clinician is perhaps likely to become more confident in his exam, more aware of subtle abnormalities, or better able to detect changes as a condition worsens.

With so much diagnostic capability available from modern imaging modalities, it is natural that clinicians are becoming more dependent on them for help with diagnosis. What they may fail to appreciate is that abnormal physical findings may provide the first evidence of disease that might otherwise go undetected (even with imaging); without these clues, conditions may progress to a more advanced and less treatable state [60]. Nystagmus, asterixis, skin lesions from melanoma or vasculitic rashes, optic papillitis, and rebound tenderness are just a few examples of conditions that can only be detected by visual inspection or palpation [30]. One hospitalist recorded findings on physical examination that were pivotal in detecting conditions that impacted the immediate hospital course and patient outcome in 26% of cases admitted to his service over a one-month period [61]. Another author recalls his failure as a young medical student to recognize the significance of a pulse differential in a patient with chest pain; the aortic dissection went undiagnosed until a day later when the patient's condition worsened, by which time his course was irreversible [31].

Perceived limitations in the physical exam may have more to do with uncertainty, poor training, and lack of confidence than the value of the exam itself. Unfortunately, the clinical work environment and workflow pressures do little to encourage better exams. With the growth of the electronic health record and institution of duty hour restrictions for interns (both meant to improve patient safety!), internal medicine residents now spend only 12% of their clinical time with patients and 40% on computer-related tasks [62]. The modern patient has been aptly described as an *iPatient*—a virtual avatar more so than a real person [63]. It may take a determined refocus to bring the physical examination back to its appropriate place in diagnosis [30].

A number of technological advancements have improved the bedside exam. Portable point-of-care ultrasound has dramatically changed the diagnostic capability of bedside clinicians. With limited training, a number of physical conditions can rapidly and accurately be detected, including ascites, pleural effusions, pericardial effusions, cholelithiasis, and hydronephrosis. In addition, ultrasound can even be used to assess physiological parameters, determine fluid status, estimate cardiac function, and predict fluid responsiveness in shock-like states [64]. The use of a PanOptic™ ophthalmoscope may improve the ability of novices to visualize the retina [65]. A device using quantitative video-oculography (under development) may help with the assessment of eye movements to improve the diagnosis of vertebrobasilar stroke in patients with acute vertigo or dizziness [66]. Development of newer, more innovative diagnostic aids may augment basic skills and help revive the bedside assessment.

The failure to detect a significant physical finding is unlikely to be discovered in a retrospective review of the medical record, since you can't prove what was unknown but may have been detected if sufficient expertise and attention had been given to the subject. Missed physical findings, like missed diagnoses, may first manifest in the unexpected deterioration of patients or unexplained outcomes. It is impossible to prove whether a finding might have been present and detectable earlier, but one can't help but wonder what might have been possible if a more masterful clinician had examined the patient.

Old-school clinicians argue that there is more to be gained at the bedside than a simple checklist of signs and symptoms. The history should provide a narrative account that gives context to symptoms [67]; the patient's story is richer and more meaningful when grounded in a relationship with a caring and empathetic clinician [68]. The history and physical exam may be both diagnostic and therapeutic [69,70]. Time at the bedside done well can engender trust, improve communication, and, in the long run, contribute to improved diagnosis [71].

Radiology

The last few decades have seen an explosion in imaging capabilities: 3-D multidetector row computed tomography (CT), positron emission tomography (PET), and magnetic resonance imaging (MRI) are now widely available in most centers. All imaging requires visual interpretation ultimately limited by human perception, and thus have some irreducible rate of error. Error rates in radiology are widely quoted as about 3%–4%; however, these rates are typical of a general practice that sees mostly normal images. Rates rise to 30% if a sample of mostly abnormal images is reviewed [72,73]. In fact, if a collection of abnormal films is reviewed a second time, an error rate of 30% is still found with another independent reviewer, and the errors are not necessarily the same [74]. Most errors in interpretation are due to flawed perception; if the abnormality is pointed out, it becomes obvious after the fact. However, errors in imaging can arise for a number of reasons as outlined and classified by Kim and Mansfield, as shown in Table 10.2 [74]. While the most common error is simply failing to see what is there (42% of errors in general radiology), the second most common error is failing to detect a second (or even third) abnormality (22%), a problem described as *search satisfying* (also referred to by some as *search satisficing)*—when the examiner abandons the search, apparently satisfied once the first abnormality is detected [75]. The accuracy for general skeletal abnormalities is about 78% on plain radiography; this falls to 40% for second and third abnormalities in the same study [75].

Perceptual errors are difficult to understand and explain. Findings missed by CT and MRI tend to occur most often in the first or last image of a series, those that may be quickly surveyed and dismissed as the examiner scrolls through images [74]. When radiologists were asked to perform a lung-nodule detection task by CT in which the image of a gorilla had been embedded, experienced radiologists failed to see the gorilla, even though it was much larger than a typical nodule [76]. Eye-tracking studies verified that the radiologists had even looked directly at the gorilla image. When given an opportunity to look again specifically for the gorilla after completing their screen for a lung-nodule, the radiologists were able to visualize it. Experts are apparently not immune to the phenomenon of *inattentional blindness* [77]. There seems to an irreducible human fallibility in perception that limits accuracy in visual diagnosis.

Clinical Laboratory

Routine clinical laboratory tests include common and basic examinations of blood and serum, such as chemistry panels, hematology, serology, and coagulation

TABLE 10.2

Classification of Errors in Radiology and Their Frequency Distribution

Type	Cause	Explanation	Errors (%)
1	Complacency	A finding is appreciated but attributed to the wrong cause	0.9
2	Faulty reasoning	A finding is appreciated and interpreted as abnormal but is attributed to the wrong cause (true-positive finding misclassified)	9.0
3	Lack of knowledge	A finding is seen but is attributed to the wrong cause because of lack of knowledge	3.0
4	Underreading	A finding is present on the image, but is missed	42.0
5	Poor communication	An abnormality is identified and interpreted correctly but the message doesn't reach the clinician	0.0
6	Technique	A finding is missed because of the limitations of the examination or technique	2.0
7	Prior examination	A finding is missed because of failure to consult prior radiologic studies or reports	5.0
8	History	A finding is missed because of inaccurate or incomplete clinical history	2.0
9	Location	A finding is missed because the location of the lesion is outside the area of interest on an image	7.0
10	Satisfaction of search	A finding is missed because of failure to continue to search after the first abnormality is identified	22.0
11	Complication	Complication from a procedure	0.5
12	Satisfaction of report	A finding was missed because of overreliance on the radiology report from a previous examination	6.0

Source: From Kim, Y.W., Mansfield, L.T., *AJR Am J Roentgenol.*, 202(3), 465–70, 2014; Bruno, M.A. et al., *Radiographics.*, 35(6), 1668–76, 2015.

studies. These studies are performed countless times each day in processes that are tightly controlled and regulated. Laboratory medicine is highly automated and very accurate. The *total testing process* (TTP) involves much more than the few steps taken when the test is performed in the clinical lab. The TTP involves numerous steps or phases as described below and in Table 10.3 [78,79].

- *Pre pre-analytic*: The decision to test the selection and ordering of the test; the collection, identification, labeling and transport of the appropriate specimen to the lab

TABLE 10.3

Laboratory Errors by Stage of Total Testing Process (TTP) and the Distribution. Frequency of All Errors (%)

Stage of Total Testing Process (TTP)	Error Rate (%)	Distribution Frequency of Errors (%)
Pre pre-analytic	—	46–68.2
Inappropriate test request		
Order entry		
Patient/specimen misidentification		
Sample collected from infusion route		
Sample collection problem (hemolysis, clotting, inadequate volume)		
Inappropriate container		
Handling, storage, and transportation		
Pre-analytic	0.913	3.0–5.3
Sorting and Routing		
Pour-off		
Aliquoting, pipetting and labeling		
Centrifugation (time and/or speed)		
Analytic	0.002	7.0–13
Equipment malfunction		
Sample mix-ups		
Interference (endogenous or exogenous)		
Undetected failure in quality control		
Post-analytic	0.0715	12.5–20
Erroneous validation of analytical data		
Failure in reporting or addressing the report		
Excessive turnaround time		
Improper data entry, manual transcription error		
Failure, delay in reporting critical values		
Post post-analytic	—	25–45.5
Delayed, missed reaction to laboratory reporting		
Incorrect interpretation		
Inappropriate, inadequate follow-up plan		
Failure to order appropriate consultation		

Source: Adapted from Plebani, M., *Ann Clin Biochem.*, 47, 101–10, 2010.

- *Pre-analytic*: The acceptance of the specimen by the lab and the processing of the specimen that precedes testing, including centrifuging, diluting, and aliquoting the specimen
- *Analytic*: The actual test, now largely automated
- *Post-analytic*: The reporting of results to the clinician
- *Post post-analytic*: The interpretation and use of the test in decision making and communicating the result to the patient

The total error rate for all phases of work for the clinical lab is remarkably low, occurring in approximately 1.9% of all tests [79]. The analytic phase, where the test is actually performed, has an admirable error rate of only 0.002% [79]. The majority of all clinical lab errors (60%–70%) occur in the steps taken outside the physical realm of the lab and mostly outside the control of lab personnel [78]. Stat labs, in which the pre-analytic process is simplified, have an error rate less of than 1% [80]. In most cases of diagnostic errors, clinically significant errors resulting from laboratory testing problems occur not from errors generated by the lab itself, but rather in decisions made in selecting and using tests, interpreting test results, informing the patient of actionable results, and integrating the results into management decisions—tasks that ultimately depend upon the reasoning, organizational, and relational skills of healthcare providers.

Anatomical Pathology

Anatomical pathology refers to the examination of tissue, through either cytology (aspirates or fine needle biopsy) or histology (solid tumors, biopsies). While the clinical laboratory has automated processes for testing, tissue samples in anatomical pathology require more specialized processing, and ultimately require visual examination for the detection of abnormalities. Like clinical laboratory tests, anatomical pathology has both pre- and post-analytic phases, but unlike the clinical lab, the analytic phase of tissue examination has more variability and potential for error [81]. The most common source of error in anatomical pathology is from misinterpretation, either from errors in the visual interpretation of findings, or because of inadequate specimen collection or poor specimen preparation. Perception may be affected by fatigue, excessive workload, and ambient light; perceptual errors may also be attributed to the influence of heuristics and cognitive bias [82]. When studies to compare tumors obtain both cytology and histology specimens, there is an overall diagnostic discrepancy rate of 11.8% [83]. Second reviews of pathology specimens reveal an overall error rate of 6.7% across a wide range of tissue types, with one in six of these likely resulting in harm [84]. A review of published error rates on second reviews found a wide variation in error rates depending on the site and tumor type, ranging from a low of 1.3% (prostate) [85] to a high of 60% (thyroid cytology) [86,87]. Significant treatment decisions often depend on surgical or anatomical pathology, so the tolerance for error is especially low. Many recommend a routine mandatory review of all specimens in which cancer is suspected or whenever a major intervention is contemplated based on the result [88].

While the interpretative analytical phase of anatomical pathology is challenging enough, a process flow map for biopsy specimens reveals additional complexity. Ten major steps are required from arrival to the lab until the specimen reaches the pathologist, each with its own failure rate or potential for problems [89]. When detailed further, between two and three hundred individual smaller steps have been identified from the time the test is ordered, specimen acquired, specimen processed, study interpreted and reported, and results finally acted on [85]. Dermatopathology biopsy specimens may pass through the hands of twenty different people stationed in different workplaces before results are obtained [90]. It's a wonder that error rates aren't even higher.

The communication of pathology results between the pathologist and the clinical team typically occurs in the form of a pathology report. The language, terminology, and formatting of pathology reports may be confusing to clinicians. When surgeons were given a test to evaluate how well they understood pathology reports, 30% demonstrated a basic misunderstanding of the results [91].

Communication and Coordination of Care

Within each specialized area, a determined focus on accuracy can optimize results. Individuals can study, practice, and hone their clinical and cognitive skills. Labs and imaging centers can and have achieved excellence in interpretation. However, much of diagnostic error occurs during the exchange of information and coordination of care between different silos, in a sort of no-man's-land. Ambiguities often exist about who assumes care and when: who is responsible for relaying, receiving, interpreting, and integrating different types of information into the final diagnosis? These struggles are discussed in more detail in Chapter 11, and are likely very relevant to many diagnostic failures.

The State of Medical Knowledge

The accuracy of diagnosis depends on clinical expertise, reliable evidence and the appropriate use of diagnostic tests. Evidence-based medicine (EBM) was developed and introduced in the 1980s as a means of collecting, synthesizing, and analyzing data about clinical questions. The EBM approach has since dominated the evaluation of clinical research and has been used to develop clinical practice guidelines. Although widely accepted and highly respected, critics have challenged its assumptions and noted its flaws and limitations [92–94]. The evidence in EBM is viewed in a hierarchical scheme, with randomized control trials (RCTs) given the greatest weight, followed by cohort studies, then observational studies and case reports. Expert opinion is the least valued. Quantitative studies are valued over qualitative methods. RCTs are widely thought to provide the best and most objective evidence. However, in the EBM paradigm, large RCTs can almost silence voices that are considered less authoritative and lower in the hierarchy of evidence. Evidence from RCTs can be tainted by bias, or by conflicts of interest. Biases can creep into the study design and analysis in many forms, including optimism, overconfidence, confirmation, anchoring, search satisficing, groupthink, and "think within the box" [92]. Conflicts of interest may be financial or simply ideological. Since bias is unconscious, and conflicts of interest often subtle, these flaws may be unintentional and even unavoidable. And they may be difficult to detect. Critics of EBM suggest that this danger can be mitigated by leveling the hierarchy of evidence to weight other voices on a par with RCTs, or at least encourage debate from opposing views that might detect and compensate for bias.

Another concern expressed by critics, and acknowledged by EBM advocates, is a *reductionist bias*, that is, using methods to address risk and likelihood for disease using data from a population that may differ from the unique characteristics of any given patient [93]. Blind allegiance to practice guidelines based on EBM may simplify clinical work, but may also contribute to some diagnostic error.

A second limitation in the current state of diagnostics is poor penetration of existing evidence into clinical practice [95]. A review of U.S. adults found that only about half of patients received routine care recommended by national guidelines [96]. There are numerous barriers to the use of existing guidelines [95]. There is a natural inertia to change and reluctance to trust new standards. Even for highly motivated clinicians, a major barrier is simply keeping up with the sheer volume of material [95,97]. One author noted that in a single day, his medical unit cared for 18 patients with a total of 44 diagnoses; the available guidelines for his patients totaled 3679 pages [97]!

Another obstacle to diagnostic work is the lack of understanding of basic mathematical principles needed to use diagnostic tests [98,99]. Although most clinicians have some rudimentary understanding of diagnostic testing, most don't apply that knowledge at the bedside. A survey of 300 practicing physicians found that although some articulated an understanding of sensitivity and specificity, only 3% reported using Bayesian methods in their daily work, and only 1% used receiver operator curves (ROCs) and likelihood ratios [100].

Not all data is easy to apply in actual clinical work. Feinstein has urged the redefinition of how we measure the clinical utility of diagnostic tests, focusing on "diagnostic efficacy" (how the test performs in patients *with* symptoms) versus "accuracy" (how the test performs in a known population with a previously defined disease state) [101]. Convenient tools embedded within clinical documents might improve the use of data. When clinicians were given test performance characteristics of clinical data, they made more accurate predictions of posttest probabilities when the data was presented in a visual format compared with text alone [102]. Estimates of the risk of disease are often inferred from probability data; however, physicians (and laypeople) perform better when information is represented in natural frequencies [103]. Some improvement in diagnosis may be gained simply by putting data in a form that is easier to understand *in vivo*, at the point of patient contact.

Strategies and Solutions to Address Diagnostic Error

Although medicine uses scientific principles and avails itself of modern technology, ultimately we deal with biological systems that have intrinsic variability and unpredictability, a fact that makes diagnostic work difficult, imprecise, and error-prone. Diagnosis is complex; the proposed work to improve diagnosis is formidable. Each of the categories discussed in this chapter is a potential source of error. Each one offers a focus for improvement efforts. The first step toward improvement is acknowledging the need for change. The publication of the 2015 report *Improving Diagnosis in Healthcare* is an important step that may help increase awareness and recruit resources to address the challenge [24].

A number of leading experts have summarized ideas for improving diagnosis [104–109]. Three main approaches have been described: (1) improve clinician performance, (2) improve the design of the system, and (3) engage the patient; the suggested ideas are summarized in Tables 10.4–10.6. A fourth could reasonably be suggested: (4) improve the interface between clinicians, healthcare systems, and patients using help from experts in human factors engineering [108]. Finally, we need to improve the science of diagnosis and take advantage of technological advancements in non-medical areas.

Optimize the Performance of Clinicians

We can hope changes in education and training (both in content and methodology) will make clinicians better diagnosticians; ideas for improvement

TABLE 10.4

Ideas to Improve Diagnostic Performance of Clinicians

General Goal	Suggested Methods
Improve knowledge and expertise	Improve clinical experience: increase number of patients during training, use simulated cases and standardized patients, use online courses with cases, improve bedside mentoring.
	Teach biostatistics.
	Provide feedback for calibration.
	Improve teamwork and communication skills during training.
	Encourage dialog with specialists in the laboratory and radiology.
	Teach and practice awareness of personal limitations and potential for error.
Improve reasoning skills	Add training in cognition, metacognition, decision making, and reasoning.
Provide cognitive aids	Provide access to specialists, second opinions, and consultants.
	Integrate clinical decision support tools within the medical record, including differential diagnosis generators, checklists, and guidelines.
	Improve data display on computers to make data easier to view and interpret.

Source: Adapted from McDonald, K.M., Matesic, B., Contopoulos-Ioannidis, D.G., Lonhart, J., Schmidt, E., Pineda, N., Ioannidis, J.P., *Ann Intern Med.*,158(5 Pt 2), 381–9, 2013.

TABLE 10.5

Ideas to Improve System Support for Diagnosis

General Goal	Suggested Methods
Clinical laboratory	Simplify and streamline processes (fewer steps lead to fewer errors).
	Automate, when possible.
	Reduce chance of misidentified specimens (barcoding).
	Eliminate handwritten results (where possible use computer reports).
	Use real-time automated paging to notify clinicians of critical lab abnormalities.
	Use information technology support to track and follow abnormal test results.
Anatomical pathology	Use second opinions and mandatory reviews.
	Standardize processes; standardized diagnostic criteria.
	Use checklists.
	Use concise terminology and formatting reports that are unambiguous and easily understood by clinicians.
Radiology	Develop computer-aided diagnostic systems to improve visual recognition of signals.
	Improve image processing techniques to optimize lesion detection (e.g., pulmonary nodules).
	Develop checklists for common misses.
	Apply eye-tracking technology to alert radiologists of sites where prolonged dwell time suggests something aberrant.
Improve utilization of specialists	Improve the range of specialists available for medical teams, including pharmacists and librarians.
Enhance technology	Use ultrasound to improve diagnostic sampling for biopsies.
	Encourage study and development of improved diagnostic tests, such as cap-fitted and high-resolution colonoscopy, for example.
	Build an infrastructure that facilitates the flow of information between clinicians and between providers and their patients.
	Develop and implement effective clinical decision support.
Improve environment	Provide an environment with minimal distractions and allow sufficient time for productive work; optimize shift scheduling.
	Encourage usability testing for medical equipment.
Create a culture of learning	Develop methods to monitor the diagnostic process and related errors.
	Reward quality.
	Give individuals feedback.
	Facilitate communication between clinicians and departments.
	Create structures for teams and teamwork.

Source: Adapted from McDonald, K.M. et al., *Ann Intern Med.*, 158(5 Pt 2), 381–9, 2013.

TABLE 10.6

How Patients Can Improve Diagnosis

General Goal	Suggested Methods
Educate themselves	Read and know about your diagnosis. Ask questions. Ask why a test is needed, and how to get the result. Understand what to expect and when to worry if you don't improve. Ask your doctor how you can learn more about your diagnosis.
Prepare for visits	Provide a one-page summary of symptoms and timeline of illness. Keep an accurate copy of medications. Whenever possible, bring an advocate with you to help remind you of questions and listen with you.
Have a copy of all your medical record	Keep copies of tests and reports.
Monitor your health	When appropriate, keep logs of your condition (such as home blood pressure or glucose monitoring) and bring them to each doctor's visit. Ensure follow-up for every test; don't assume no news is good news.
Speak up	Verbalize your questions and concerns. Ask what unfamiliar terms mean. Ask your doctor how sure they are of the diagnosis and if you could benefit from a second opinion. If you seek a second opinion, be sure to bring all medical records with you.

Source: Adapted from McDonald, K.M., Bryce, C.L., Graber, M.L., *BMJ Qual Saf.*, 22, 8833–ii39, 2013.

are summarized in Table 10.4. The field of cognitive psychology provides insight into how we think: about how we collect data, how our minds perceive visual images, and how we make decisions. There is a growing imperative to educate physicians about strategies to mitigate cognitive biases (see Chapter 15), and increased awareness of the need for feedback on diagnostic performance. Clinicians rely on diagnostic tests to make decisions; they need to be educated on the operating characteristics of diagnostic tests and given appropriate decision support. These efforts will likely be tedious and slow, and will probably produce small and incremental change. Systems can offer support to make information more accessible at the right time, in the right format. Teamwork training can improve how we function together and in groups to help each other achieve more accurate diagnoses.

Improve Systems of Care

Our current healthcare system serves an antiquated model of care that is physician-centric, where all information flows to and between physicians. We have excellence in specialty silos, but hazy boundaries between those silos where healthcare information is lost and responsibility for patients left to chance. We are largely blind to or even indifferent to diagnostic failures, too often attributing them to difficult diseases or even patient noncompliance. Alternative models of care have been proposed that are patient-centric, promote cross-disciplinary collaboration, actively track error, and provide innovative tools for diagnostic support [24]. Those who are entrenched in the current system may lack enough creative vision and imagination to redesign how diagnosis take place; insight may come from nontraditional sources, such as philosophers, cognitive scientists, psychologists, human factors engineers, business people, economists, mathematicians, statisticians, and yes, patients themselves.

Principles from non-medical industries have helped improve common processes. The simplification, standardization, and automation of routine processes have improved common laboratory testing. Interpretive studies, such as imaging and anatomical pathology, may benefit further from computerized aids, such as the computer-aided detection of pulmonary nodules [110,111]. The growth of information technology has tremendous potential for improving and streamlining clinical care, although development has been slow. Embedding clinical decision support into routine tasks, tracking and monitoring lab tests, and designing electronic methods of informing clinicians of critical lab results are just a few of many potential applications. A few ideas for the system support of diagnosis are described in more detail in Table 10.5 and discussed in more detail in Chapter 16 (Diagnostic Support from Information Technology).

Engage and Empower Patients

Patients need (and want) to be involved in the diagnostic process, after all, they have the most to gain or lose in the process [112–114]. Patient stories reveal recurring themes of how diagnostic errors occur, and their stories are quite compelling [113]. Patients who have experienced catastrophic diagnostic error often relate that they tried to warn others but were dismissed or ignored. Many were simply not believed, their opinions or concerns thought to be uninformed or not relevant, or clinicians were too busy or preoccupied with seemingly more important tasks [113]. Change will require a paradigm shift in our model of care to give more respect to patients and their concerns and give them more responsibility and ownership for their care; this focus should be given as much priority as the taking of vital signs. An engaged patient who is informed can contribute actively and compensate for many imperfections in their process of care. Guidelines for patients, summarized in Table 10.6, give practical advice to help patients themselves take ownership in driving the diagnostic process toward a safe conclusion [114,115].

We should not expect patients to take more responsibility for their care without guidance. Physicians can help by flattening the authority gradient and prioritizing communication with patients. Healthcare teams should have someone designated available to help patients as they navigate their illness. Systems can be better designed to feed test results and educational materials directly to patients and allow them to play a larger role in tracking their test results and managing their healthcare information.

Innovation and Technological Advancements

We need breakthrough advancements in science and technology and innovative changes in the design of our systems of care. Efforts targeted directly at diagnostic error may not achieve the greatest gains. Spinoffs from space exploration have led to thousands of useful products, including infrared ear thermometers, a microvolt T-wave alternans test useful for detecting people at increased risk for sudden death cardiac events, and a MicroMed DeBakey Ventricular Assist Device small enough to be used in children [116,117]. There are likely many other technologies we can look to for improvement.

Conclusion

As the challenges and difficulties with diagnosis have been acknowledged, the next decade offers hope for improvement. The medical community and the public at large now seem more aware of the limitations of diagnosis and our potential for improvement. The diagnosis of biological systems is a remarkable and sophisticated process, and the risk of failure and misdiagnosis has probably been underestimated and underappreciated. Increasing awareness of diagnostic error has highlighted the need to think more about our clinical care and how we can use both the art and science of medicine to optimize our diagnostic endeavors.

SUMMARY POINTS

- Diagnosis is an imperfect science.
- The actual incidence of error in diagnosis is unknown, but is certainly far more common than we have acknowledged in our recent past.
- Although we have many sophisticated tools, many diagnoses can be detected by a well-done history and physical examination, the foundation of all diagnostic work.
- Errors in diagnostic reasoning may be due to inadequate knowledge, or flaws in reasoning.

- Techniques that rely on visual interpretation are subject to human limitations in perception as well as cognitive biases that influence what we "see."
- Diagnostic work relies on complex processes that are error-prone.
- Failure in the diagnostic process can occur simply due to a lack of coordination in care, and inadequate communication.
- There are five main strategies to target diagnostic error:
  - Optimize clinician performance
  - Design better systems
  - Engage patients
  - Improve the interface between patients, providers, and the health-care system
  - Advance the science and technology of diagnosis through innovation

References

1. Kohn LT, Corrigan J, Donaldson MS. *To Err is Human: Building a Safer Health System*. Washington DC: National Academy Press, 2000.
2. Schiff GD, Kim S, Krosnjar N, Wisniewski MF, Bult J, Fogelfeld L, McNutt RA. Missed hypothyroidism diagnosis uncovered by linking laboratory and pharmacy data. *Arch Intern Med*. 2005;165(5):574–7.
3. Schiff GD, Kim S, Abrams R, Cosby K, Lambert B, Elstein AS, Hasler S, Krosnjar N, Odwazny R, Wisniewski MF, McNutt RA. Diagnosing diagnosis errors: Lessons from a multi-institutional collaborative project. In: Henrikson K, Battles JB, Marks ES, Lewin DI (eds). *Advances in Patient Safety: From Research to Implementation (Volume 2: Concepts and Methodology)*. Rockville (MD): Agency for Healthcare Research and Quality (US): 2005 Feb. Advances in Patient Safety.
4. Graber M. Diagnostic errors in medicine: A case of neglect. *Jt Comm J Qual Patient Saf*. 2005;31(2):106–13.
5. Graber ML, Franklin N, Gordon R. Diagnostic error in internal medicine. *Arch Intern Med*. 2005;165(13):1493–9.
6. Cosby KS, Roberts R, Palivos L, Ross C, Schaider J, Sherman S, Nasr I, Couture E, Lee M, Schabowski S, Ahmad I, Scott RD 2nd. Characteristics of patient care management problems identified in emergency department morbidity and mortality investigations during 15 years. *Ann Emerg Med*. 2008;51(3):251–61.
7. Croskerry P, Campbell S. Comment on The Canadian Adverse Events Study: The incidence of adverse events among hospital patients in Canada. *CMAJ*. 2004;171(8):833; author reply 834.
8. Graber ML, Trowbridge R, Myers JS, Umscheid CA, Strull W, Kanter MH. The next organizational challenge: Finding and addressing diagnostic error. *Jt Comm J Qual Patient Saf*. 2014;40(3):102–10.

9. Wachter RM. Why diagnostic errors don't get any respect—and what can be done about them. *Health Aff (Millwood)*. 2010;29(9):1605–10.
10. Graber ML, Wachter RM, Cassel CK. Bringing diagnosis into the quality and safety equations. *JAMA*. 2012;308(12):1211–2.
11. Singh H, Graber ML. Improving diagnosis in health care: The next imperative for patient safety. *New Engl J Med*. 2015;373(26):2493–5.
12. Newman-Toker DE, Pronovost PJ. Diagnostic errors: The next frontier for patient safety. *JAMA*. 2009;301(10):1060–2.
13. Zwaan L, Singh H. The challenges in defining and measuring diagnostic error. *Diagnosis*. 2015;2(2):97–103.
14. Singh H, Meyer AND, Thomas EJ. The frequency of diagnostic errors in outpatient care: estimations from three large observational studies involving U.S. adult populations. *BMJ Qual Saf*. 2014;23(9):727–31.
15. Singh H, Sittig DF. Advancing the science of measurement of diagnostic errors in healthcare: The Safer Dx framework. *BMJ Qual Saf*. 2015;24(2):103–10.
16. Leape LL, Berwick DM, Bates DW. Counting deaths due to medical errors. *JAMA*. 2002;288(19):2404–5.
17. Shojania KG, Burton EC, McDonald KM, Goldman L. Changes in rates of autopsy-detected diagnostic errors over time: A systematic review. *JAMA*. 2003;289(3):2849–2856.
18. Winters B, Custer J, Galvagno SM Jr, Colantuoni E, Kapoor SG, Lee H, Goode V, Robinson K, Nakhasi A, Pronovost P, Newman-Toker D. Diagnostic errors in the intensive care unit: A systematic review of autopsy studies. *BMJ Qual Saf*. 2012;21(11):894–902.
19. Custer JW, Winters BD, Goode V, Robinson KA, Yang T, Pronovost PJ, Newman-Toker DE. Diagnostic errors in the pediatric and neonatal ICU: A systematic review. *Pediatri Crit Care Med*. 2015;16:29–36.
20. Gandhi TK, Kachalia A, Thomas EJ, Puopolo AL, Yoon C, Brennan TA, Studdert DM. Missed and delayed diagnoses in the ambulatory setting: A study of closed malpractice claims. *Ann Intern Med*. 2006;145(7):488–96.
21. Blendon RJ, DesRoches CM, Brodie M, Benson JM, Rosen AB, Schneider E, Altman DE, Zapert K, Herrmann MJ, Steffenson AE. Views of practicing physicians and the public on medical errors. *N Engl J Med*. 2002;347(24):1933–40.
22. Berner ES, Graber ML. Overconfidence as a cause of diagnostic error in medicine. *Am J Med*. 2008;121(5A):S2–S23.
23. Saber Tehrani AS, Lee H, Mathews SC, Shore A, Makary MA, Pronovost PJ, Newman-Toker DE. 25-year summary of U.S. malpractice claims for diagnostic errors 1986–2010: An analysis from the National Practitioner Data Bank. *BMJ Qual Saf*. 2013;22:672–80.
24. National Academies of Science, Engineering and Medicine. *Improving Diagnosis in Healthcare*. Washington, DC: National Academy Press, 2015.
25. Teutsch C. Patient–doctor communication. *Med Clin North Am*. 2003;87(5):1145–55.
26. Hobus PP, Schmidt HG, Boshuizen HP, Patel VL. Contextual factors in the activation of first diagnostic hypotheses: Expert-novice differences. *Med Educ*. 1987;21(6):471–6.
27. Stiell A, Forster AJ, Stiell IG, van Walraven C. Prevalence of information gaps in the emergency department and the effect on patient outcomes. *CMAJ*. 2003; 169(10):1023–8.

28. Heuer JF, Gruschka D, Crozier TA, Bleckmann A, Plock E, Moerer O, Quintel M, Roessler M. Accuracy of prehospital diagnoses by emergency physicians: Comparison with discharge diagnosis. *Eur J Emerg Med*. 2012;19(5):292–6.

29. Crombie DL. Diagnostic process. *J Coll Gen Pract*. 1963;6:579–89.

30. Shattner A. Revitalizing the history and clinical examination. *Am J Med*. 2012;125(4):e1–e3.

31. Jauhar S. The demise of the physical exam. *N Engl J Med*. 2006;354(6):548–51.

32. Fred HL. Hyposkillia: Deficiency of clinical skills. *Texas Heart Inst J* 2005;32(3):255–7.

33. Patel K. Is clinical examination dead? *BMJ*. 2013;346:f3442.

34. Verghese A, Horwitz RI. In praise of the physical examination. *BMJ*. 2009;339:b5448.

35. Schattner A. The clinical encounter revisited. *Am J Med*. 2014;127(4):268–74.

36. Paley L, Zornitzki T, Cohen J, Friedman J, Kozak N, Schattner A. Utility of clinical examination in the diagnosis of emergency department patients admitted to the department of medicine of an academic hospital. *Arch Intern Med*. 2011;171(15):1394–6.

37. Edmonds ZV, Mower WR, Lovato LM, Lomeli R. The reliability of vital sign measurements. *Ann Emerg Med*. 2002;39(3):233–7.

38. Roubsanthisuk W, Wongsurin U, Saravich S, Buranakitjaroen, P. Blood pressure determination by traditionally trained personnel is less reliable and tends to overestimate the severity of moderate to severe hypertension. *Blood Press Monit*. 2007;12(2):61–8.

39. Sebo P, Pechere-Bertschi A, Herrman FR, Haller DM, Bovier P. Blood pressure measurements are unreliable to diagnose hypertension in primary care. *J Hypertens*. 2014;32(3):509–17.

40. Pickering TG, James GD, Boddie C, Harshfield GA, Blank S, Laragh JH. How common is white coat hypertension? *JAMA*. 1988;259(2):225–8.

41. Ebell MH. Evaluation of the patient with suspected deep venous thrombosis. *J Fam Pract*. 2001;50(2):167–71.

42. Cummings S, Papadakis M, Melnick J, Gooding GA, Tierney Jr LM. The predictive value of physical examinations for ascites. *West J Med*. 1985;142(5):633–6.

43. Hansen M, Sindrup SH, Christensen PB, Olsen NK, Kristensen O, Friis ML. Interobserver variation in the evaluation of neurological signs: Observer dependent factors. *Acta Neurol Scand*. 1994;90(3):145–9.

44. Joshua AM, Celermajer DS, Stockler MR. Beauty is in the eye of the examiner: Reaching agreement about physical signs and their value. *Int Med J*. 2005;35(3):178–87.

45. Metlay JP, Kapoor WN, Fine MJ. Does this patient have community-acquired pneumonia? Diagnosing pneumonia by history and physical examination. *JAMA*. 1997;278(17):1440–5.

46. Breum BM, Rud B, Kirkegaard T, Nordentoft T. Accuracy of abdominal auscultation for bowel obstruction. *World J Gastroenterol*. 2015;21(34):10018–24.

47. Grover SA, Barkun AN, Sackett DL. The rational clinical examination. Does this patient have splenomegaly? *JAMA*. 1993;270(18):2218–21.

48. Grayburn PA, Smith MD, Handshoe R, Friedman BJ, DeMaria AN. Detection of aortic insufficiency by standard echocardiography, pulsed Doppler echocardiography, and auscultation. *Ann Intern Med*. 1986;104(5):599–605.

49. Barton MB, Harris R, Fletcher SW. The rational clinical examination: Does this patient have breast cancer? The screening clinical breast examination: Should it be done? How? *JAMA*. 1999;282(13):1270–80.

50. Patel R, Bushnell DL, Sobotka PA. Implications of an audible third heart sound in evaluating cardiac function. *West J Med*. 1993;158(6):606–9.

51. Adedeji OA, McAdam WA. Murphy's sign, acute cholecystitis and elderly people. *J R Coll Surg Edinb*. 1996;41(2):88–9.

52. Kuschner SH, Ebramzadeh E, Johnson D, Brien WW, Sherman R. Tinel's sign and Phalen's test in carpal tunnel syndrome. *Orthopedics*. 1992;15(11):1297–302.

53. Badgett RG, Tanaka DJ, Hunt DK. Can moderate chronic obstructive pulmonary disease be diagnosed by historical and physical findings alone? *Am J Med*. 1993;94(2):188–96.

54. Marantz PR, Kaplan MC, Alderman MH. Clinical diagnosis of congestive heart failure in patients with acute dyspnea. *Chest*. 1990;97(4):776–81.

55. Criqui MH, Fronek A, Klauber MR, Barrett-Connor E, Gabriel S. The sensitivity, specificity, and predictive value of traditional clinical evaluation of peripheral arterial disease: Results from noninvasive testing in a defined population. *Circulation*. 1985;71(3):516–22.

56. Thomas KE, Hasbun R, Jekel J, Quagliarello VJ. The diagnostic accuracy of Kernig's sign, Brudzinski's sign, and nuchal rigidity in adults with suspected meningitis. *Clin Infect Dis*. 2002;35(1):46–52.

57. Verghese A, Charlton B, Kassirer JP, Ramsey M, Ioannidis JPA. Inadequacies of physical examination as a cause of medical errors and adverse events: A collection of vignettes. *Am J Med*. 2015;128(2):1322–24.

58. Barkun AN, Camus M, Green L, Meagher T, Coupal L, De Stempel J, Grover SA. The bedside assessment of splenic enlargement. *Am J Med*. 1991;91(5):512–8.

59. Alvarado A. A practical score for the early diagnosis of acute appendicitis. *Ann Emerg Med*. 1986;15(5):557–64.

60. Grais IM. Little things can be big diagnostic clues. *Tex Heart Inst J*. 2011;38(6):617–9.

61. Reilly BM. Physical examination in the care of medical inpatients: An observational study. *Lancet*. 2003;362(9390):1100–5.

62. Feddock CA. The lost art of clinical skills. *Am J Med*. 2007 Apr;120(4):374–8.

63. Verghese A. Culture shock: Patient as icon, icon as patient. *N Engl J Med*. 2008;359(26):2748–51.

64. Bailitz J. A Problem-based approach to resuscitation of acute illness or injury. In: Cosby KS, Kendall JL (eds). *Practical Guide to Emergency Ultrasound*. 2nd ed. Philadelphia PA: Lippincott Williams & Wilkins; 2014.

65. Petrushkin H, Barsam A, Mavrakakis M, Parfitt A, Jaye P. Optic disc assessment in the emergency department: A comparative study between the PanOptic and direct ophthalmoscopes. *Emerg Med J*; 2012;29(12):1007–8.

66. Newman-Toker DE, Saber Tehrani AS, Mantokoudis G, Pula JH, Guede CI, Kerber KA, Blitz A, Ying SH, Hsieh YH, Rothman RE, Hanley DF, Zee DS, Kattah JC. Quantitative video-oculography to help diagnosis of stroke in acute vertigo and dizziness: Toward an ECG for the eye. *Stroke*. 2013;44(4):1158–61.

67. Sanders L. *Every Patient Tells a Story: Medical Myths and the Art of Diagnosis*. New York: Broadway Books, 2009.

68. Hurwitz B. Narrative and the practice of medicine. *Lancet*. 2000 Dec 16;356(9247):2086–9.

69. Richardson B. Clinical examination is essential to reduce overdiagnosis and overtreatment. *BMJ*. 2014;348:g2920.
70. Bleakley A, Marshall RJ. The embodiment of lyricism in medicine and Homer. *Med Humanit*. 2012; 38(1):50–4.
71. Swendiman RA. Deep listening. *Acad Med*. 2014;89(6):950.
72. Berlin L. Radiologic errors, past, present and future. *Diagnosis*. 2014;1(1):79–84.
73. Bruno MA, Walker EA, Abujudeh HH. Understanding and confronting our mistakes: The epidemiology of error in radiology and strategies for error reduction. *Radiographics*. 2015;35(6):1668–76.
74. Kim YW, Mansfield LT. Fool me twice: Delayed diagnoses in radiology with emphasis on perpetuated errors. *AJR Am J Roentgenol*. 2014;202(3):465–70.
75. Ashman CJ, Yu JS, Wolfman D. Satisfaction of search in osteoradiology. *AJR Am J Roentgenol*. 2000;175(2):541–4.
76. Drew T, Vo ML, Wolfe JM. The invisible gorilla strikes again: Sustained inattentional blindness in expert observers. *Psychol Sci*. 2013;24(9):1848–53.
77. Eitam B, Shoval R, Yeshurun Y. Seeing without knowing: Task relevance dissociates between visual awareness and recognition. *Ann N Y Acad Sci*. 2015;1339:125–37.
78. Plebani M, Sciacovelli L, Aita A, Pelloso M, Chiozza ML. Performance criteria and quality indicators for the pre-analytical phase. *Clin Chem Lab Med*. 2015;53(6):943–8.
79. Plebani M. The detection and prevention of errors in laboratory medicine. *Ann Clin Biochem*. 2010;47(Pt 2):101–10.
80. Carraro P, Plebani M. Errors in a stat laboratory: Types and frequencies 10 years later. *Clin Chem*. 2007;53(7):1338–42.
81. Hollensead SC, Lockwood WB, Elin RJ. Errors in pathology and laboratory medicine: Consequences and prevention. *J Surg Oncol*. 2004;88(3):161–81.
82. Crowley RS, Legowski E, Medvedeva O, Reitmeyer K, Tseytlin E, Castine M, Jukic D, Mello-Thoms C. Automated detection of heuristics and biases among pathologists in a computer-based system. *Adv in Health Sci Educ*. 2013;18(3):343–63.
83. Raab SS, Grzybicki DM, Janosky JE, Zarbo RJ, Meier FA, Jensen C, Geyer SJ. Clinical impact and frequency of anatomic pathology errors in cancer diagnoses. *Cancer*. 2005;104(10):2205–13.
84. Raab SS, Nakhleh RE, Ruby SG. Patient safety in anatomic pathology: Measuring discrepancy frequencies and causes. *Arch Pathol Lab Med*. 2005;129:459–66.
85. Epstein JI, Walsh PC, Sanfilippo F. Clinical and cost impact of second-opinion pathology. Review of prostate biopsies prior to radical prostatectomy. *Am J Surg Pathol*. 1996;20(7):851–7.
86. Baloch ZW, Hendreen S, Gupta PK, LiVolsi VA, Mandel SJ, Weber R, Fraker D. Interinstitutional review of thyroid fine-needle aspirations: Impact on clinical management of thyroid nodules. *Diagn Cytopathol*. 2001 Oct;25(4):231–4.
87. Raab SS, Grzybicki DM. Quality in cancer diagnosis. *CA Cancer J Clin*. 2010;60(3):139–165.
88. Kronz JD, Westra WH, Epstein JI. Mandatory second opinion surgical pathology at a large referral hospital. *Cancer*. 1999;86(11):2426–35.
89. Raab SS, Grzybicki DM, Condel JL, Stewart WR, Turcsanyi BD, Mahood LK, Becich MJ. Effect of Lean method implementation in the histopathology section of an anatomical pathology laboratory. *J Clin Pathol*. 2008;61(11):1193–9.

90. Wyers W. Confusion: Specimen mix-up in dermatopathology and measures to prevent and detect it. *Dermatol Pract Concept*. 2014;4(1):27–42.
91. Powsner SM, Costa J, Homer RJ. Clinicians are from Mars and pathologists are from Venus. *Arch Pathol Lab Med*. 2000;124(7):1040–6.
92. Seshia SS, Makhinson M, Philips DF, Young GB. Evidence-informed person-centered healthcare (part I): Do "cognitive biases plus" at organizational levels influence quality of evidence? *J Eval Clin Pract*. 2014;20(6):734–47.
93. Seshia SS, Makhinson M, Young GB. Evidence-informed person-centred health care (part II): Are "cognitive biases plus" undermining the EBM paradigm responsible for undermining the quality of evidence? *J Eval Clin Pract*. 2014;20(6):748–58.
94. Greenhalgh T, Howick J, Maskrey N, Evidence Based Medicine Renaissance Group. Evidence based medicine: A movement in crisis? *BMJ*. 2014;348:g3725.
95. Grimshaw JM, Eccles MP, Lavis JN. Hill SJ, Squires JE. Knowledge translation of research findings. *Implement Sci*. 2012;7(50):1–17.
96. McGlynn EA, Asch SM, Adams J, Keesey J, Hicks J, DeCristofaro A, Kerr EA. The quality of health care delivered to adults in the United States. *New Engl J Med*. 2003;348(26):2635–45.
97. Allen D, Harkins KJ. Too much guidance? *Lancet*. 2005;365(9472):1768.
98. Gigerenzer G, Gaissmaier W, Kurz-Milcke E, Schwartz LM, Woloship S. Helping doctors and patients make sense of health statistics. *Psychol Sci*. 2008;8(2):53–96.
99. Whiting PF, Davenport C, Jameson C, Burke M, Sterne JA, Hyde C, Ben-Shlomo Y. How well do health professionals interpret diagnostic information? A systematic review. *BMJ Open*. 2015;5:e008155.
100. Reid MC, Lane DA, Feinstein AR. Academic calculations versus clinical judgments: Practicing physicians' use of quantitative measures of test accuracy. *Am J Med*. 1998;104(4):374–80.
101. Feinstein AR. Misguided efforts and future challenges for research on "diagnostic tests." *J Epidemiol Community Health*. 2002;56(5):330–2.
102. Ben-Schlomo Y, Collin SM, Quekett J, Sterne JAC, Whiting P. Presentation of diagnostic information to doctors may change their interpretation and clinical management: A web-based randomised controlled trial. *PLoSOne*. 2015;10(7):e0128673.
103. Hoffrage U, Gigerenzer G. Using natural frequencies to improve diagnostic inferences. *Acad Med*. 1998;73:538–40.
104. McDonald KM, Matesic B, Contopoulos-Ioannidis DG, Lonhart J, Schmidt E, Pineda N, Ioannidis JP. Patient safety strategies targeted at diagnostic errors: A systematic review. *Ann Intern Med*. 2013; 158 (5 Pt 2): 381–9.
105. Singh H, Graber ML, Kissam SM, Sorensen AV, Lenfestey NF, Tant EM, Henriksen K, LaBresh KA. System-related interventions to reduce diagnostic errors: A narrative review. *BMJ Qual Saf*. 2012;21(2):160–70.
106. Graber ML, Kissam S, Payne VL, Meyer AN, Sorensen A, Lenfestey N, Tant E, Henriksen K, Labresh K, Singh H. Cognitive interventions to reduce diagnostic error: A narrative review. *BMJ Qual Saf*. 2012;21(7):535–57.
107. El-Kareh R, Hasan O, Schiff GD. Use of health information technology to reduce diagnostic errors. *BMJ Qual Saf*. 2013;22 Suppl 2:ii40–ii51.
108. Henriksen K, Brady J. The pursuit of better diagnostic performance: A human factors perspective. *BMJ Qual Saf*. 2013;22 Suppl 2:ii1–ii5.
109. Trowbridge RL, Dhaliwal G, Cosby KS. Educational agenda for diagnostic error reduction. *BMJ Qual Saf*. 2013;22 Suppl 2:ii28–ii32.

110. Zeng JY, Ye HH, Yang SX, Jin RC, Huang QL, Wei YC, Huang SG, Wang BQ, Ye JZ, Qin JY. Clinical application of a novel computer-aided detection system based on three-dimensional CT images on a pulmonary nodule. *Int J Clin Exp Med*. 2015;8(9):16077–16082.
111. Jacobs C, van Rikxoort EM, Murpy K, Prokop M, Schaefer-Proko CM, van Ginneken B. Computer-aided detection of pulmonary nodules: A comparative study using the public LIDC/IDRI database. *Eur Radiol*. 2015 Oct 6. (in press)
112. Graedon T, Graedon J. Let patients help with diagnosis. *Diagnosis*. 2014;1(1):49–51.
113. Haskell HW. What's in a story? Lessons from patients who have suffered diagnostic failure. *Diagnosis*. 2014;1(1):53–4.
114. McDonald KM, Bryce CL, Graber ML. The patient is in: patient involvement strategies for diagnostic error mitigation. *BMJ Qual Saf*. 2013 Oct;22 Suppl 2:ii33–ii39.
115. The Joint Commission: Facts about Speak Up Initiatives. Available at: http://www.jointcommission.org/assets/1/18/Facts_Speak_Up.pdf. Accessed December 5, 2015.
116. NASA. NASA Technology Spinoffs. Improving technology on earth. Available at:https://www.nasa.gov/centers/johnson/pdf/167752main_FS_Spinoffs508c.pdf. Accessed Dec 6, 2015.
117. NASA. NASA Technology Transfer Program Spinoff. Bring NASA technology down to earth. Available at: https://spinoff.nasa.gov. Accessed December 6, 2015.

11

The Role of the Healthcare System in Diagnostic Success or Failure

Karen Cosby

CONTENTS

Diagnosis Relies on Both Cognitive and System Processes

We have described the act of making a diagnosis as mostly a cognitive process. The mental process of diagnosis occurs in the mind of the clinician and is largely unseen. Like Sherlock Holmes, or the more recent television character Dr. House, the doctor acts like a detective, asks a series of questions (the "history"), then pokes and prods ("examines the patient"), and finally declares a diagnosis in a somewhat mystical manner, as if by superior intellect, good training, or even just good luck. The relationship between doctor and patient is historically romanticized by the famous Norman Rockwell painting that portrays a caring family doctor and his patient in a comfortable office setting, imbued with a sense of comfort and compassion. Although many admire the skill set possessed by the fictional character Dr. House and others may long for the traditional relationship between doctor and patient illustrated by Rockwell, neither image accurately reflects medical diagnosis today. Advancements in biomedical science allow us to recognize a growing variety and number of conditions, offer more treatment options, improve quality of life, and prolong survival. However, with that progress comes greater complexity. While in some cases a simple office visit with a single experienced provider will secure a likely diagnosis and treatment strategy, in many others, diagnosis relies on an increasingly elaborate coordination of specialists, complex processes, and technical procedures occurring over time and in places remote to the visit with one's primary physician. Thus, we need a system designed to support diagnosis.

What Is "The System?"

The doctor and patient are at the center of the diagnostic process, but they often rely on services and procedures to establish a diagnosis. Common needs include laboratory testing, imaging (such as x-ray, computed tomography (CT) scans, ultrasound, echocardiography), procedures (such as biopsies and endoscopies), and specialty consultations. Each test requires a series of actions, typically beginning with patient identification, determination of the clinical question to be addressed, selection of the correct test, performance of the test, interpretation of the result, communication with the ordering doctor, follow up with the patient and further diagnostic reasoning to integrate the results with the diagnostic impression. Diagnosis becomes a complex practice with an elaborate network of people and processes; the "system" is the network required to complete these steps. The system includes factors proximate to the patient (local resources such as people, supplies, equipment), or even remote (including public policy and healthcare financing that may determine access to care and resources for diagnosis). Care in emergency departments has been described as dependent on local system factors (the microsystem), wider hospital resources (the macrosystem), and distant factors that may influence the ability to care for patients, as illustrated in Figure 11.1 [1]. Flaws or gaps in any part of this system can create risk for diagnostic errors.

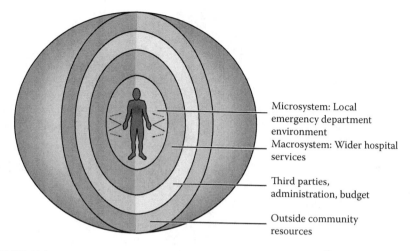

Microsystem: Local emergency department environment

Macrosystem: Wider hospital services

Third parties, administration, budget

Outside community resources

FIGURE 11.1

A model of the multiple layers of system factors that contribute to medical error (and diagnosis). (Reprinted from *Annals of Emergency Medicine*, 42(6), Cosby, K., A framework for classifying factors that contribute to error in the emergency department, 815–23, Copyright 2003, with permission from Elsevier [1].)

Sydney Dekker has described "the system" as

> "... a dynamic and complex whole, interacting as a structured functional unit to achieve goals (e.g. treating patients). ...The behavior of a system reflects the linkages and interactions among the components that make up the entire system. All medicine is practiced within a system...These system design factors can help or hinder medical professionals from doing their job..." [2]

A complete listing of system factors is far too detailed to provide here. Each type of task has requirements. Each practice environment has its own requisite equipment, supplies, processes and procedures. The coordination of care within each environment and communication between them is necessary to provide robust, reliable care.

System Flaws Contribute to Diagnostic Failure

The diagnostic thought process may be primarily cerebral, but it relies on a healthy clinician in a supportive environment with access to necessary resources and accurate data. System flaws have been identified as significant contributing factors to diagnosis error in a variety of healthcare settings and specialties [3–7]. Some of the evidence of the impact of system flaws on diagnostic accuracy is summarized in Figure 11.2.

- 65% of diagnostic errors in internal medicine have system factors as a major contributing factor (2)

- 65% of diagnostic errors in emergency medicine were attributed in part to teamwork failures, and another 41% involved other system flaws (3)

- 75% of diagnostic errors by medical trainees across a wide variety of specialties are due in part to system factors (4)

FIGURE 11.2
System factors that are common contributing factors to diagnostic error. (From Dekker, S.W. and Leveson, N.G., *BMJ Qual Saf.*, 24(1), 7–9, 2015; Graber, M. et al., *Arch Intern Med.*, 165(13), 1493–99, 2005 and Cosby, K.S. et al., *Ann Emerg Med.*, 51(3), 251–61, 2008 [2–4].)

Historically, clinicians have been slow to acknowledge potential system contributions to diagnostic error. The traditional professional model promotes individual accountability, and pointing a finger at the system seems to some like relinquishing ownership of diagnosis, or even sharing a portion of our professional identity with a system for which we do not have primary responsibility [8]. But the recognition of system contributions to diagnosis error brings with it acknowledgment that if the system could cause failure, perhaps an improved system design might improve diagnosis. This revelation has given new insight into potential strategies for optimizing the diagnostic process. Rather than decreasing accountability, appreciation of system factors has provided support for the medical work product and invited clinicians to contribute to the meaningful design of their workspace and workflow, and even recruit outsiders like human factors engineers into the medical realm [2].

Diagnosis Is a Process, Not Just an Endpoint

Diagnosis might be better understood as a series of tangible steps primarily involved in collecting data to substantiate or refute diagnostic possibilities. The process can be outlined or mapped into phases of work as shown in Figure 11.3. By examining each phase of work, we can identify processes that need specific design.

Access to Care and Patient Involvement

The patient must first recognize the need to seek care. Timely recognition of some conditions may be challenging; public health measures have focused on educating the public on the need to seek immediate care for signs of a stroke in order to make the diagnosis quickly enough to offer effective treatment.

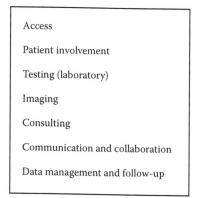

Access

Patient involvement

Testing (laboratory)

Imaging

Consulting

Communication and collaboration

Data management and follow-up

FIGURE 11.3
Phases of diagnostic work.

Unrecognized hypertension is another example of a common diagnosis that is often neglected or delayed due to either a lack of patient awareness or inability to access routine care. Public health measures to educate the public and provide routine diagnostic screening can address this limitation. Once patients decide to seek care, they need access. A patient who lacks insurance may avoid care, or select only those tests they feel they can afford. A variety of specific patient factors may further complicate diagnostic efforts. Patients may be afraid to seek care if they feel they will be judged harshly or treated poorly; for example, obese patients recognize that they are discriminated against and tend not to seek care when they should, fearing embarrassment or humiliation by the healthcare system [9]. Other patient factors may impact the ability to communicate and engage the system, such as language and cultural barriers, cognitive impairment, and psychiatric illness. Patient factors have been identified as common contributing factors to diagnostic error in emergency medicine [1,4]. Although patient factors are not necessarily an intrinsic part of the system, systems can (and probably should) find ways to compensate for these difficulties, such as by providing interpreter services and employing social workers and patient advocates for help as the failure to design for these stakeholders is seen by some as a key system flaw. Access to care implies that there is a fully functioning healthcare center with ready and expert staff. Adequate staffing includes well-trained, qualified, licensed individuals with sufficient equipment and supplies.

Testing

Comprehensive healthcare settings require testing facilities. The laboratory is a good example of a common diagnostic center that is used by most diagnosticians on a daily basis that has evolved its own system. Laboratory testing itself is mapped into phases. The need for a test begins with test selection and ordering. Once the test is ordered, the pre-analytic phase involves identification of the correct patient, sample procurement, sample labeling, transportation to the lab site, and receipt by the lab. The analytic phase involves performing the test itself; it requires functioning and calibrated equipment, reagents, and technical expertise. Once the test is completed, the process moves to the post-analytic phase in which the test result is recorded and communicated to the ordering physician. Lastly, the physician interprets the result and applies it in his decision making. At each phase in this testing procedure, well-designed processes are needed to ensure accuracy. Individual labs have highly refined processes for the analytic phase of work, involving calibration of equipment and quality assurance. This phase is highly reliable and accurate. The pre-analytic and post-analytic phases are the source of most laboratory errors, probably because they cross domains of care and share responsibility with those outside the laboratory itself [10,11].

Common tests such as hematology and chemistry are provided by largely automated systems with little potential for error. However, the clinical pathology examination of tissue is highly dependent on specimen processing and subjective interpretations. There is a surprising discrepancy rate in the interpretations of tissue for cancer, high enough for most systems to require mandatory second reviews of new cancer cases and/or review of cases in multidisciplinary conferences. Major discrepancies in final pathology reports (defined as impacting treatment and prognosis) have been noted in as many as 18% of endocrine and thyroid biopsies, 23% of endometrial cancers, 28% of liver biopsies, and 7.8% of breast cancers [12]. (See case described in Box 11.1.)

Imaging

Similar to the laboratory, imaging centers have their own systems and processes. Just like the laboratory, the order for an image begins with a clinical question, a decision to image, and a selection of the most appropriate test and technique. The correct patient must be identified and prepared (proper pre-imaging testing for renal function, pregnancy status, establishment of an intravenous line if needed), and transported to the site. Once in the imaging center, the patient must be identified; the image is then acquired and transmitted to the radiology suite. The formal process of interpretation

---

BOX 11.1   DIAGNOSIS ERROR FROM THE LABORATORY

A patient presents with a new diagnosis of a brain mass. A preliminary biopsy reports pathology consistent with glioblastoma. A second biopsy report performed later reports the cell type as likely metastatic adenocarcinoma. An investigation reveals that specimens from different patients were misidentified, leading to diagnostic errors in two patients.

BOX 11.2   ERROR CAUSED BY INFORMATION LOST IN THE HEALTHCARE SYSTEM

A 21-year-old man with asthma presents to his local ED for three weeks of chest pain and generalized fatigue. His vital signs are normal. The triage nurse notes that his oxygenation is normal and he is not wheezing. He is given a low triage acuity and a low priority compared with other, more acutely ill patients. The department is very busy and wait times are long. In order to facilitate care in the waiting patients, the ED uses standing orders for routine tests. A chest x-ray is performed from the waiting room. After a 6-hour delay, the patient gives up and goes home.

The x-ray is seen and interpreted by the radiologist who notes a new mediastinal mass. The radiologist notes this on his report and transmits the report to the picture archiving system. Once the patient leaves, however, there is no patient encounter, and no one to note the result. The finding goes unnoticed until the patient presents a month later with progressive signs and symptoms of lymphoma. Who was responsible for communicating with the patient? The radiologist? The emergency physician? The hospital administration? The quality assurance department? Or is the patient responsible for not asking about or getting the results?

---

is performed and a report is generated and sent to the ordering physician. Finally, the physician interprets and applies the information to the patient.

The process of both laboratory testing and imaging requires site-specific measures to ensure accuracy. Laboratories need to process specimens and confirm that they are adequate for the test. Radiology images must confirm that the image quality and technique are sufficient to answer the question posed by the ordering team. Factors such as ambient light and visual fatigue are relevant to the accuracy of a radiologist, whereas sample handling and specimen preparation are important factors in laboratory medicine [13]. While radiologists tend to focus on their expertise in image interpretation, and laboratory directors focus on equipment maintenance and technician certification, the accurate communication of test results is integral to both processes, and a common source of error [13,14] (See Box 11.2).

Consultation

At times, specialty consultation is necessary. Processes need to be in place to define what specialists are available, when they are available (or not), the timeframe to expect an evaluation, and the methods by which recommendations can be shared. Whenever multiple physicians and teams are involved in patient care, there must be an explicit understanding regarding who takes primary

> **BOX 11.3   WHO IS RESPONSIBLE FOR PATIENTS DURING TRANSITIONS IN CARE?**
>
> A patient hospitalized in respiratory distress is noted to have a markedly elevated white blood cell count. The result is so abnormal that the laboratory holds the results until they can be repeated and confirmed by a manual examination of the blood smear. The abnormal result is finally communicated to the emergency physician long after the patient has been moved to the intensive care unit, but wanting to be helpful, the physician agrees to notify the admitting team of the result. Busy with other sick patients, the hospitalist asks the emergency physician to order a hematology consultation. The emergency physician, already preoccupied with other duties, simply places an electronic order for a hematology consultation but does not personally call the hematologist directly. The hematologist sees that the electronic request was placed late at night, and assumes it can wait until morning. By then the patient is in septic shock and succumbs to acute myeloid leukemia, a diagnosis that needed rapid treatment for survival. Discussion at the following Morbidity and Mortality conferences centers on the question, "Who was responsible for the diagnostic delay: the emergency physician, the admitting team, or the hematologist?"

responsibility for coordinating care. Availability of specialty care needs to be consistent and known to those who depend on consultants. (See Box 11.3).

Communication and Collaboration

The responsibility for making a final diagnosis, traditionally assumed by a single physician, is now often shared between teams of physicians. In hospitals, the responsibility for diagnosis may rotate as teams change call. Responsibility may also float between specialists when a diagnosis is narrowed to within a specialist's domain. Without proper procedures for handoffs, the ultimate responsibility for ownership of the diagnosis may not be openly claimed by any single individual. The initial diagnosis by an emergency physician may change on admission to a primary care service, then be reassigned to a surgeon should an operative intervention be necessary. Over time, different physicians may contribute to diagnosis and management, the extent of their involvement changing with the needs of the patient. The careful coordination of care between providers (including thought process and diagnostic strategies) is essential to diagnostic accuracy.

Medical care tends to occur in silos that are culturally and academically distinct from each other. There are potential difficulties created by the segregation into subspecialization. Specialty-specific language may not be understood, and may even be misinterpreted by other physicians. (See case examples in Boxes 11.4 through 11.6.) Clear communication may require direct conversations to clarify understanding.

Advancements in diagnostic methods may require increasing collaboration with individuals with expertise outside the traditional medical team

---

BOX 11.4   ERROR GENERATED BY CONFUSING ANATOMICAL TERMS

A 50-year-old woman with ovarian cancer presents with two days of unilateral leg edema. Her physician orders an ultrasound of her lower extremity to rule out a deep venous thrombosis (DVT). The sonographer reports that there is clot in the superficial femoral vein. The physician explains to the patient that she does not have a DVT and will not need anticoagulation, since only the superficial vein is involved. The patient leaves and later dies of a massive pulmonary embolus. On review of the case, the radiologist notes that the nomenclature of the superficial venous system often confuses clinicians. In fact, some have recommended that the name be changed to avoid misunderstanding, or that the report explicitly describe the presence of a clot in the deep system.

BOX 11.5   MISCOMMUNICATION CAUSED BY TECHNICAL SPECIALTY LANGUAGE

A patient undergoes a surgical procedure to resect what is believed to be a relatively unusual but benign tumor. A frozen section identifies and names the tissue. The surgeon continues to operate with the understanding that the tumor is benign. The patient recovers and keeps his follow-up appointments. His condition unexpectedly deteriorates several months later. On review of the final pathology report with the pathologist, the surgeon learns that the biopsy was actually malignant. The misunderstanding came from not being familiar with the terminology used by the pathologist.

BOX 11.6   ERROR CREATED BY USE OF A NON-STANDARDIZED ABBREVIATION

A patient with dyspnea undergoes a computed tomography scan of the chest. When a "PE" is reported by the radiologist, the patient is anticoagulated. The following day it becomes evident that the radiologist used the abbreviation "PE" for pleural effusion, not pulmonary embolus.

structure. New diagnostic tests may be implemented without medical staff fully understanding how to use or interpret them, and some primary care physicians express uncertainty in their use of lab tests [15]. Newer diagnostic tests, including complex genetic testing and coagulation studies, have become so nuanced that laboratorians argue that they need to provide more support to clinicians who may not know how to order and use the growing list of test options. The College of American Pathologists has advocated for a model of care delivery that promotes and reimburses clinical pathologists for consultations for test selection and interpretation. In addition, some teams have incorporated medical librarians into their bedside rounds to research questions. In order to improve communication and ensure that the medical team has accurate and timely information about the patient, other teams have added family members and social workers to their bedside rounds.

Data Management and Patient Follow-Up

Communication of test results and patient follow-up are two of the most essential but precarious aspects of diagnosis. Healthcare systems and medical practices often lack effective processes to reliably communicate and follow-up test results [16,17]. In the United Kingdom, the Royal College of General Practitioners' report on *Delayed Diagnosis of Cancer* noted that the ordering, tracking, and managing of test results was a prime cause of diagnostic delay in cancer [18]. Similarly, U.S. studies report that "missed opportunities" for earlier diagnosis of lung and colorectal cancer occur in about a third of cases, largely due to failure to note and follow up abnormal test results [19,20]. One U.S. study of colorectal cancer found that two-thirds of cases had a delay of six months or more due to system factors, including scheduling delays and abnormal findings lost to follow-up [21]. A review of 56 cases of missed and delayed diagnoses of breast and colorectal cancer found that half were due to logistical breakdowns in tracking screening tests and arranging timely follow-ups [22]. A survey of pediatricians reported that failure to follow up on diagnostic laboratory test results contributed to 39% of diagnostic delays [23]. A third of abnormal Papanicolaou (Pap) smears are lost to follow-up [24]. Thirty percent of diagnosis-related malpractice cases reported by the Controlled Risk Insurance Company (CRICO) were attributed in part to failure in follow-up; another 20% had a failure or delay in reporting findings, and 15% had delays in scheduling or performing tests [25]. Patients may move from hospital to hospital and doctor to doctor, each time making it more difficult to establish a long-term relationship with a single system or provider and further complicating the flow of information. Gandhi describes the loss of information and miscommunication as "fumbled handoffs" and notes that current systems do not reliably ensure that abnormal test results are received and acted on [26].

At the same time, some physicians suffer from information overload. It is estimated that primary care physicians handle 930 lab and 60 pathology/

radiology reports each week [17] and most are dissatisfied with existing result management systems [16]. In contrast, emergency physicians often suffer from a lack of information, referred to as an *information gap*, resulting from patient factors (language, altered mental status, lack of medical records, and so on) that severely limit their information about patients' medical histories [27].

Within a given system, there is often a problem with sharing information. While individuals may function well within their specialty silo, coordination between silos is often neglected. Just who owns the responsibility for patient data and patient follow-up? The answer to this may vary by local standards, but it is sometimes left undefined. The traditional role of the single primary care doctor or group practice is not universal. If we don't know who owns the data, and who is ultimately responsible for the patient, how can we provide reliable care?

These phases of diagnosis involve a series of complex tasks, each with potential for contributing to diagnostic error; examples are given in Table 11.1.

Who Owns the System? Who Is Responsible for the Design of the System?

To complicate matters further, it can be difficult to determine just who "owns" the system. Medicine is learned as an individual skill; professional expertise for providers is based on individual excellence in clinical skills. Whether someone practices in an office, a clinic, or a hospital, they enter an environment dependent on services they often don't control. Much of the complexity of medical practice evolved without much forethought or intentional design. There are of course business managers, hospital administrators, group contractors, and malpractice insurers all contributing to the management and maintenance of practice settings; they have assorted purposes. But just who owns safety? Who helps create a safe, reliable system ensuring that the diagnostic process is timely and accurate? Increasingly we are realizing that everyone must contribute to an understanding of how system factors influence the ability to care for patients, and actively contribute to the design (or redesign) of diagnostic work.

Human Factors

Human factors engineering focuses on optimizing human performance by designing the environment and task to accommodate human fallibility.

TABLE 11.1

Diagnostic Error Examples for Major Phases of Evaluation

Domain	Diagnostic Error Example
Access to care	A 50-year-old man lost his job and health insurance. He develops chest pain and is afraid to seek care because of the financial expense. He suffers a cardiac arrest from undiagnosed heart disease.
Patient involvement	A 30-year-old man presents with heart failure from undiagnosed hypertension. He knew he had a family history of hypertension but never sought screening.
Laboratory testing	A laboratory has a history of frequent hemolyzed specimens with hyperkalemia. Tired of repeated false positives, a skeptical and ill-rested on-call physician fails to evaluate a patient.
Imaging	A radiologist amends a radiology report to note a new lung mass. The ordering physician views the report, but does not scroll all the way through the document to see the amendment attached to the end of the initial report. The diagnosis is missed.
Consultation	A patient has an abnormal mammogram that notes the presence of a smooth regular mass that is likely a fibroadenoma. There is a separate area of stippled calcifications concerning for malignancy. A surgical consultation request is completed by an office assistant who lists the reason for consultation as "likely fibroadenoma" but fails to mention the calcifications. The surgeon reassures the patient that nothing more than follow-up is indicated. The patient returns a year later with metastatic disease.
Communication	An elderly patient relates in detail a recent fall and residual wrist pain to her home nurse. No one mentions this to the physician who fails to diagnose the fracture.
Coordination of care	A patient is referred to a tertiary center with a hand-written note mentioning hypercalcemia. The crinkled note is lost and the patient seizes and dies.
Data management	A surgeon has the practice of reviewing final pathology reports during the postoperative visit. A young patient fails to follow up after an apparently uncomplicated appendectomy. The pathology report showing a carcinoid is lost to follow-up.

An effective system is designed to take human factors into account. A good deal of progress in patient safety has come from the human factor approach, including methods for investigating and analyzing adverse events, usability testing of medical equipment, analysis of medication errors, and teamwork training. Human factors can help improve diagnosis by helping design processes to get necessary information to clinicians in a timely way and in a us able form. The development of information technology to support clinical decision making is still in early development, but is considered one of the most promising solutions to diagnostic error [28]. Clinicians tend to have a set perspective on their practice and environment. Human factors engineers and safety experts outside the traditional discipline of medicine may bring fresh insights and new solutions to the task of diagnosis [1,29].

System Awareness

There are important implications to the recognition of the system contributions to diagnostic error.

1. Although diagnostic error occurs at the level of the provider, it is essential that clinicians understand how system flaws may impact them and their ability to do their work. (See Box 11.7.) You can't improve something you don't know about. System awareness is the first step in navigating system flaws.

2. Safe systems need to be designed intentionally and actively maintained. No system will ever guarantee perfection due to the dynamic nature of disease and the variable processes typical of physiological systems. However, good design can overcome many natural limitations in clinical practice.

3. There is a tendency to recognize harm only after the event; this reactive approach inevitably leads to a patchwork approach to fixing flaws. Good design requires a systematic proactive approach.

The System as Solution for Diagnostic Error

While this chapter focuses on system flaws as a source of diagnostic error, many will suggest that system design may provide potential solutions for common types of diagnostic errors. We may never be able to fully compensate for human fallibility, but well-designed systems can optimize human functioning and create processes that make success more likely. Improvements in system design are on the horizon. These improvements will be more effective if designed with the input of frontline clinicians and the patients who rely on them.

BOX 11.7 SYSTEM-RELATED DIAGNOSTIC ERROR

A 56 year-old male was admitted to the intensive care unit (ICU) with an 8-day history of increasing cough, fever, shortness of breath, hypoxemia, and pulmonary infiltrates. The admitting diagnosis was pneumonia and the patient was started on intravenous broad-spectrum antibiotic coverage. A nephrology consultation was requested to investigate a slightly elevated creatinine level, and the possibility of Wegener's granulomatosis or a related multi system vasculitis was raised, with the recommendation to obtain cANCA and related serologies. The tests were ordered, but the patient expired of a massive pulmonary hemorrhage three days later. An autopsy was consistent with vasculitis and the cANCA eventually returned strongly positive, consistent with the diagnosis of Wegener's granulomatosis. On investigation, it was learned that the cANCA serum was in the laboratory freezer awaiting the providers to fill out a special "send-out" test request form. None of the team members were aware of the requirement to fill out a send-out test request form.

Analysis: This case illustrates many different failures that contributed to the fatal and potentially preventable outcome in this case. Over 85% of patients with Wegener's granulomatosis respond to appropriate chemotherapy if treated in time.

- *Over-reliance on printed policy: Providers were not aware of the laboratory's policies.*

- *Failure to appreciate the immediacy of testing needs: The laboratory failed to appreciate the need for a very rapid turnaround time on vasculitis testing.*

- *Communication failure: The laboratory failed to communicate the need for the form to be completed; the care team failed to indicate the urgency of their need and to follow up on the missing test results. Clinical providers and the laboratory had no established communication pathway.*

- *Inadequate supervision: Supervising physicians could have been more assertive in investigating the missing test.*

SUMMARY POINTS

- System factors play a significant role in diagnostic errors.
- Awareness of system factors can help patients and clinicians better navigate the complicated processes in their diagnostic evaluations.
- Patients, clinicians, and safety experts all have much to contribute to necessary system design.

References

1. Cosby KS. A framework for classifying factors that contribute to error in the emergency department. *Ann Emerg Med*. 2003;42(6):815–23.
2. Dekker SW, Leveson NG. The systems approach to medicine: Controversy and misconceptions. *BMJ Qual Saf*. 2015;24(1):7–9.
3. Graber M, Franklin N, Gordon R. Diagnostic error in internal medicine. *Arch Intern Med*. 2005;165(13):1493–99.
4. Cosby KS, Roberts R, Palivos L, Ross C, Schaider J, Sherman S, Nasr I, Couture E, Lee M, Schabowski S, Ahmad I, Scott RD 2nd. Characteristics of patient care management problems identified in emergency department morbidity and mortality investigations during 15 years. *Ann Emerg Med*. 2008;51(3):251–61.
5. Singh H, Thomas EJ, Petersen LA, Studdert DM. Medical errors involving trainees: A study of closed malpractice claims from 5 insurers. *Arch Intern Med*. 2007;167(19):2030–36.
6. Thammasitboon S, Thammasitboon S, Singhal G. System-related factors contributing to diagnostic errors. *Curr Prob Pediatr Adolesc Health Care*. 2013; 43(9):242–47.
7. Berenson RA, Upadhyah DK, Kaye DR. *Placing Diagnosis Errors on the Policy Agenda*. Washington, DC: Urban Institute. Princeton, NJ: Robert Wood Johnson Foundation; 2014.
8. Levitt P. When medical errors kill. American hospitals have embraced a systems solution that doesn't solve the problem. *Los Angeles Times*. March 15, 2014.
9. Puhl RM, Heuer CA. The stigma of obesity: A review and update. *Obesity (Silver Spring)*. 2009;17:941–64.
10. Carraro P, Plebani M. Errors in a stat laboratory: Types and frequencies 10 years later. *Clin Chem*. 2007;53(7):1338–42.
11. Bonini P, Plebani M, Ceriotti F, Rubbli F. Errors in laboratory medicine. *Clin Chem*. 2002;48(5):691–98.
12. Dahl J. Quality, assurance, diagnosis, treatment and patient care. *Pt Saf Qual Healthcare*. 2006. Available at: www.psqh.com/marapr06/pathologist.html. Accessed August 1, 2016.
13. Lee CS, Nagy PG, Weaver SJ, Newman-Toker DE. Cognitive and system factors contributing to diagnostic errors in radiology. *AJR Am J Roentgenol*. 2013; 201(3):611–17.
14. Brook OR, O'Connell AM, Thornton E, Eisenberg RL, Mendiratta-Lala M, Kruskal JB. Anatomy and pathophysiology of errors occurring in clinical radiology practice. *Radiographics*. 2010;30(5):1401–10.
15. Hickner J, Thompson PJ, Wilkinson T, Epner P, Sheehan M, Pollock AM, Le J, Duke CC, Jackson BR, Taylor JR. Primary care physicians' challenges in ordering clinical laboratory results and interpreting results. *J Am Board Fam Med*. 2014;27(2):268–74.
16. Poon EG, Gandhi TK, Sequist TD, Murf HJ, Karson AS, Bates DW. "I wish I'd seen this test earlier!": Dissatisfaction with test result management systems in primary care. *Arch Intern Med*. 2004;164(20):2223–28.
17. Poon EG, Wang SJ, Gandhi TK, Bates DW, Kuperman GJ. Design and implementation of a comprehensive outpatient Results Manager. *J Biomed Inform*. 2003;36(1–2):80–91.

18. Royal College of General Practitioners Report on Delayed Diagnosis of Cancer, March 2010. National Patient Safety Agency. Available at: www.nris.npsa.nhs. uk/EasySiteWeb/getresource.axd?AssetID=69895. Accessed February 6, 2017.

19. Singh H, Hirani K, Kadiyala H, Rudomiotov O, Davis T, Khan MM, Wahls TL. Characteristics and predictors of missed opportunities in lung cancer diagnosis: An electronic health record-based study. *J Clin Oncol*. 2010;28(20):3307–15.

20. Singh H, Daci K, Petersen LA, Collins C, Peterson NJ, Shethia A, El-Serag HB. Missed opportunities to initiate endoscopic evaluation for colorectal cancer diagnosis. *Am J Gastroenterol*. 2009;104(10):2543–54.

21. Wahls TL, Peleg I. Patient- and system-related barriers for the earlier detection of colorectal cancer. *BMC Fam Pract*. 2009;10:65.

22. Poon EG, Kachalia A, Puopolo AL, Gandhi TK, Studdert DM. Cognitive errors and logistical breakdowns contributing to missed and delayed diagnosis of breast and colorectal cancers: A process analysis of closed malpractice claims. *J Gen Intern Med*. 2012;27(11):1416–23.

23. Singh H, Thomas EJ, Wilson L, Kelly PA, Pietz K, Elkeeb D, Singhal G. Errors of diagnosis in pediatric practice: A multi-site study. *Pediatrics*. 2010;126(1):70–79.

24. Marcus AC, Crane LA, Kaplan CP, reading AE, Savage E, Gunning J, Bernstein G, Berek JS. Improving adherence to screening follow-up among women with abnormal Pap smears: Results from a large clinic-based trial of three intervention strategies. *Med Care*. 1992;30(3):216–30.

25. Schaefer M. Reducing office practice risks. Forum. Risk Management Foundation of the Harvard Medical Institutions 2000;20(2):1–21. Available at: https://www.rmf.harvard.edu/~/media/Files/_Global/KC/Forums/2000/ ForumFeb2000.pd. Accessed March 9, 2015.

26. Gandhi TK. Fumbled handoffs: One ball dropped after another. *Ann Intern Med*. 2005;142(5):352–58.

27. Stiell A, Forster AJ, Stiell IG, van Walraven P. Prevalence of information gaps in the emergency department and the effect on patient outcome. *CMAJ* 2003;169(10):1023–28.

28. El Kareh R, Hasan O, Schiff GD. Use of health information technology to reduce diagnostic errors. *BMJ Qual Saf*. 2013;22 Suppl 2:ii40–ii51.

29. Henriksen K, Brady J. The pursuit of better diagnostic performance: A human factors perspective. *BMJ Qual Saf*. 2013;22 Suppl 2:ii1–ii5.

12

Do Teams Make Better Diagnoses?

Karen Cosby

CONTENTS

Introduction

We tend to view the work of diagnosis as a solo task with images of the doctor sitting, thinking, and deciding. In academia, we envision a gray-haired professor encircled by captivated young trainees. After all, diagnosis is important work requiring expertise and leadership. The doctor is the leader; the surgeon is the captain of the ship.

A more realistic model for diagnosis, suited to work as it actually happens today, is one that includes cross-discipline collaboration and teamwork. Admittedly, no single person possesses all the necessary expertise. Knowledge, skills, and expertise are shared across the spectrum of medicine. Diagnosis is no longer the domain of a single person but rather the joint work product of many individuals. Diagnosis is also best seen as a process that is incremental, occurring over time and space, involving multiple testing modalities and consultations. Whether we realize it or not, our skill in working and collaborating with other professionals may be a significant factor in our diagnostic success. The effective use of teams and teamwork principles has been suggested as one strategy to improve diagnosis.

Everyone loves the concept of teams and teamwork. The idea of teamwork conjures up images of the *esprit de corps* of a group of individuals working (even racing) toward a goal, supporting one other, and sacrificing for the good of the whole. But does a teamwork model fit the work of diagnosis, a process that is largely intangible and unseen? Before arguing for teamwork in diagnosis, we must first examine the work of diagnosis.

The Anatomy of a Diagnosis: How Are Decisions and Diagnoses Made?

In all parts of life, decisions must be made, and relevant models for decision making described in non-medical disciplines provide an interesting comparison for diagnostic work (Table 12.1). Business and management models

TABLE 12.1

Comparison of the Classical Model for Decision Making with the Process of Making a Diagnosis

Classical Model for Decision Making	Process of Making a Diagnosis
Identify a problem	Identify a symptom or syndrome
Gather information	Obtain a history, examine the patient
Analyze the situation	Consider a differential diagnosis, test
Consider options	Interpret the tests
Select preferred alternatives	Decide on most likely diagnosis(es)
Act, decide	Treat
Evaluate the outcome	Evaluate response to treatment, outcome

for decision making closely resemble the diagnostic process; in fact, diagnosis is a kind of decision – the choosing of a diagnostic label that forms the foundation on which all other clinical actions are made [1,2].

The process of diagnosis is well described in Kassirer and Kopelman's classic text as a hypothetico-deductive process, beginning with hypothesis generation and refinement, followed by testing, causal reasoning, and verifying, as discussed in more detail in Chapter 2 [3]. Their scheme is similar to other analytical models of reasoning, and can be compared with the classical rational model for decision making [2]. When all information is known and there is a high level of certainty and unlimited time to consider the options, a rational deliberate process can be used. Given unlimited time and resources, this thorough and methodical model is highly effective and likely very accurate. However, real-life decision making typically has limitations, such as imperfect information, time limits, and uncertainty. Alternative models better describe decisions made under these constraints [4,5]. A clinical situation may demand a decision despite less than perfect certainty and incomplete information. In such moments, rather than be paralyzed with indecision and inaction, people can set "bounded" expectations, choosing to make a "good enough" decision with less than perfect information (the *bounded rationality* model). In more extreme situations, circumstances may arise when rapid, almost split-second decisions must be made in spite of high risk and great uncertainty. In police, fire, and healthcare emergencies, some decisions must be made in the absence of complete understanding, almost like a reflexive action that relies largely on experience and intuition (*intuitive* model) [6,7]. Lastly, some circumstances are unique and require innovation. The *creative* model for decision making fits a situation in which no expert exists and a totally novel solution is needed [8].

TABLE 12.2

Context-Dependent Models for Decision Making

Model	Characteristics
Rational	• Problem is clear and unambiguous • High level (if not perfect) certainty • All relevant information is known • No time constraints • Slow and deliberate
Bounded rationality	• Uncertainty cannot be removed • Information is never complete • Time does not allow opportunity to consider all alternatives
Intuitive	• Time pressure precludes thorough assessment of all alternatives • There is urgency to act • Relies on experience • Sometimes requires reaction that relies on unconscious pattern recognition
Creative	• Novel problem • Requires innovative solution • No one has expertise

These four different models of decision making are described in Table 12.2 [8]. No single model fits all types of diagnostic work; in fact, there is a role for each, and for blends of each model in different situations. The diagnostic workup for a newly recognized cancer requires expertise across multiple specialties and is a situation ideally suited to a formal, thorough, rational approach. The evaluation of acute chest pain in the emergency department (ED) is not. The goal of the evaluation of chest pain in an ED is not focused so much on what is causing the pain as much as determining that it is not something dangerous or life threatening; in that context, the bounded rationality model is better suited. In an instance of cardiac arrest, there is no time to ponder; considered diagnoses require rapid action to quickly detect any process that may be potentially reversible. Although a methodical process for such cases can be rehearsed, the keen eye of an experienced clinician may be able to quickly recognize the problem with rapid recognition and intuition gained over time (the intuitive model). Finally, a novel problem requires a creative solution. On rare occasions, a patient may present with a unique set of circumstances and problems; this requires an innovative strategy, one that must be created as a one-of-a-kind solution.

Each of these models can be further informed by the dual process model described in earlier chapters. System 2 analytic processing is required for the rational decision-making model, and System 1 decision making is required for the intuitive model. The bounded rationality model lies somewhere in between, toggling between System 1 and System 2; as uncertainty increases and time available for thinking decreases, there is an increasing reliance on heuristics and intuition.

Who Should Make the Decision (or Diagnosis)?

The need for individual decision making, as opposed to a preference for group effort, is similarly dependent on setting and context (Table 12.3) [9]. Situations that favor a single individual decision maker include fast-paced, time-dependent problems that must be addressed urgently. Whenever there is a clear and unambiguous expert, most would favor his or her advice over a group of less experienced individuals. If a decision point arises that is routine and easy to resolve, it may be simply more efficient to allow an individual to decide, especially if there is nothing likely to be gained from a group effort. A group of experts may contribute to better decisions when problems are complex, such as when no single person has broad expertise, when important problems arise that have significant implications for long-term outcomes, and when there is time for deliberation.

These two examples are extremes; many situations exist that lie somewhere in between, and the choice of an individual leader versus group effort is often left to chance or made without much thought. Often frontline healthcare work is not even designed to provide an opportunity for group effort in diagnosis.

TABLE 12.3

Advantages and Disadvantages of Individual versus Group Decisions

Individual Decides		Group Decides	
Pro	Con	Pro	Con
Fast!	Limits generation of ideas	Generates diversity of ideas	Slower
Best individual in group may outperform the group	Must identify who has expertise, determine who decides	Encourages group buy-in with the plan	Beware of *Groupthink*
Accountability is easy to determine		Team must practice cooperation, can be made to feel more cohesive in the process	

Who Is on the Diagnostic Team?

Diagnosis is often viewed as an individual act performed at a single point in time. But many diagnoses involve numerous acts by a variety of people, some not routinely acknowledged or appreciated for the role they play. Take the example of a patient who has a myocardial infarction (MI). A bystander calls a 911 operator who verifies the location and notifies a local emergency medical first responder crew. A paramedic stabilizes and transports the patient to safety. The emergency medicine physician recognizes ST-segment elevation on an electrocardiogram (EKG) and triggers the cardiac catheterization (cath) lab. The on-call cath team rushes to prepare the lab and an interventional cardiologist diagnoses an occlusion of the right coronary artery. The patient is monitored and treated in a coronary care unit by nurses; a cardiologist and hospitalist continue medical care, identifying and managing cardiovascular risk factors until the time of discharge to the care of the primary care doctor. Diagnosis is suspected, confirmed, and verified in different steps. Individuals at each step are all diagnosticians in their own right. Each of them recognizes relevant diagnoses ("the chest pain on First and Elm street," "acute chest pain," "ST-segment elevation acute coronary syndrome," "a stenosis of the right coronary artery," the "post-cath patient," "hypertension" and "hypercholesterolemia"). Each individual is necessary for the diagnosis to be made and managed in different moments in time. The all-too-common failure to recognize the role that each individual plays in diagnosis is not only shortsighted but also limits our understanding of diagnosis.

Sometimes diagnosis occurs in a series of steps with different participants acting sequentially, but all are necessary to the ultimate diagnosis.

Participants need not necessarily recognize their role within the diagnostic process or be aware of their interdependency. However, their work will likely be more accurate, their actions more timely, and their team transitions more effective when they do.

Structure and Concept of Medical Teams

Recent initiatives strongly advocate for improving teamwork and instituting team training in medicine [10–12]. But before assuming that teamwork can improve diagnosis, we need to determine just who the team is, what we mean by teamwork, and how that might impact diagnostic work. A team is roughly defined as a group of individuals all working together toward a shared mission with common and interdependent goals. That is, they need each other. Teamwork principles argue that individual skills and excellence are necessary but insufficient for optimal results. Important team traits include interdependence and interconnectivity. That is, each participant has a role, understands and respects the other roles, and has a sense of awareness of and dependence on other team members [13].

Diagnostic work takes place in assorted and highly varied settings. No single team model adequately describes diagnostic work. Some medical teams are fairly well-structured and clearly defined. A surgeon may choose someone as a first assistant, and may arrange to have a favorite scrub nurse or circulating nurse, and may even be fortunate enough to have a preferred choice of anesthesiologist. Such teams can be familiar with one another and aware of each other's strengths and limitations. However, a team arranged for an emergency case with on-call individuals may not even know each other's names and be largely unaware of their relative skill levels. Some fluid teams are dynamic and assembled on the spur of the moment, such as code teams or rapid response teams. Some unit-specific teams may train together; others may have never met one another before being placed together. Some diagnostic work takes place in private offices remote from anything resembling teams. For many settings, the concept of teams and teamwork may seem altogether foreign.

The concept of teamwork is simpler and more obvious in treatment centers. Treatments tend to be more standardized, and treatment units more specialized. Infusion centers for chemotherapy, intensive care units, dialysis centers, burn units and other specialty care centers all develop around certain types of problems and common patterns of treatment. It is natural to consider how teamwork around these common processes can be developed. In contrast, diagnosis is a more abstract concept, and is a more varied and less visible process. The need for teamwork with diagnostic work is more easily understood through examples of misdiagnosis.

Team Work Failure Can Give Rise to Diagnostic Failure

At best, diagnosis is uncertain and difficult. But when systems lack team structure and processes, the path to diagnosis is made all the more difficult.

Examples of a Flawed Diagnostic Process: Doctors Acting as Individuals

A 55-year-old man presents with chest pain and is suspected to have an aortic dissection. A young emergency physician new to the hospital encounters the patient. He pages the on-call cardiologist who requests that an echocardiogram be done to evaluate the aortic arch. He also contacts the on-call cardiothoracic surgeon to warn him of his concern, who refuses to see the patient until his diagnostic test of choice, an angiogram, is done by an interventional radiologist. When he calls the interventional radiologist, the on-call radiologist refuses as he thinks a computed tomography (CT) scan is an easier and preferred imaging modality. The echocardiology technician fails to respond when paged, since he is rarely called and has fallen asleep with his pager at his desk. The emergency physician contacts the intensivist to arrange a hospital bed in the intensive care unit (ICU), but the intensivist argues about the drug of choice in managing the patient's blood pressure, and refuses to help until the diagnosis is "firmly established." Finally, the emergency physician obtains a CT scan that demonstrates the dissection. The cardiothoracic surgeon is angry because too much intravenous contrast has been administered, making the angiogram he requested now too risky. The diagnosis is ultimately made, but each of the doctors disagrees about the diagnostic pathway, and each has differing ideas about who should direct the workup for the patient. Each of the clinicians is an expert in his own right, but each has a different idea for optimal diagnosis and management. Each of them dogmatically believes that he is demanding what is best. They fail to recognize that they are dependent on each other to make and manage the diagnosis of aortic dissection.

When controversy and disagreements surround diagnostic workups, the ability to rapidly and efficiently reach a diagnosis is affected. Worse still, the frontline clinician facing such challenges may be reluctant to encounter that difficulty again, and may even raise his threshold for testing in the future. Aortic dissections are relatively uncommon, and the diagnosis is often delayed or even missed. Impediments in the diagnostic pathway only add to the challenge.

Salas describes three aspects of teamwork behavior that are lacking in this example [14]:

1. Team behavior that allows information exchange
2. Team cognition, a shared mental model
3. Team attitudes that reflect cohesion

Individuals can be "correct" yet fail to deliver quality care if they are interdependent while simultaneously at odds with each other. In this example, the individuals did not see themselves as team members but rather individual experts. Without team training, an individual may be rigid and intolerant, viewing himself as the ultimate authority that should dictate all decisions.

Example of a Failed Diagnostic Process: System Failure

A patient has a spinal tumor and travels far from home to have surgery by a renowned expert, who assures him that the tumor is benign. He is discharged home to the care of his usual primary care provider. The final pathology report is completed three weeks later and copies of the pathology report are sent to both his surgeon and his primary care doctor. The surgeon's office assistant files the report in the patient's records, but with no follow-up visit scheduled, the surgeon never sees it. The primary care doctor reviews the report, but is unfamiliar with some of the language and terms used and assumes all is well. A few months later, the patient has recurrent pain, undergoes a second procedure, and is found to have a rare cancer. When the first pathology report is reviewed, it becomes clear that the diagnosis was missed. An investigation of this incident revealed a flaw in the system to review and track abnormal pathology reports by both the surgeon and his hospital. Additionally, there was flawed communication between the pathologist and the primary care physician attributed to overly technical language and unfamiliar terms used in the pathology report.

Communication has been identified as the most critical teamwork competency for healthcare providers [14,15]. Failure to follow up test results and failed communication between physicians are common sources of diagnostic failures. Communication is not only dependent on individuals seeking information exchange; the reliable flow of information depends on institutional processes that facilitate the flow of information between individuals, even between different settings (inpatient, ambulatory clinics, offices, other hospitals, and so on).

Concepts of Teamwork Are New to the Culture of Healthcare

Teamwork has gradually improved healthcare [16,17]. We have yet to prove that teamwork *per se* results in improved diagnoses. However, the evidence we have makes it clear that some diagnostic failures can be attributed to a lack of team structure and the absence of traits that characterize highly functioning teams, including effective communication, attitudes and actions that reflect awareness of our interdependence with other people and processes, and shared mental models. Historically, clinicians have been trained in silos and primarily rewarded for individual excellence. They compete with one another for training positions. They establish hierarchical relationships and have "turf battles." A less competitive environment that fosters a spirit of cooperation and expectation of teamwork might actually be threatening to those who are accustomed to autonomy and authority in making decisions. But ultimately, clinicians are dependent on both a system of care and a network of others to achieve their goals and can benefit from a change in culture that fosters teamwork.

For any one moment in time, individuals may function as solo diagnosticians, but over time, most rely on many others and depend on teamwork principles to be successful, whether they realize it or not. Teamwork is a relatively new concept in medicine, particularly with regard to diagnosis. Teamwork principles that are necessary for timely and accurate diagnoses include communication and coordination of care over the continuum of the patient experience.

Are More Opinions Better?

There is an assumption that more equals better. But on a practical level, just how many opinions matter, and how many can we afford?

Role of Second Opinions

Patients are encouraged to get second opinions before surgery or before planning a major or controversial treatment plan. Sometimes patients seek second opinions when they do not understand their diagnosis, doubt their diagnosis, receive conflicting reports, or just want clarity about their diagnosis [18]. Perhaps some simply hope that a second opinion might give them a better result and a more favorable prognosis. A number of options are available for patients who seek second opinions. Most insurers will cover second opinions within their own networks of care, or patients can venture out to academic centers and online services that offer opinion services, such as Johns Hopkins and the Cleveland Clinic [19,20]. A Harvard group of physicians has organized a private second opinion program known as Best Doctors, Inc. [21]. Second opinions are valued: they give a chance for a fresh look at the evidence, add time to consider more options, and provide one more opportunity to catch flaws in clinical reasoning. The American Medical Association has a policy on second opinions [22]. Are these consultations and tests worthwhile? We simply do not have enough data to know. Results from Best Doctors, Inc. report that patient-initiated consultations for cases with diagnostic concerns result in a change in diagnosis in 14.8% of cases [23]. Some of these are just refined and perhaps better clarified diagnoses, although 20.9% of the revised diagnoses were considered to have moderate to major clinical impacts. Although this study gives a glimpse of diagnostic questions that patients may identify, we do not know the final outcomes. Was the second diagnosis correct for the patient? Did the patient pursue a third diagnosis? What was the eventual outcome? Several questions remain.

Institutions often have mandatory second reads for tests that are known to have high discordant rates. Second opinions are widely used in pathology. Discrepancies in diagnosis with second reads have been reported for 12.9% of lymphomas [24], 56% of dermatopathology specimens (22% of which were classified as major) [25], 26% of fine needle aspirations for thyroid nodules [26], and 18% of urothelial cancers [27]. In general radiology, 2%–26% of images have discrepancies with second reads [28]. In abdominal/pelvic CT scans, radiologists disagree with each other 26% of the time, and with themselves in 32% of cases [29]. These variations in performance are due to numerous factors: quality of specimens and staining, image technique, and perhaps changes in perception affected by fatigue, workload, and adequacy of search (just how thoroughly did the pathologist or radiologist search?). In fact, there is no guarantee that the second read is in fact the correct one. Ideally, a discrepancy between the first read and the second one should at least mandate a third. If a second opinion sometimes catches flaws, should a third or fourth be sought? How much can be gained by adding more experts, and how many is practical?

Role of Expert Consensus and Specialty Guidelines

If two opinions are good, are more even better? At the point of patient care, there are natural limitations to the number of voices and opinions that can be considered. Behind the scenes of clinical work, in conference rooms, libraries, and networking sites, groups of experts can conduct studies, collect data, analyze the literature, and generate guidelines to support diagnostic work. The number of experts who contribute to the literature and produce clinical rules is truly unlimited. In the end, however, diagnosis still relies on front-line individuals proximate to the patient, who collect the facts, frame a diagnostic question, and use their experience and judgment to apply the evidence to any given case.

Might Clinicians Make Better Diagnoses Working in Pairs or Groups?

Much of our discussion has involved using more people working separately, independent of one another. But what if they worked together? Beyond the obvious change in clinical practice models, would diagnosis be improved if clinicians worked in groups simultaneously? One study observed the success of medical students who were placed in pairs and asked to solve a diagnostic problem [30]. The students in pairs performed better than those acting alone: they were more accurate, took longer but planned a more efficient workup, and were more confident in their results. However, their confidence was not well-calibrated (their confidence was greater with wrong diagnoses).

Groups offer many advantages: they allow for a diverse set of members who may focus on different aspects of care; collectively, they increase the amount of time spent with the patient and potentially collect more data. Working in groups may help clinicians think aloud and better crystallize their thoughts. Theoretically, groups might improve the range of novel ideas considered [31]. In practice, however, groups often fail to live up to their potential, largely because of social interactions that may constrain individuals from fully participating in generating ideas and sharing knowledge. Groups can improve performance above that of the average individual, but are often inferior to the most superior member, and often fall short of their overall potential [32]. Certainly, the current organization of work in medicine seldom allows clinicians to work side by side to solve diagnostic challenges, and workplace pressures to be efficient hardly seem to support the idea.

Models of Teamwork for Medicine

While teamwork is not a routine part of all medical settings, some areas already incorporate team structure in diagnosis.

Oncology

The diagnosis and management of cancer is one of the best examples of multidisciplinary diagnostic work [33]. The diagnosis of cancer requires the expertise of many specialists. Medical oncologists rely on the coordinated efforts of pathologists, radiologists, surgeons, radiation therapists, nutritionists, social workers, palliative care teams, and ethicists [34–36]. Oncology teams often meet at tumor boards and multidisciplinary conferences where representatives from different disciplines gather to review, discuss, and debate diagnosis. Their joint effort allows the team to better assess the type and stage of tumors, and helps oncologists plan their treatment strategy. They may ask if the assessment of the tumor fits the overall clinical picture: does the pathologist's interpretation match the degree of invasiveness of the disease noted by the surgeon, or do the findings by the radiologist match the clinical course noted by the oncologist? This type of collaborative approach leads to a team-based diagnosis.

Specialty Guidelines

Another model of team care is evidenced by the work of the American Heart Association to improve time to diagnosis of acute ischemic events (both acute coronary syndromes and stroke) and facilitate the rapid detection of conditions that benefit from time-dependent thrombolysis. These team structures require institutional commitment as well as cross-discipline collaboration to adopt and implement standardized protocols. Other standardized processes in resuscitation (advanced cardiac life support, advanced trauma life support, pediatric advanced life support) have provided structure for the rapid identification and care of cardiac arrests and trauma. Now considered routine, they have defined effective team structures and standardized processes that unquestionably improve diagnosis and treatment.

Growing Medical Team

The concept of medical teams is evolving. New voices are offering help, many that would not have been considered even a decade ago. Medical rounds commonly include pharmacists who help recognize common drug interactions and adverse side effects [37]. Librarians are available in some hospitals to identify questions and find evidence to expand on diagnostic considerations [38]. Increasingly, clinical pathologists are expanding their work to bedside consultations to assist with test ordering and interpretation [39–41]. Medical ethicists and palliative care specialists are beginning to join rounds to add additional perspectives on diagnostic and treatment decisions [34].

Patient-Centered Diagnosis

While we aim to redefine how diagnostic work takes place, a new player has been introduced to the process—the patient himself [42,43]. The Institute of Medicine report *Crossing the Quality Chasm* has argued for a model of care that is more patient-centric, although to date most of the emphasis has been placed on treatment [11]. The idea of a patient participating and contributing to a differential diagnosis turns the traditional model of diagnosis on its head. Patients who have experienced a diagnostic failure offer much to this conversation [44]. If a diagnosis hinges on a patient's description of his or her symptoms, might an educated, informed, and engaged patient be better able to provide necessary details? Ferguson has described and fostered the e-patient movement—encouraging patients to be equipped, enabled, empowered, and engaged and use the Internet to search for their own answers [45]. Many a doctor will groan when approached by a patient who has prediagnosed himself after a Google search, and most would argue that Google is not an ideal resource for diagnostic queries. Yet, engaged patients may be helpful advocates who can take the time to check resources that physicians may not have time for. Patient advocacy groups suggest that patients can be trained to participate in diagnosis; one even emphasized the need for patient engagement with the expression "nothing about me without me" [46]. Guidelines have been suggested to help patients participate, including preparing for their doctor visit and bringing a short one-page summary of symptoms with a time line, asking if any of their symptoms do not fit the leading diagnosis, and understanding the expected course of their illness (so they can return if they do not respond as expected) [44–47]. Portals have been designed to provide patients with access to their medical records and appointment schedules. Most poignantly, Helen Haskell of Mothers Against Medical Error suggests that "diagnosis is born in a relationship, and it forms the basis for everything else that happens in healthcare" [48]. The relationship between provider and patient may be the most important link in the diagnostic team.

Do We Need Teamwork, or Culture Change?

Many have argued that teamwork is one solution for some diagnostic failures, but simply assigning people to teams will likely do little to achieve better results. The argument for teamwork is based on the recognition that a lack of teamwork and poor task design often lead to diagnostic errors [10]. Good, effective, and highly functioning teams can address many of the problems that manifest in healthcare, but teams tend to reflect the culture that gives rise to them, and may reproduce the same flaws that exist in the underlying organization.

Potential problems with team structure include the following:

- Teamwork implies effective, timely, and accurate communication. However, many communication failures are the result of a complex dynamic in relationships between people. If we fail to address the cultural issues that impact how individuals communicate in healthcare, much of the improvement we seek in team structures will fail [49].

- Individuals tend to contribute less to a project as more people are added to the task, a concept described by Latane as *social loafing* [50]. When German workers were asked to pull a rope, the effort they expended decreased with the successive addition of more workers, a phenomenon known as the *Ringelman effect* [50].

- Teams may lead to the diffusion of responsibility, such that no one feels, or takes, responsibility for team actions. This tendency has been observed outside the field of medicine where bystanders are less inclined to intervene in an emergency if others are present [51]. This tendency may be compounded further in medicine when members are uncertain of their roles, or fearful of taking charge when there are others with greater authority or seniority.

- Physicians work in hierarchical settings, and the role of any given team member may be ambiguous [49,52]. The willingness to take charge, make a decision, or speak for the team may be influenced by *authority gradients*, fears of criticism or reprisal, or uncertainty [53].

- Teams can develop an illusion of invulnerability and become overly confident.
- Even good teams can suffer from *collective rationalization* that can prevent them from recognizing weaknesses or flaws in their decisions.
- Very dominant teams may have unquestioned belief in their group. Strong groups may give the appearance of unanimity, when in fact, less dominant members feel discouraged from questioning or challenging the team. *Groupthink* may sway the team members to buy into a decision simply because it seems to reflect the majority opinion. Unquestioned belief in *Groupthink* can have devastating and even fatal consequences.

Team training is now integrated into many medical schools and the healthcare community is widely adopting strategies to teach and promote teamwork. However, we should use caution to ensure that the teams we build are healthy and effective. And ultimately, we need to understand that the change we seek is not necessarily one of team structure, but rather a change in attitudes and culture.

Conclusion

We can describe diagnosis as a decision that is made in a single moment by a single person, or a decision made by a number of individuals acting in sequence, or a process that occurs over space and time depending on numerous people and processes, or as a real-time face-to-face collaborative process. These differing images reflect the varied settings in which diagnosis takes place. Individual excellence in cognition is essential for all diagnostic work. But some diagnostic work also requires interpersonal communication skills and professional collaboration. Diagnostic excellence within healthcare organizations also requires institutional commitment to reliable processes for the timely and accurate flow of information, support for second opinions, and structures for team development. The concept of multidisciplinary medical teams has become rather expansive and open to a number of new voices and opinions. Regardless of setting, a culture of safe and accurate diagnosis is one that engages the patient, the only constant throughout the diagnostic journey.

Is diagnosis a team effort or an individual skill? The answer is both. No amount of team structure can compensate for poor decision-making skills or a lack of individual expertise. However, we can also argue that individual skills are insufficient for diagnosis in the complex medical system that clinicians and patients must navigate. Future work in building team structures and support for diagnosis looks promising, but we should recognize that it will require a significant investment in resources to ensure that we form reliable and effective teams.

SUMMARY POINTS

- Diagnosis has traditionally been considered the work of individual clinicians. In fact, diagnosis often requires the collaboration of multiple experts over space and time.
- Methods for diagnosis may vary depending on the urgency of the condition and the degree of certainty. Four models of diagnosis describe different processes for these circumstances:
  - Rational deliberate: When certainty is high and there is time to thoughtfully consider possibilities.
  - Bounded rationality: When circumstances constrain either available information and/or time for considering the options.
  - Intuitive: When the need for action precludes rigorous thought, or when pattern recognition is strong.
  - Creative: For novel problems for which expertise is generally lacking.
- Features of health teams include interdependence and interconnectivity.
- Ineffective communication contributes to diagnostic failures.
- Diagnosis can be improved by developing healthy teams and improving teamwork skills, but both require changes in the culture of healthcare institutions.

References

1. Li B. The classical model of decision making has been accepted as not providing an accurate account of how people typically make decisions. *Int J Business and Management*. 2008;3(6):151–54.
2. Club Managers Association of America. Team decision making. Available at: https://www.cmaa.org/bmiteam/decision/page3.asp. Accessed November 14, 2015.
3. Kassirer JP, Kopelman RI. *Learning Clinical Reasoning*. Baltimore, MD: Williams & Wilkins; 1991.
4. Jones BD. Bounded rationality. *Annu Rev Polit Sci*. 1999;2:297–321.
5. Club Managers Association of America. The behavioral theory of decision making. Available at: https://www.cmaa.org/bmitam/decision/page4.asp. Accessed November 14, 2015.
6. Kahneman D. *Thinking, Fast and Slow*. New York: Farrar, Straus and Giroux; 2011.

7. Khatri N, Ag HA. The role of intuition in strategic decision making. *Human Relations*. 2000;53(1):57–86.

8. Carpenter M, Bauer T, Erdogan B. *Principles of Management v1.1*. Nyack, New York: Flat World Knowledge; 2010.

9. Club Managers Association of America. Deciding alone or using a team. Available at: https://www.cmaa.org/bmiteam/decision/page7.asp. Accessed November 14, 2015.

10. Institute of Medicine. *Improving Diagnosis in Health Care*. Quality Chasm Series. Washington, DC: National Academy Press; 2015.

11. Institute of Medicine. *Crossing the Quality Chasm: A New Health Care System for the 21st Century*. Washington, DC: National Academy Press; 2001.

12. Salas E, DiazGranados D, Weaver SJ, King H. Does team training work? Principles for health care. *Acad Emerg Med*. 2008;15(11):1002–1009.

13. Lerner S, Magrane D, Friedman E. Teaching teamwork in medical education. *Mt Sinai J Med*. 2009;76(4):318–29.

14. Weaver SJ, Feitosa J, Salas E, Seddon R, Vozenilek JA. The theoretical drivers and models of team performance and effectiveness for patient safety. In: Salas E, Frush K, editors. *Improving Patient Safety through Teamwork and Teamwork Training*. Oxford: Oxford University Press; 2013, p. 4.

15. JCAHO. Root causes: A failure to communicate—Identifying and overcoming communication barriers. *Jt Comm Perspect Patient Safety*. 2002;2(9):4–5.

16. Schmutz J, Manser T. Do team processes really have an effect on clinical performance? A systematic literature review. *Br J Anaesth*. 2013 Apr;110(4):529–44.

17. Weaver SJ, Dy SM, Rosen MA. Team-training in healthcare: A narrative synthesis of the literature. *BMJ Qual Saf*. 2014 May;23(5):359–72.

18. Payne VL, Singh H, Meyer AN, Levy L, Harrison D, Graber ML. Patient-initiated second opinions: Systematic review of characteristics and impact on diagnosis, treatment, and satisfaction. *Mayo Clin Proc*. 2014 May;89(5):687–96.

19. Johns Hopkins medical second opinion program. Available at: www.hopkinsmedicine.org/second_opinion/. Accessed November 15, 2015.

20. Cleveland Clinic. MyConsult online medical second opinion. Available at: www.my.clevelandclinic.org/online-services/myconsult. Accessed November 15, 2015.

21. Best Doctors, Inc. Available at: www.bestdoctors.org. Accessed November 15, 2015.

22. AMA policy on Second Opinions. Opinion 8.041-Second opinions. Available at: https://web.archive.org/web/20160511115339/http://www.ama-assn.org/ama/pub/physician-resources/medical-ethics/code-medical-ethics/opinion8041.page. Accessed February 15, 2017.

23. Meyer AN, Singh H, Graber ML. Evaluation of outcomes from a national patient-initiated second opinion program. *Am J Med*. 2015;128(10):1138e25–e33.

24. Bowen JM, Perry AM, Laurini JA, Smith LM, Klinetobe K, Bast M, Vose JM, Aoun P, Fu K, Greiner TC, Chan WC, Armitage JO, Weisenburger DD. Lymphoma diagnosis at an academic centre: Rate or revision and impact on patient care. *Br J Haematol*. 2014;166(2):202–208.

25. Gaudi S, Zarandona JM, Raab SS, English JC 3rd, Jukic DM. Discrepancies in dermatopathology diagnoses: The role of second review policies and dermatopathology fellowship training. *J Am Acad Dermatol*. 2013;68(1):119–28.

26. Park JH, Kim HK, Kang SW, Jeong JJ, Nam KH, Chung WY, Park CS. Second opinion in thyroid fine-needle aspiration biopsy by the Bethesda system. *Endocr J*. 2012;59(3):205–12.

27. Coblentz TR, Mills SE, Theodorescu D. Impact of second opinion pathology in the definitive management of patients with bladder cancer. *Cancer.* 2001;91(7):1284–90.

28. Goddard P, Leslie A, Jones A, Wakeley C, Kabala J. Error in radiology. *Br J Radiol.* 2001;74(886):949–51.

29. Abujudeh HH, Boland GW, Kaewlai R, Rabiner P, Halpern EF, Gazelle GS, Thrall JH. Abdominal and pelvic computed tomography (CT) interpretation: Discrepancy rates among experienced radiologists. *Eur Radiol.* 2010;20(8):1952–57.

30. Hautz WE, Kammer JE, Schauber SK, Spies CD, Gaissmaier W. Diagnostic performance by medical students working individually or in teams. *JAMA.* 2015;313(3):303–4.

31. Christensen C, Larson JR Jr. Abbott A, Ardolino A, Franz T, Pfeiffer C. Decision making of clinical teams: Communication patterns and diagnostic error. *Med Decis Making.* 2000;20(1):45–50.

32. Hill GW. Group versus individual performance: Are N + 1 heads better than one? *Psychol Bul.* 1982;91(3):517–39.

33. Taplin SH, Weaver S, Chollette V, Marks LB, Jacobs A, Schiff G, Stricker CT, Bruinooge SS, Salas E. Teams and teamwork during a cancer diagnosis: Interdependency within and between teams. *J Oncol Pract.* 2015;11(3):231–38.

34. Le Divenah A, David S, Bertrand D, Chatel T, Viallards ML. Multidisciplinary consultation meetings: Decision-making in palliative chemotherapy. *Sante Publique.* 2013;25(2):129–35.

35. Alcantara SB, Reed W, Willis K, Lee W, Brennan P, Lewis S. Radiologist participation in multi-disciplinary teams in breast cancer improves reflective practice, decision making and isolation. *Eur J Cancer Care (Engl).* 2013;23(5):616–23.

36. European Partnership Action Against Cancer consensus group. Policy Statement on multidisciplinary cancer care. *Eur J Cancer.* 2014;50(3):475–80.

37. Kucukarsian SN, Peters M, Miynarek M, Nafziger DA. Pharmacists on rounding teams reduce preventable adverse drug events in hospital general medicine units. *Arch Intern Med.* 2003;167(17):2014–18.

38. Aitken EM, Powelson SE, Reaume RD, Ghali WA. Involving clinical librarians at the point of care: Results of a controlled intervention. *Acad Med.* 2011;86(2):1508–12.

39. Marques MB, Anastasi J, Ashwood E, Baron B, Fitzgerald R, Fung M, Krasowski M, Laposata M, Nester T, Rinder HM. Academy of Clinical Laboratory Physicians and Scientists. The clinical pathologist as consultant. *Am J Clin Pathol.* 2011;135(1):11–12.

40. Laposata M. Putting the patient first: Using the expertise of laboratory professionals to produce rapid and accurate diagnoses. *Lab Med.* 2014;45(1):4–5.

41. Govern P. Diagnostic management efforts thrive on teamwork. Available at: www.news.vanderbilt.edu/2013/03/diagnostic-management-efforts-thrive-on-teamwork/. Accessed November 14, 2015.

42. McDonald KM. The diagnostic field's players and interactions: From the inside out. *Diagnosis.* 2014;1(1):55–58.

43. Millenson ML. Telltale signs of patient-centered diagnosis. *Diagnosis.* 2014;1(1):59–61.

44. Graedon T, Graedon J. Let patients help with diagnosis. *Diagnosis.* 2014;1(1):49–51.

45. Ferguson T and the e-Patient Scholars Working Group. e-Patients and how they can help us heal health care. Available at: e-patients.net/e-Patients-White-Paper.pdf. Accessed November 15, 2015.

46. Delbanco T, Berwick DM, Boufford JI, Edgman-Levitan S, Ollenschlager G, Plamping D, Rockefeller RG. Healthcare in a land called Peoplepower: Nothing about me without me. *Health Expect*. 2001;4(3):144–50.

47. McDonald KM, Bryce CL, Graber ML. The patient is in: Patient involvement strategies for diagnostic error mitigation. *BMJ Qual Saf*. 2013;22 Suppl 2:ii33–ii39.

48. Haskell HW. What's in a story? Lessons from patients who have suffered diagnostic failure. *Diagnosis*. 2014;1(1):53–54.

49. Sutcliffe KM, Lewton E, Rosenthal MM. Communication failures: An insidious contributor to medical mishaps. *Acad Med*. 2004;79(2):186–94.

50. Latane B, Williams K, Harkins S. Many hands make light the work: The causes and consequences of social loafing. *J Pers Soc Psychol*. 1979;37(6):822–32.

51. Darley JM, Latane B. Bystander intervention in emergencies: Diffusion of responsibility. *J Pers Soc Psychol*. 1968;8(4):377–83.

52. Dayton E, Henriksen K. Communication failure: Basic components, contributing factors, and the call for structure. *Jt Comm J Qual Patient Saf*. 2007;33(1):34–47.

53. Cosby KS, Croskerry P. Profiles in safety: Authority gradients in medical error. *Acad Emerg Med*. 2004;11(2):1341–45.

13

How Much Diagnosis Can We Afford?

Karen Cosby and Mark L. Graber

CONTENTS

Introduction

The cost of medical care is a serious concern in the industrialized world, threatening the financial security of individuals, businesses, and even countries. The United States outspends all other Organization for Economic

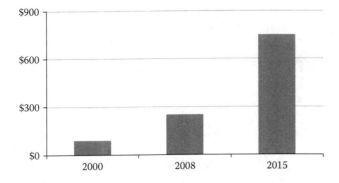

FIGURE 13.1
The rising costs of medical diagnosis in the United States in billions of dollars, 2000–2015 projected. (Adapted from Feldman, L., *Manag Care.*, 18(5), 43–45, 2009 [2].)

Cooperation and Development (OECD) countries on healthcare, but has lower life expectancy and poorer outcomes for many key measures. Much of this difference in cost can be attributed to greater utilization of expensive medical technology, including imaging and pharmaceuticals that may not all generate better outcomes [1]. Costs relating to diagnosis make up just a small fraction of this expense, around 10% according to recent estimates. Even this small fraction, however, amounts to the staggering sum of $250 billion per year in the United States [2]. Even more disturbing is the fact that these costs have increased at an annual rate of 14% over the last decade, such that by 2015 they are projected to be $750 billion per year (see Figure 13.1) [2].

Diagnosis is at the heart of all medical care, and is the basis for all subsequent interventions. Accurate and timely diagnosis is essential to all other healthcare decisions. Do we get our money's worth? What makes up these costs? Why are costs increasing so rapidly? Can we control cost without compromising quality? Will measures to control cost force us to limit options, or discourage the growth of industries that support healthcare and drive innovation? These are difficult questions.

The escalating costs of diagnosis reflect a host of factors, including the ever-increasing complexity of medical care, new and better tests, increased utilization of expensive imaging, better access to medical care, increased adherence to recommendations for screening tests, and inevitably, some waste. It all adds up. Consider the costs of a simple chemistry test, like measuring serum calcium. Run as part of a multi-channel analyzer on 5 microliters of blood, the cost for the reagent is about 10 cents. The autoanalyzer can run over 1000 tests per hour! The final charge might well be $25, the difference reflecting the lab's overhead for its clinical and administrative staff, the amortized cost of the $5 million instrument, and a healthy "profit" for the clinical lab, one of the few departments in the hospital that actually makes money for the institution.

Cost of Medical Imaging and Advancements in Technology

The ever-escalating expense of medical imaging accounts for many of the rising costs of diagnostics and total healthcare. According to the U.S. Government Accountability Office (GAO), Medicare payments for imaging services doubled between 2000 and 2006, a growth rate more rapid than that of any other healthcare expenditure (see Figure 13.2) [3]. Much of the increase has been attributed to the newer and most expensive imaging modalities: computed tomography scans (CT), magnetic resonance imaging (MRI), positron emission tomography (PET), and related nuclear imaging. These have increased at an annual rate of 17% [4]. There are over 7000 MRI imaging facilities in the United States, twice as many per capita as any other industrialized nation, and the United States orders two to three times as many advanced imaging studies as most other countries (seen in Table 13.1) [5,6]. Whether or not this pace of growth will continue or plateau simply cannot be known. At some point, the sensitivity of advanced imaging becomes clinically irrelevant as we become capable of detecting subtle and clinically insignificant abnormalities that do not impact treatment. However, there is always the promise of new developments on the horizon that will likely continue to expand our capability to better characterize disease and feed the appetite for testing.

Large regional variations in these costs suggest that at least some imaging studies may be ordered unnecessarily or inappropriately. Expenses for imaging ranged from a low of $62 per beneficiary in Vermont to $472 in Florida, an eightfold difference [3]. Siegel, chief of nuclear medicine imaging

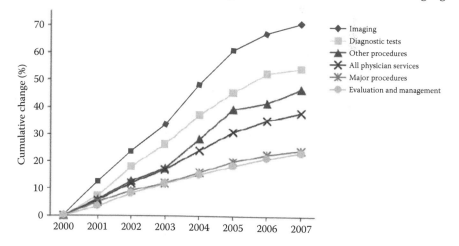

FIGURE 13.2

The rate of change for use of imaging and diagnostic tests is rising faster than any other component of healthcare services. (From Iglehart, J.K., *N Engl J Med*, Health insurers and medical-imaging policy: A work in progress, 360, 1031, Copyright 2009, Massachusetts Medical Society. Reprinted with permission from Massachusetts Medical Society [3].)

TABLE 13.1

Comparison of the Use of Diagnostic Imaging by Country, 2014 Data

Country	CT Exams/1000 population	MRI Exams/1000 population
Australia	115.5	35.3
Canada	148.5	54.9
Denmark	150.5	75
France	187.9	95.5
The Netherlands	79.5	51.2
United States	254.7	109.5

Source: OECD Health Statistics 2016. Available at: http://www.oecd.org/els/health-systems/health-data.htm. Accessed August 5, 2016.

at Washington University in St Louis, recently estimated that 10%–15% of the requests he gets "raise questions" about their appropriateness [7]. This component may be addressed in the future through better dissemination and use of guidelines. The American College of Radiology maintains an extensive online set of appropriateness guidelines to optimize test use; however, most physicians are unaware of the resource, and integration into clinical practice has been limited thus far [8,9]. Adding decision support algorithms into the ordering process has been suggested as one way of encouraging their use.

The unanswered question is whether the increased cost of imaging improves healthcare value and by how much. Assessing value, however, is a challenge; the costs are always clear, but the benefits, if any, may play out over time, and are more difficult to translate into a dollar value. The Cost-Effectiveness Analysis (CEA) Registry, maintained at the Center for the Evaluation of Value and Risk at Tufts Medical Center in Boston, is a database of over 4000 cost–utility analyses on a wide variety of diseases [10,11]. Only 6% of its analyses focus on diagnostic modalities, but the number is growing all the time. In 2003, the Medicare Modernization Act authorized the Agency for Healthcare Research and Quality (AHRQ) to conduct research on comparative clinical effectiveness. The American Recovery and Reimbursement Act of 2009 has since allocated over $1.1 billion to fund cost-effectiveness (CE) research, ensuring that the science of evaluation will continue to mature as more high-quality studies are funded [12]. A sampling of cost-effectiveness analyses related to medical imaging is presented in Table 13.2 [13–17].

Cost-Effectiveness (CE) Analysis

CE analysis is the accepted approach to judging the relative value of healthcare costs [12,17–22]. These analyses are complex but critically important to determining which tests are worthwhile and which are not. Each analysis typically compares an intervention state to a control condition (such as no intervention), constructing a "C/E" ratio: the difference in the cost measured

TABLE 13.2

Examples of Cost-Effectiveness (CE) Analysis

Test	Cost-Benefit
CT imaging for suspected appendicitis	CT scans eliminated unnecessary admissions, expedited surgery in the patients who needed it, and saved an average of $447 per patient [13].
PET scanning for cancer	Cost-effective for evaluating pulmonary nodules and staging non-small cell lung cancer. Clinically effective for diagnosis and staging of other cancer types, though cost-effectiveness in these settings is not conclusive [14].
CT imaging for headache	Only 1%–2% of scans reveal critical new findings at a cost of $50,078/significant case detected [15].
CT coronary angiography to triage patients with chest pain in the emergency department	CT angiography was cost-saving for women and cost-effective for men at a cost of $5400/QALY [16].
Stress electrocardiography in men with atypical chest pain	Compared to not testing, the procedure cost was $57,700 per QALY saved [17].

in dollars divided by the difference in the outcomes or effectiveness measured in Quality Adjusted Life Years (QALY). As an example, compared to no screening at all, screening for cervical cancer every 3 years costs $11,830 per life year saved, an intervention that would be considered highly cost-effective [21].

The QALY is an artificial measure used by healthcare economists to more precisely estimate effectiveness. These analyses hinge on determining the net benefit of an intervention, and the outcome of most interest is whether a specific intervention prolongs life. Does mammography at age 40 prolong life for a population of women screened? Does screening for abdominal aortic aneurysms prolong life? The outcome in these studies, measured in years of life, seems simple enough, but what if a disease impacts quality of life but not necessarily duration, or reduces both quality and duration? The QALY can be adjusted by a health utility value between 0 and 1 that adjusts for quality of life. This scale adjusts how one's current quality of life compares to perfect health (scored as 1) on a scale where death is rated 0. A patient who recently had a stroke and has not recovered speech or the ability to use their right leg might estimate that their quality of life is only half what it used to be, corresponding to a utility factor of 0.5. Over time, a chronic condition may worsen and the utility value may change. The calculation of QALY in such an example would weight years of life with health utility for a final QALY calculation. Figure 13.3 demonstrates an example calculation for QALY when a condition affects both quality of life and longevity [22]. A wide variety of approaches are available to help patients estimate their own quality of life, such as the Health Utility Index, the Quality of Well-Being Scale, and the Health and Activity Limitation Index. For research studies where population averages are needed, standards have been developed that can be used in place of individual estimates.

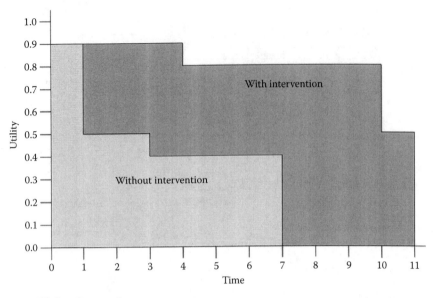

Without Intervention:

    Cost = $0

    QALY = (0.9 * 1) + (0.5 * 2) + (0.4 * 4) = 3.5

With Intervention:

    Cost = $54,000

    QALY = (0.9 * 4) + (0.8 * 6) + (0.5 * 1) = 8.9

$$CE = \frac{(\$54,000 - \$0)}{(8.9 - 3.5 \text{ QALY})} = \$10,000/\text{QALY}$$

FIGURE 13.3
Demonstration of the calculation of cost-effectiveness of an intervention when condition affects quality of life as well as longevity. (Adapted from Whitehead, A.J. and Ali, S., *Br Med Bull.*, 96, 5–21, 2010 [22].)

    CE analysis provides an estimate based on clinical studies that can be used by health economists and policy makers to compare different strategies in care. There is no absolute standard for an acceptable CE ratio and assessments about desirable CE ratios are inevitably influenced by context, setting, funding (public or private), and resources [23,24]. Interventions that are both more effective and less costly than existing standards of care are obviously worthy of support. In general, U.S. interventions that cost <$50,000-$60,000/QALY are considered "cost-effective" [24]. Canadian guidelines are most likely to adopt interventions that cost <$20,000/QALY, whereas interventions with CE ratios from $20,000-$100,000/QALY are commonly recommended for adoption [25].

Categories of Diagnostic Work and Their Costs

In order to better understand and assess diagnosis, it is helpful to describe three different categories of diagnosis as viewed through three different lenses. Diagnosis may:

1. Screen for (as yet) unrecognized conditions, such as occult cancer or hypertension, or
2. Test for inherent risks or tendencies toward future illness (by molecular or genetic analyses), or
3. Evaluate the source of symptoms (such as chest pain)

Each of these involves distinctly different thought processes and analyses. The costs and benefits of diagnosis can also be assessed through different perspectives:

1. The population at large (societal needs), versus
2. The individual, or
3. The scientific community

These differing perspectives can at times be at odds with each other. From a societal perspective, the cost of investigating a rare or unclassified disease for which there is no cure is hardly justified; yet on a personal level, the desire to find a better understanding of a rare degenerative condition may be quite compelling, regardless of the cost if it involves a young child for whom no one has given hope or even a diagnostic label. From a scientific perspective, there is the argument that we cannot advance the state of medical knowledge without investigating and studying rare conditions. There is a cost of diagnosis, a cost of treatment, and a cost of failure. The economics of healthcare is complicated: it involves objective statistics and facts of course, but priorities and choices are also influenced by cultural, humanitarian, and philosophical values that may change relative to the perspective from which you view the matter.

Routine Screening Tests

The best screening tests detect disease at an early stage, allowing for early and effective treatment. When applied to large populations, however, any test will yield false positive results. Complications or problems that result from investigating these will offset the benefit gained from early detection in the patients with true positive results. For example, CT scanning of the lungs can be used as a sensitive screening test for lung cancer, but will also detect small indeterminate nodules that are likely benign. Any analysis of potential benefit will have to consider the cost and potential harm required to investigate other abnormalities.

The process by which screening tests are evaluated is an excellent example of how CE analysis works in practice. The United States Preventive Service Task Force (USPSTF) is considered the leading organization in this effort [26]. This independent body of experts uses existing peer-reviewed evidence to make and disseminate recommendations for preventive care [27]. The task force determines the strength of evidence that can either support (recommend) a particular screen, or declare it "medically unnecessary" (not recommend). If there is insufficient evidence to determine effectiveness, they may make no recommendation. To be recommended, tests must be sensitive, specific and accurate; there must be evidence that early intervention is effective; and the cost-benefit analysis, taking into account disease prevalence and the costs of screening, must be favorable. The tests recommended by the USPSTF all meet the criteria that the net benefits outweigh the known costs and risks. An example of USPSTF guidelines includes common and widely used recommendations for routine cancer surveillance and common diseases; a few examples are shown in Table 13.3 [27]. For example, the USPSTF recommends screening for hypertension in adults over 18 years of age; they recommend against PSA-based prostate cancer screening; they make no recommendation (indeterminate strength of evidence) for routine screening for open angle glaucoma in adults. The USPSTF lists a number of services as "not medically necessary," including total body scans, heavy metal screens, and repeated ultrasound examinations during pregnancy. Such tests consume healthcare resources without the net benefits expected of nationally recommended screening. However useful these recommendations may be, they are often slow to be acknowledged and implemented in practice, a common and well-recognized challenge with knowledge translation [28]. There are many examples throughout medicine in which there is a failure to use evidence from research to alter clinical behavior. For example, although evidence has long been available to indicate that routine chest x-rays create more harm than good, the practice is still widespread [29].

TABLE 13.3

Recommendations of the U.S. Preventive Services Task Force Regarding Common Screening Tests for Disease Detection

Screening Recommended	Screening Not Recommended, or Evidence Indeterminate
• Breast cancer screening in women over 50 • Cervical cancer screening in women • Screening for hypertension, lipid disorders, and diabetes in men and women • Colorectal cancer screening in men and women over 50 • Osteoporosis screening in women over 65 or over 60 if at high risk	• Lung cancer screening, even in smokers • Screening for coronary artery disease • Whole body imaging to detect cancer • Screening for prostate cancer • Genetic screening for Alzheimer's dementia

Source: http://www.uspreventiveservicestaskforce.org.

Controversies about appropriateness of screening can create a political firestorm, which happened when new federal guidelines for mammograms were announced in 2009 that raised the screening age for breast cancer from 40 to 50 [30]. Patients, advocacy groups, and politicians clashed over decisions in which individuals' personal interests conflicted with population-based recommendations [30–33]. The types of risks patients and their physicians are willing to take (the risk of unnecessary biopsies versus the risk of delayed diagnosis) and the price they are willing to pay may vary by individual, and is not always predictable or even rational.

Genetic Screening: Molecular Diagnostics and the New Era of Medical Genomics

Since the completion of the Human Genome Project (HGP) in 2001, a rapid explosion in accessible technology has helped characterize genomes associated with disease, making it possible to detect many conditions before they even manifest symptoms. Potential parents can be screened before conception to determine the risk for hereditary disease in their offspring. In assisted reproduction, selected embryos can be tested for a number of genetic conditions before implantation; a few examples are shown in Table 13.4 [34]. Prenatal genetic screening is widely available using both invasive and non-invasive methods. Newborns are now routinely screened at birth for dozens of genetic disorders, typically including inborn errors of metabolism, cystic fibrosis, and hemoglobinopathies [34,35].

It is now feasible to obtain a full genomic analysis of an individual. However, much of the information may not be useful, and might even be misleading. A number of common conditions have been found to have a genetic basis, however, their ultimate expression is influenced by many factors and the usefulness of testing is not yet established [36]. Awareness of susceptibility for some hereditary cancer syndromes (such as hereditary non-polyposis colorectal cancer, familial adenomatous polyposis, and hereditary breast/

TABLE 13.4
Monogenetic Diseases Detectable by Preimplantation Tests

Inheritance Pattern	Condition
Dominant	• Huntington disease • Myotonic dystrophy • Charcot-Marie-tooth disease
Recessive	• Thalassemia • Cystic fibrosis • Spinal muscular atrophy (Werdnig-Hoffman) • Sickle cell disease
Sex-linked	• Fragile X syndrome • Duchenne muscular dystrophy • Hemophilia

Source: Adapted from Bodurtha, J. and Straus, J.F. 3rd. *N Engl J Med.*, 366(1), 64–73, 2012.

ovarian cancer linked to BRCA1/2 genes) may help individuals modify their lifestyle, screen more often, and potentially treat certain cancers more aggressively [37]. For other conditions, there is too much complexity to make sound recommendations based on genetics alone (such as type 2 diabetes and coronary artery disease) [36,38].

The capability to detect genotypic abnormalities is increasing at a rapid rate. More than 3000 diseases can now be identified from their genetic footprint and over 600 laboratories offer genetic testing internationally. The U.S. National Institutes of Health (NIH) supports a resource center at the University of Washington in Seattle that boasts over 54,000 tests for inherited disorders [39]. Although the costs of genetic testing may run into thousands of dollars, this is in the same cost range as CT and MRI, and in certain circumstances, genetic testing may provide diagnostic information at a lower cost than competing methods, sometimes with less risk.

Advancements in molecular genetics have had implications for all types of diagnosis, from screening, confirming diagnosis, and guiding specific targeted treatments (see Table 13.5) [40–44]. As the science of molecular genetics continues to grow, frontline clinicians will be expected to provide knowledgeable advice on inherited diseases, and will soon be expected to be "genetically literate" [38]. However, only a few of these tests are currently judged to be cost-effective. The criteria to perform these evaluations are now well-defined, however, more work must be done to determine how the information they generate should be integrated into care [38,45].

TABLE 13.5
Use of Genetic Testing for Diagnosis

How Test Can Be Used	Example Disease or Condition
Establish a diagnosis	Huntington disease: Genetic testing is 99% sensitive and 100% specific in detecting the Huntington gene abnormality, and is especially helpful in patients without a clear-cut family history [40].
Confirm a diagnosis	Cystic fibrosis: In cases where sweat testing is inconclusive, genetic testing for CFTR variants can confirm the diagnosis [41].
Predictive testing	Hereditary non-polyposis colorectal cancer: MLH1 and MSH2 are genes involved in genetic repair, and are markers for this disease. Of all patients with colon cancer, however, this predisposition accounts for only 1%–5% of cases. Screen-positive patients could undergo a colonoscopy to detect cancer at an earlier age. The test costs $1400–2700 [42,43].
Personal decision making	Breast cancer: BRCA genetic abnormalities are identified in 3.3% of Caucasian women with breast cancer. Screen-positive patients could undergo more intense surveillance for cancer. The test costs $300–5000 [44].

"Doctor, Why Do I Have Pain?": Diagnosing Symptomatic Disease

While screening recommendations are based on population risks at large, symptom-based diagnosis is highly individualized, and diagnostic pathways are often variable between clinicians and different settings. Because there are so many potential diagnoses and so many varied presentations of disease, it is difficult to standardize diagnosis, although there is a growing body of evidence to guide clinicians.

In complaint-driven diagnosis, each patient encounter is unique. An otherwise healthy 20-year-old who develops chest pain after shoveling snow is likely to just have a musculoskeletal strain, but could have a myocardial infarction (MI), an aortic dissection, pericarditis, or a pneumothorax. Missing any of these diagnoses can be fatal. Missed MIs and aortic disasters are among the leading causes of missed diagnosis (and malpractice claims) in almost every healthcare setting. The determination of what tests to order is not only based on population strategies; all that matters in a specific moment of time is the symptoms of the patient in front of you. How much certainty do we need to be safe? How do we decide when to test, and when not to test? This question is the medical version of the signal-detection problem faced by the military when calibrating the settings of the radar system. If set too high, the screen will be filled with signals that can overwhelm the analyst, and a large bird may be mistaken for an incoming missile. If set too low, the screen will be mostly empty, and the missile might look like a bird, or be missed altogether.

Evidence-based practice, combined with the use of epidemiologic approaches to decision making, offers an approach to answering these questions. The basic idea is to find the sweet spot—the perfect balance point where you find the most missiles while minimizing false alarms. This point can be determined with sufficient data on the characteristics of the test in question, but often a judgment call must be made. What risks are we willing to take? How much do we test? What cost is acceptable? The U.S. system of healthcare, and the professional ethos of healthcare providers, is largely unforgiving of any imperfection in the detection of disease, lending an unrealistic goal of perfect sensitivity that leads to many costly evaluations.

A growing number of clinical guidelines and decision rules are available to improve utilization of tests so that they are at least evidence-based and cost-effective (although they don't remove uncertainty, and don't necessarily achieve perfection). Acute ankle injuries are common and x-rays are often ordered for those injuries symptomatic enough to bring a patient to an emergency department (ED). The yield, however, on these studies is low—fractures are detected in only 10%–15% of cases. The Ottawa Ankle Rule is a validated instrument that reliably predicts which injuries need imaging [46]. In validation studies, it detected every single malleolar and midfoot fracture and eliminated the need for x-rays in 30%–35% of patients [47]. In follow-up studies of some 2,342 patients presenting with ankle injuries, use of the rule detected every fracture, decreased the time spent in the ED from 116 to 80 minutes, and reduced total costs by more than half [47]. In an ideal world, one might imagine that every ED would adopt the new rule immediately and enthusiastically. Similar to many innovations in medicine, however, diffusion is slow, adoption is less than universal, and initial effects wane. As an example, a randomized controlled trial in France confirmed the extremely high sensitivity of the rule, but only a 20% reduction in x-rays ordered was recorded, and usage tended to decrease over time [48]. Despite that, healthcare providers have an overall favorable view of clinical decision rules. Emergency providers in particular rate clinical decision rules positively when they fit within their workflow, help them make diagnoses, and save time [49].

Factors That Influence Decisions to Test

Sometimes it is simply easier to test. Patients have expectations and often demand a test to prove the presence (or at least be convinced of the absence) of disease. Physicians too are sometimes anxious since they are driven to reach unrealistic standards for both accuracy and efficiency. There are many factors that drive testing independent of diagnostic utility.

The Cost of "Thinking" (or Not)

The core bedside skills clinicians use rely on listening and interpreting patients' perceptions and descriptions of their symptoms, and examining them for physical findings to refute or support diagnostic considerations. Sometimes the history and physical ("H&P") is itself sufficient to establish a diagnosis. The unseen task that is taken for granted but too often neglected is the thinking process: the synthesis of information, the generation of a differential diagnosis, and the development of a considered workup. It is the foundation on which all other decisions are made. The relative cost of the patient encounter is negligible, yet all subsequent testing and imaging, the interpretation of the results, and the final clinical integration of the results rest on that assessment. Many lament the apparent "lost art" of medicine, and mistakes in diagnosis due to inadequate physical examinations [50–54]. Trainees have been heard to say that, "It's easier to order a CT than to spend time figuring out if I really need it." Quality measures from emergency medicine practice focus on "throughput" and "time to electrocardiogram (EKG) for MI" but give little attention to accuracy (and quality of thinking). Although electronic health records improve the exchange of information, the quality of information entered may suffer from too little engagement with the subject matter himself: the patient! Once testing is done, clinicians have been noted to rely more heavily on imaging reports or laboratory data, even when it contradicts physical findings [55]. Time spent with the patient and in communication with consultants is an under-recognized but cost-efficient tool available for diagnosis, and is largely unreimbursed. Kassirer has argued that much of the behavior that drives over-testing is discomfort with uncertainty and our need to verify diagnoses with duplicate testing [56].

Defensive Medicine

Extensive literature on the topic of defensive medicine portrays it as one of the major afflictions of modern medicine, an epidemic that wastes precious healthcare resources. Recent surveys of practicing physicians found that nine out of ten agreed with the statement that "Doctors order more tests and procedures than patients need to protect themselves from malpractice suits" [57,58]. An analysis in 2008 by PricewaterhouseCoopers estimated that almost half of healthcare spending in this country is wasted, and defensive medicine is the leading contributor [59], accounting for some $60 billion/year [60]. According to data from the Massachusetts Medical Society, defensive medicine accounted for 20% of tests and radiological studies, 28% of subspecialty consultations, and 13% of hospital admissions [61].

Doctors have good reason to be fearful. Diagnostic errors account for the majority of successful malpractice cases, have the largest payments, and are associated with the worst patient outcomes, including death and major disability [62]. Physicians have strong personal and professional grief when outcomes are poor. One author notes, "there is a strong personal incentive—judicial, societal, professional—to be as close to 100% right as medical science allows anyone possibly to be" [63]. This drive to perfection is ultimately costly.

Over-Diagnosis

One test tends to beget another. Clinicians can be tempted to pursue further evaluation of spurious abnormalities on tests, feeling pressure to explain or account for every finding. False positive tests and the detection of *incidentalomas* may confound workups and lead to more testing or unnecessary invasive procedures [64]. Kassirer describes the "stubborn pursuit for diagnostic certainty" as a driving force for inappropriate and sometimes excessive workups [56]. Over-testing has obvious economic consequences, but there are also clinical consequences. The inappropriate use of CT scans causes unnecessary exposure to radiation and potential contrast reactions (anaphylaxis, renal damage). One author describes an amusing acronym emphasizing the consternation faced by "victims of medical imaging technology (VOMIT)" [65]. The overzealous use of technology is widely recognized, so much so that a partnership between clinicians and patients now exists to better define and control appropriate medical interventions. *Choosing Wisely* is a partnership between providers and patients that aims to identify tests and procedures that have little or no value [66,67].

The Art of Doing Nothing: The Decision Not to Test

There are times when doing less is best, akin to the wise advice to "Take an aspirin and call me in the morning." In some settings, individual clinicians who have established relationships with patients, known to have good access and reliable telephone follow-up, may accept a wait-and-see attitude, knowing that they will be available to reassess and respond if a patient does not improve as expected. Rather than order tests, some may allow time and course to declare itself.

Sometimes the likelihood of disease is high enough and the benefit of therapy clear enough (and the potential risk of adverse events with treatment small enough) that a treatment threshold is reached, and no formal diagnostic testing is needed. An exudative pharyngitis with fever and cervical adenopathy is often taken as sufficient proof that even a negative streptococcal screen would not change the decision to treat for streptococcal pharyngitis. Typical manifestations of a very likely condition (e.g., fever and myalgias during an influenza epidemic in an immunocompetent patient) may not require anything more than a history and exam to secure a reasonable diagnosis. At another extreme, there may be times when there is little to be gained by intervention, so testing may be deferred even when the likelihood of disease is significant. For instance, patients with limited life expectancy may forego cancer screening. The decision to test for all of these conditions requires a thoughtful assessment of the risks and benefits of testing and the potential benefit of intervention, and a consideration of the unique characteristics of the patient and his relationship with the clinician. These decisions involve some degree of diagnostic reasoning, but with little to no financial cost. They are examples of thoughtful, deliberate diagnoses that are both safe and cost-conscious.

Fear of the Unknown Diagnosis

No one wishes to over-test or over-diagnose. Given a known outcome, most would choose a simple and direct path, limiting cost and risk to the patient. However, symptom-based diagnosis has an unknown end point. A missed diagnosis may cost the patient their life or result in disability. For professionals, there is an ever-present threat of the unknown, the "what if?" that must be managed. Risk-adverse personalities may just prefer to "play it safe" without fully recognizing the additional risks they impose with excessive testing. When a serious diagnosis is missed, a patient suffers. But clinicians may be a *second victim* [68–70] in this scenario. Wu has noted that clinicians may suffer their own personal anguish, including self-doubt, guilt, anger, frustration, fear, and depression, when a patient is harmed from their mistake. Clinicians need a better science and better tools to achieve optimal diagnoses.

Driving Cost-Conscious Diagnosis

Optimal diagnosis begins with an accurate diagnosis. Cost-effective diagnosis achieves accuracy with an efficient use of resources and avoids the pitfalls of unnecessary testing. The goal of marrying accuracy with efficiency achieves optimal outcomes in a financially responsible manner. Research in better defining which conditions merit testing, and which do not, may make significant contributions to improving the value of the diagnostic experience. Newman-Toker illustrates this concept effectively in defining clinical parameters detected by basic bedside maneuvers to determine which patients with dizziness benefit from early imaging with MRI and which patients do not benefit at all from imaging or hospitalization [71]. Similarly, Green et al. demonstrated an economic benefit to early diagnosis of celiac disease that helped avoid unnecessary ED visits and repeated imaging studies [72]. Further work on better classifying disease presentations and syndromes may offer much improvement. The integration of clinical decision support, evidence-based clinical guidelines, and appropriateness tools may also be useful. The goal may be to improve the cost of healthcare and diagnosis, but ultimately improving the science of diagnosis may improve both quality and cost.

SUMMARY POINTS

- The price of healthcare expenditure for diagnosis does not necessarily equate to quality of care or outcome.
- Excess use of advanced imaging technology accounts for much of the high cost of diagnostics.
- Quality Adjusted Life Year (QALY) is a useful measure of cost-effectiveness that can be used to assess and compare diagnostic tests.
- The American College of Radiology publishes appropriateness guidelines to improve the use of imaging resources and minimize over-testing.
- The U.S. Preventive Service Task Force (USPSTF) makes recommendations for routine screening tests, noting tests that are useful and recommended, versus those that are not.
- The growing field of molecular diagnostics can characterize personal genomes, but guidelines for their use have lagged behind their development.
- Ultimately, decisions about testing can be guided by evidence, but are also influenced by provider and patient preferences, personal and societal values, and risk aversion.

References

1. Squires D, Anderson, C. U.S. Health care from a global perspective: Spending, use of services, prices, and health in 13 countries. Available at: http://www.commonwealthfund.org/publications/issue-briefs/2015/oct/us-health-care-from-a-global-perspective. Accessed October 31, 2015.
2. Feldman L. Managing the cost of diagnosis. *Manag Care*. 2009;18(5):43–45.
3. Iglehart JK. Health insurers and medical-imaging policy: A work in progress. *N Engl J Med*. 2009;360(10):1030–37.
4. Government Accountability Office. Report to congressional requestors: Medicare Part B imaging services: Rapid spending growth and shift to physician offices indicate need for CMS to consider additional management practices. June 2008. Available at: www.gao.gov/new.items/do8452.pdf. Accessed October 31, 2015.
5. Squires DA. Explaining high health care spending in the United States: An international comparison of supply, utilization, prices, and quality. *Issue Brief (Commonw Fund)*. 2012;10:1–14.
6. OECD Health Statistics 2016. Available at: http://www.oecd.org/els/health-systems/health-data.htm. Accessed August 5, 2016.
7. Rowan K. Rising costs of medical imaging spur debate. *J Natl Cancer Inst*. 2008;100(23):1665–67.
8. Sheng AY, Castro A, Lewiss RE. Awareness, utilization, and education of the ACR appropriateness criteria: A review and future directions. *J Am Coll Radiol*. 2016;13(2):131–6.
9. American College of Radiology. ACR Appropriateness Criteria®. Available at: http://www.acr.org/ac. Accessed October 31, 2015.
10. Thorat T, Cangelosi M, Neumann PJ. Skills of the trade: The Tufts Cost-Effectiveness Analysis Registry. *J Benefit-Cost Analysis*. 2012;3(1):1–9.
11. Cost-Effectiveness Analysis Registry. Available at: https://research.tufts-nemc.org/cear4/. Accessed October 31, 2015.
12. Manchikanti L, Falco FJ, Boswell MV, Hirsch JA. Facts, fallacies, and politics of comparative effectiveness research: Part I. Basic considerations. *Pain Physician*. 2010;13(1):E23–54. Review.
13. Rao PM, Rhea JT, Novelline RA, Mostafavi AA, McCabe CJ. Effect of computed tomography of the appendix on treatment of patients and use of hospital resources. *N Engl J Med*. 1998;338(3):141–46.
14. Buck AK, Hermann K, Stargardt T, Dechow T, Krause BJ, Schreyogg J. Economic evaluation of PET and PET/CT in oncology: Evidence and methodologic approaches. *J Nucl Med*. 2010;51(3):401–12.
15. Jordan YJ, Lightfoote JB, Jordan JE. Computed tomography imaging in the management of headache in the emergency department: Cost efficacy and policy implications. *J Natl Med Assoc*. 2009;101(4):331–35.
16. Ladapo JA, Hoffman U, Bamberg F, Nagurney JT, Cutler DM, Weinstein MC, Gazelle GS. Cost-effectiveness of coronary MDCT in the triage of patients with acute chest pain. *AJR Am J Roentgenol*. 2008;191(2):455–63.
17. Kuntz KM, Fleischmann KE, Hunink MG, Douglas PS. Cost-effectiveness of diagnostic strategies for patients with chest pain. *Ann Intern Med*. 1999;130(9):709–18.

18. Russell LB, Gold MR, Siegel JE, Daniels N, Weinstein MC. The role of cost-effective analysis in health and medicine. Panel on Cost-Effectiveness in Health and Medicine. *JAMA*. 1996; 276(14):1172–77.

19. Gold MR, Siegel JE, Russell LB, Weinstein MC, editors. *Cost-Effectiveness in Health and Medicine*. New York: Oxford University Press; 1996.

20. Cohen DJ, Reynolds MR. Interpreting the results of cost-effectiveness studies. *J Am Coll Cardiol*. 2008;52(25):2119–26.

21. Essselen KM, Feldman S. Cost-effectiveness of cervical cancer prevention. *Clin Obstet Gynecol*. 2013; 56(1):55–64.

22. Whitehead SJ, Ali S. Health outcome in economic evaluation: The QALY and utilities. *Br Med Bull*. 2010;96:5–21.

23. Azimi NA, Welch HG. The effectiveness of cost-effectiveness analysis in containing costs. *J Gen Intern Med*. 1998;13(10):664–69.

24. Owens DK. Interpretation of cost-effectiveness analyses. *J Gen Intern Med*. 1998 Oct;13(10):716–17.

25. Laupacis A, Feeny D, Detsky AS, Tugwell PX. How attractive does a new technology have to be to warrant adoption and utilization? Tentative guidelines for using clinical and economic evaluations. *CMAJ*. 1992; 146(4):473–81.

26. Moyer V, Bibbins-Domingo K. The U.S. Preventive Services Task Force: What is it and what does it do? *N C Med J*. 2015;76(4):238–42.

27. U.S. Preventive Services Task Force. Available at: http://www.uspreventiveservicestaskforce.org/Page/Name/home. Accessed October 31, 2015.

28. Straus SE, Tetroe J, Graham I. Defining knowledge translation. *CMAJ*. 2009 Aug;181(3–4):165–68.

29. Mauri D, Kamposioras K, Proiskos A, Xilomenos A, Peponi C, Dambrosio M, Zacharias G, Koukourakis P, Pentheroudakis G, Pavlidis N. Old habits die hard: Chest radiography for screening purposes in primary care. *Am J Managed Care*. 2006;(11):650–56.

30. US Preventive Services Task Force. Screening for breast cancer: U.S. Preventive Services Task Force recommendation statement. *Ann Intern Med*. 2009;151(10):716–26.

31. American Cancer Society (2009) Press release: American Cancer Society responds to changes to USPSTF mammography guidelines. Available at: http://pressroom.cancer.org/index.php?s=43&item=201. Accessed January 28, 2013.

32. McCarthy M. U.S. panel reaffirms controversial 2009 mammography recommendations. *BMJ*. 2015;350:h2174.

33. Wang AT, Fan J, Van Houten HK, Tilburt JC, Stout NK, Montori VM, Shah ND. Impact of the 2009 U.S. Preventive Services Task Force Guidelines on screening mammography rates on women in their 40s. *PLoS One*. 2014;9(3):e91399.

34. Bodurtha J, Strauss JF 3rd. Genomics and perinatal care. *N Engl J Med*. 2012;366(1):64–73.

35. Francescatto L, Katsanis N. Newborn screening and the era of medical genomics. *Semin Perinatol*. 2015;39(8):617–22.

36. Janssens AC, van Duijn CM. Genome-based prediction of common diseases: Advances and prospects. *Hum Mol Genet*. 2008;17(R2):R166–73.

37. Fostira F, Thodi G, Konstantopoulou I, Sandaltzopouos R, Yannoukakos D. Hereditary cancer syndromes. *J BUON*. 2007;12 Suppl 1:s13–22.

38. Emery J, Hayflick S. The challenge of integrating genetic medicine into primary care. *Br Med J*. 2001;322(7293):1027–30.

39. GENE tests. Available at: https://www.genetests.org. Accessed October 31, 2015.

40. Batepe M, Xin W. Huntington disease: Molecular diagnostics approach. *Curr Protoc Hum Genet*. 2015;87(9):26.

41. Dequeker E, Stuhrmann M, Morris MA, Casals T, Castellani C, Claustres M, Cuppens H, des Georges M, Ferec C, Macek M, Pignatti PF, Scheffer H, Schwartz M, Witt M, Schwarz M, Girodon E. Best practice guidelines for molecular genetic diagnosis of cystic fibrosis and CFTR-related disorders: Updated European recommendations. *Eur J Hum Genet*. 2009;17(1):51–65.

42. Strafford JC. Genetic testing for lynch syndrome, an inherited cancer of the bowel, endometrium, and ovary. *Rev Obstet Gynecol*. 2012;5(1):42–49.

43. U Conn Health. Division of Medical Genetics. Hereditary NonPolyposis Colorectal Cancer (HNPCC). Available at: www.humangenetics.uconn.edu/hereditary/info_hnpcc.html. Accessed October 31, 2015.

44. Organization. Available at: www.breastcancer.org. Accessed October 31, 2015.

45. Higashi MK, Veenstra DL. Managed care in the genomics era: Assessing the cost effectiveness of genetic tests. *Am J Manag Care*. 2003;9(7):493–500.

46. Stiell IG, Greenberg GH, McKnight D, Nair RC, McDowell I, Reardon M, Stewart JP, Maloney J. Decision rules for the use of radiography in acute ankle injuries: Refinement and prospective validation. *JAMA*. 1993;269(9):1127–32.

47. Stiell IG, McKnight RD, Greenberg GH, McDowell I, Nair RC, Wells GA, Johns C, Worthington JR. Implementation of the Ottawa ankle rules. *JAMA*. 1994;271(11):827–32.

48. Auleley GR, Ravaud P, Giraudeau B, Kerboull L, Nizard R, Massin P, Garreau de Loubresse C, Vallee C, Durieux P. Implementation of the Ottawa ankle rules in France: A multicenter randomized controlled trial. *JAMA*. 1997;277(24):1935–39.

49. Richardson S, Khan S, McCullagh L, Kline M, Mann D, McGinn T. Healthcare provider perceptions of clinical prediction rules. *BMJ Open*. 2015;5(9):e008461.

50. Natt B, Szerlip HM. The lost art of the history and physical. *Am J Med Sci*. 2014; 348(5):423–25.

51. Verghese A, Charlton B, Kassirer JP, Ramsey M, Ioannides JP. Inadequacies of physical examination as a cause of medical errors and adverse events: A collection of vignettes. *Am J Med*. 2015;128(12):1322–24.

52. Jauhar S. The demise of the physical exam. *N Engl J Med*. 2006;354(6):548–51.

53. Dalvi B. The "Lost" frontier of clinical medicine: Have we reached a point of no return? *Ann Pediatr Cardiol*. 2009;2(1):1–2.

54. Fred HL. Hyposkillia: Deficiency of clinical skills. *Tex Heart Inst J*. 2005; 32(3):255–57.

55. Marks LB. "Error bars" in medical imaging: Stealth and treacherous. *Radiology*. 2015;277(2):318–28.

56. Kassirer JP. Our stubborn quest for diagnostic certainty: A cause of excessive testing. *N Engl J Med*. 1989;320(22):1489–91.

57. Bishop TF, Federman AD, Keyhani S. Physicians' views on defensive medicine: A national survey. *Arch Intern Med*. 2010;170(12):1081–83.

58. Kanzaria HK, Hoffman JR, Probst MA, Caloyeras JP, Berry SH, Brook RH. Emergency physician perceptions of medically unnecessary advanced diagnostic imaging. *Acad Emerg Med*. 2015;22(4):390–98.

59. The price of excess: Identifying waste in healthcare spending. Available at: http://www.oss.net/dynamaster/file_archive/080509/59f26a38c114f2295757b b6be522128a/The%20Price%20of%20Excess%20-%20Identifying%20Waste%20 in%20Healthcare%20Spending%20-%20PWC.pdf. Accessed February 15, 2017.
60. Kessler D, McClellan M. Do doctors practice defensive medicine? *Quart J Economics*. 1996;111(2):353–90.
61. Massachusetts Medical Society. Investigation of defensive medicine in Massachussets. Nov. 2008. Available at: www.massmed.org/defensive medicine. Accessed October 30, 2015.
62. Saber Tehrani AS, Lee H, Mathews SC, Shore A, Makary MA, Pronovost PJ, Newman-Toker DE. 25-Year summary of U.S. malpractice claims for diagnostic errors 1986–2000: An analysis from the National Practitioner Data Bank. *BMJ Qual Saf*. 2013;22(8):672–80.
63. Kassirer J. Correspondence: Our stubborn quest for diagnostic certainty. *N Engl J Med*. 1989 Nov 2;321(8):1272–73.
64. Chojniak R. Incidentalomas: Managing risks. *Radiol Bras*. 2015;48(4):IX–X.
65. Hayward R. VOMIT (victims of modern imaging technology): An acronym for our times. *BMJ*. 2003;326:1273.
66. Morden NE, Colla CH, Sequist TD, Rosenthal MB. Choosing wisely: The politics and economics of labeling low-value services. *N Engl J Med*. 2014;370(7):589–92.
67. Choosing Wisely. An initiative of the ABIM Foundation. Available at: http://www.choosingwisely.org. Accessed October 31, 2015.
68. Wu AW. Medical error: The second victim. *West J Med*. 2000;172(6):358–59.
69. Levinson W, Dunn PM, Portland O. A piece of my mind: Coping with fallibility. *JAMA*. 1989; 261(15):2252.
70. Seys D, Wu AW, van Gerven E, Vieugeis A, Euwema M, Panella A, Scott SD, Conway J, Sermeus W, Vanhaecht K. Health care professionals as second victims after adverse events: A systematic review. *Eval Health Prof*. 2013;36(2):135–62.
71. Newman-Toker DE, McDonald KM, Meltzer DO. How much diagnostic safety can we afford, and how should we decide? A health economics perspective. *BMJ Qual Saf*. 2013;22 Suppl 2:ii11–ii20.
72. Green PH, Neugut AI, Naiyer AJ, Edwards ZC, Gabinelle S, Chinburapa V. Economic benefits of increased diagnosis of celiac disease in a national managed care population in the United States. *J Insur Med*. 2008; 40(3–4):218–28.

Section V

The Fix

14

Medical Education and the Diagnostic Process

Pat Croskerry

CONTENTS

Introduction

It is an extremely rewarding and satisfying experience for medical educators to have the opportunity to watch medically naïve individuals develop and mature into qualified physicians through the undergraduate years of medical education. Cognitively speaking, it is the busiest of times for the medical learner. At no other time will they have to process and memorize so much information in such a short period of time. The assumption is made that educators have done all they can along the way to facilitate the process of the individual becoming a competent physician.

Two Types of Knowledge

Essentially, there are two tasks that need to be accomplished—the learner must acquire two types of knowledge: *knowing that* (declarative) and *knowing how* (procedural) knowledge. The first involves imparting basic medical knowledge. This is usually, but not necessarily, built on a foundation of premed-acquired knowledge in the basic sciences. Generally, most medical schools impart this knowledge well, albeit on the basis of performance in National Board Examinations in which candidates are assessed on a body of medical knowledge defined by the examiners. Many time-weathered approaches have been retained in teaching medicine and it is a process that continues to evolve as new ideas are introduced to successive generations of learners. As Lucey notes, from a pedagogical perspective at least, current medical education programs are superb: "Objectives-based teaching has yielded to competency-based learning. Passive lectures have given way to interactive small-group learning experiences. Simulation-based training in communication, procedural skills and complex decision making has replaced practice on patients" [1]. These processes, she notes, aim at producing the "personally expert sovereign physician"—one who has mastered a significant body of biomedical science along the road toward becoming "autonomous, independent, and authoritative" [1].

The second process, procedural knowledge, involves training in thinking, reasoning, and making decisions based on that knowledge, referred to as clinical reasoning and decision making. Historically, medical educators have performed less well in this area. Much of the time, no explicit strategies have been used. It seems that an implicit assumption has been made that medical students would acquire these skills passively or informally, through coaching by instructors, tutors, mentors, and by experienced clinician exemplars. Few physicians reading this chapter will have had a formal course in clinical reasoning. This is not to say that no effort has been made to teach medical students clinical reasoning and decision making, however, now and in the past, approaches have been generally confined to courses such as *Clinical Epidemiology and Biostatistics*, or *Introduction to Medical Decision Making*, often given in the first or second year of the undergraduate curriculum. Such courses are typically aimed at developing an understanding of research methodology and statistics, the skills required to interpret study results, and the ability to determine the impact the study would be expected to have on clinical practice. Evidence-based medicine (EBM) is strongly emphasized. Instruction is given in the various tools of biostatistics: sensitivity, specificity, Bayes' theorem, likelihood ratios, number needed to treat, number needed to harm, and others [2]. Equipped with such knowledge, students would be expected to be able to conduct effective clinical appraisals of studies in the medical literature. As noted in Chapter 8, such procedural and declarative knowledge is vital mindware. Without it, rationality may be compromised (mindware gaps).

The other compromise of rationality content occurs through contaminated mindware (Figure 8.2). This, too, involves both declarative and procedural knowledge. The declarative moiety consists of learning about a variety of cognitive vulnerabilities, principally the names of cognitive and affective biases and logical failures, as well as the terminology of this literature. But the learner also needs to acquire procedural knowledge of how these factors exert their influence and, importantly, how they impact rational thinking.

Multiple Factors Affecting Clinical Decision Making

Over the last few decades, a variety of approaches toward teaching and assessing clinical reasoning have been explored. While the main focus has been directed toward optimizing the manner in which the learner's brain processes and integrates the abundant information to which it is exposed, there has been a growing awareness of the impact of bias and other distortions of reasoning (Chapter 7), as well as a variety of other individual (Chapter 9), patient (Chapter 17), and ambient factors on decision making. The assumption has prevailed that medical decision makers are rational, well-rested, well-slept, well-fed, emotionally stable, and not subject to resource limitations or cognitive overloading; that is, decisions are assumed to be made under optimal conditions. With few exceptions, until quite recently [3–5], the impact of a variety of individual and ambient factors on clinical reasoning and decision making were not explicitly dealt

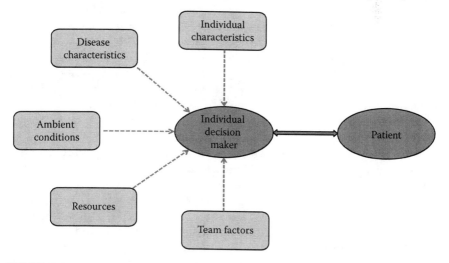

FIGURE 14.1
Major influences on clinical reasoning and decision making.

with, factors illustrated in Figure 14.1. In particular, individual cognitive and affective biases on the clinical reasoning process were not considered, even though these are probably the most powerful factors influencing a physician's reasoning, judgment, and decision making (topics dealt with at length in Chapter 7).

The treatment of biases in courses of medical decision making were usually confined to those associated with study design (e.g., sample selection); publication bias (e.g., certain studies that are more likely to be published than others; the greater likelihood of studies with positive findings being published rather than those with negative findings); other editorial biases, such as the editor choosing specific individuals who are likely to review favorably or unfavorably; English language bias; the deliberate withholding of data with findings that do not support a sponsoring pharmaceutical company's commercial interests; or studies in which a reviewer might systematically include certain studies over others.

Scientific Decision Making and Flesh and Blood Decision Making

Historically, it appears that the science of medical decision making has been perceived quite differently from the actual processes that might occur under the ambient conditions in which clinical reasoning usually takes place. Indeed, over 60% of the lead authors for the 51 articles published in the journal *Medical Decision Making* in 2004 were not medically trained, and therefore did not have exposure to the realities and responsibilities of clinical medicine. It might be argued that one doesn't have to be a chicken to know what an egg looks like, and perhaps non-medically trained researchers were just as, or even better, able to investigate the clinical decision-making process as medically trained researchers. However, many clinicians were known to despair at the intensely dry text associated with quantitative medical decision making. Others were disappointed in the apparent disinterest in heuristics and non-quantitative aspects of medical decision making. One reviewer, at that time, wryly noted that "The broader community of medical decision makers has not embraced the topic of heuristics and biases approach with sustained enthusiasm" [6]. The papers published in the journal, until fairly recently, seemed to reflect this perception. Overall, the emphasis in medical decision-making has been on the quantitative properties of the medical decision making process and not the 'flesh and blood' aspects. This was consistent with what was expected of Lucey's utopian physician [1], and to be fair to the authors of *Medical Decision Making*, the journal was not titled the Journal of Clinical Reasoning, although some may have had an implicit understanding that it was aimed in that direction.

Does Medical Decision Making Meet Normative Standards?

However, medical decision making and the clinical reasoning process are inextricably related. Thus, anything that affects medical decision making should affect clinical reasoning. Although many medical educators may be oblivious to this point, one of the strongest factors in the clinical reasoning process is the *rationality* of the decision maker (discussed in Chapter 8). Over the last few decades, hundreds of studies in cognitive psychology have been done on various aspects of rationality. As Stanovich notes, the results of these studies have established that an individual's performance in reasoning tasks falls short of what would be considered *normative* standards, that is, those that optimize the accuracy of our belief systems and the efficacy of our actions: "... people's responses sometimes deviate from the performance considered normative on many reasoning tasks. They have thinking habits that lead to suboptimal actions (instrumental rationality) and beliefs (epistemic rationality). For example, people assess probabilities incorrectly; they test hypotheses inefficiently; they violate the axioms of utility theory; they do not properly calibrate degrees of belief; their choices are affected by irrelevant context; they ignore the alternative hypothesis when evaluating data; and they display numerous other information processing biases" [7]. Thus, if human judgment and decision making is flawed, how likely is it that medical decision making is at all times rational, and if it is, at times, irrational, how can we ignore such failings in any teaching of clinical reasoning?

Increased Awareness of the Impact of Cognitive and Affective Biases

Over the last two to three decades, things have begun to change in the medical literature and we are now at a point where we can explicitly discuss the cognitive processes involved in making a diagnosis. Two books in 1991 drew attention to the cognitive calibration of the physician as diagnostician. In the first, an insightful and underappreciated work by Riegelman, two basic error types were described: errors of ignorance and errors of implementation, corresponding respectively to failures in declarative and procedural knowledge. Riegelman's emphasis was on the latter, and he detailed how multiple steps in this process may be adulterated through cognitive failure [8]. Kassirer and Kopelman's book *Learning Clinical Reasoning* in 1991 explicitly addressed a number of specific cognitive biases in clinical reasoning [9]. Their classic textbook has been updated recently, with the addition of John Wong as a third author [10]. Since the initial work of Detmer et al. on biases in surgeons in 1978 [11], the impact of cognitive and affective biases on clinical decision making has now become a focus of studies in Anesthesiology [12], Dermatology [13], Emergency Medicine [14], Medicine [15,16], Neurology [17], Neurosurgery [18], Obstetrics [19], Ophthalmology [20], Pathology [21,22], Pediatrics [23], Pediatric Psychiatry [24], Psychiatry [25], Radiology [26,27], Surgery [28,29], medical education [30], and specialty environments such as the Intensive Care Unit [31], Forensics [32], Hematology [33], and Dentistry [34]. Following on from Kassirer and Kopelman, texts now specifically address the impact of bias on clinical reasoning [3–5]. In the United Kingdom, a new initiative has emerged recently called Clinical Reasoning in Medical Education (CreME) [35], which is represented in more than half of all medical schools in the United Kingdom and which aims to share and develop resources in the clinical teaching of clinical reasoning. And yet, many medical school curricula still fail to explicitly address the topic, with even less giving it the attention it deserves.

A Model for Learning about Clinical Reasoning and Decision Making

We are beginning to understand medical decision making more fully and are now able to articulate the complexity of the processes involved. In Figure 14.2, an overall schema that was developed at Dalhousie University Medical

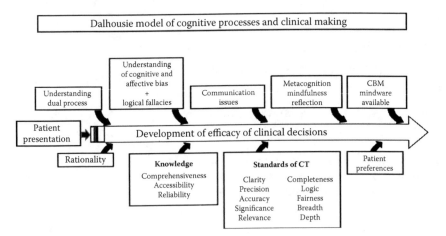

FIGURE 14.2
The Dalhousie model for development of clinical reasoning. From the initial presentation of the patient with their symptoms and signs on the left in the figure, a number of processes work to gradually refine the clinical decision. (CBM, cognitive bias mitigation; CT, critical thinking).

School illustrates some of the major factors influencing the development of clinical reasoning. The model may serve as a basis for course content in clinical reasoning, as well as providing a map of developmental stages in learning critical content in clinical decision making.

The patient's presentation of symptoms and signs begins at the extreme left in the figure. Even at this stage, various biases may begin to manifest themselves. The patient may frame their symptoms in a particular way that influences and perhaps misleads the physician (as illustrated in Box 14.1). A variety of other biases such as *Triage Cueing, Diagnosis Momentum* and

---

BOX 14.1   CLINICAL CASE EXAMPLE OF VARIOUS BIASES THAT MAY MANIFEST AT A PATIENT'S PRESENTATION TO THE EMERGENCY DEPARTMENT

A sprightly 68-year-old female presents at an emergency department, complaining of shoulder pain. She relates that she was mowing her lawn and the mower got stuck in an awkward spot and she had to twist and push hard to release it. She feels she may have strained her shoulder in the process (*framing bias*). Other than some left shoulder pain, which is now beginning to ease, she is otherwise well and feels fine. She would just like to have her shoulder checked out.

At triage she is noted to have shoulder sprain brought on by mowing (*triage cueing*). Her vital signs are all stable and she is triaged to the Minors' area, where after a brief wait she is seen by an emergency physician. It was the practice in this department that emergency physicians would work the first six hours of their shift on the main floor and then go to the Minors' area for the last two hours. The implicit assumption made by those working in the Minors' area is that they will not be dealing with complex cases. This is a manifestation of the *Geography is Destiny* bias, whereby the physician assumes, "if I am working in the Minors' area, I expect to see patients with minor conditions or illnesses."

The physician takes a brief history and proceeds to examine the patient. Her shoulder is slightly limited in range of motion but appears otherwise normal. He orders an x-ray, which shows no acute injury, but notes some osteoarthritic changes (*confirmation bias*). He orders a sling for the patient and advises her to take anti-inflammatories and rest it for the next few days. His discharge diagnosis is "osteoarthritis."

Several hours later, the patient returns to the same ED having experienced a "weak spell" associated with some nausea and vomiting. She is pale, diaphoretic, and hypotensive. An EKG at triage reveals an acute inferior myocardial infarct.

Geography is Destiny may operate at this time. From a bias standpoint, it is especially important to be vigilant at the very outset of a patient's presentation when subtle cues and biases may be particularly influential and when the decision maker may be at their most vulnerable. The processes above the line in Figure 14.2 are suggested topics to be covered during undergraduate training; they are described in detail in this book. The properties of the Dual Process model for decision making should be reviewed so that students are completely familiar with it; awareness needs to be raised about the extent of cognitive and affective biases in medical decision making, and emphasis needs to be placed on the recognition of specific affective and cognitive biases that may be particularly relevant for clinical decision making (see Appendix I: Common Cognitive and Affective Biases in Medicine). Patients or referring colleagues may also fall victim to logical fallacies that distort the presentation of information; for example, framing and verbal priming may confuse correlation with causation (*post hoc ergo propter hoc*), as in Box 14.1. Training in communication issues should also be provided during undergraduate years, with additional emphasis on some of the biases associated with intra- and inter-professional communication, such as verbal priming and others (e.g., primacy and recency, see Figure 7.1 in Chapter 7). The process of metacognition should also be reviewed—essentially encouraging the learner to think about how they think. This can be augmented with specific reflective practice and mindfulness training, the features of which may contain various cognitive bias mitigation strategies [36], which will be discussed in more detail in Chapter 15. Undergraduates should also receive specific instruction in cognitive bias mitigation using "mindware"—a term proposed by David Perkins to describe specific tools (rules, procedures or other forms of knowledge) that we have learned and memorized that can be used to optimize problem solving and decision making [37,38]. Mindware is discussed more fully in Chapter 8.

In the bottom half of the figure, there are various elements that further contribute to both declarative and procedural components of medical training. Most importantly, medical educators should be clear about what rational behavior is and how it may be attained. Characteristics of the rational clinician are described in Chapter 8. There is no substitute for comprehensive medical knowledge. With current cognitive aids, there is less need for arduous memorizing; however, students need to be aware that if they are deficient in knowledge, they must be familiar with the means to quickly access reliable sources. With all management of information, the standards of critical thinking need to be upheld to cultivate sound intellectual traits, as illustrated in Figure 14.3. Finally, patient preferences are paramount and ultimately all decisions must include the patient and consider their values and preferences.

Critical thinkers routinely
apply intellectual standards to the elements of
reasoning in order to develop intellectual traits

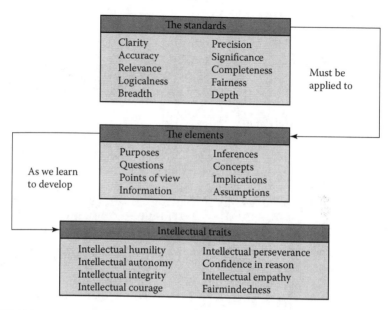

FIGURE 14.3
Application of the standards of critical thinking in the development of intellectual traits. (Reproduced with permission from Paul, R. and Elder, L. *Foundation for Critical Thinking*, Tomales, CA, 2014. http://www.criticalthinking.org/store/products/poster-standards-elements-traits/192.

Conclusion

Although Lucey's "sovereign physicians" may have been regarded as successful and highly effective in the twentieth century, in retrospect, we can be reasonably sure they were not. They would have been relatively naïve about the impact of cognitive bias on clinical reasoning, and unaware of the extent of their diagnostic failures, which have only recently been unmasked. Their failure rate was probably at least as high, or higher, as that noted in recent studies [39], (Chapter 8). What was missing from the skill set of this autonomous, independent, and authoritative figure was an awareness and understanding of the various sources of cognitive failure in the diagnostic process. Principal among them are cognitive and affective biases, and logical failures in reasoning. Yet, many medical schools do not presently teach about these important influences on clinical decision making, nor do they touch on other cognitive failings that may lead to irrational behavior (Chapter 8), arguably the most important influence on diagnostic failure.

In developing the twenty-first-century physician, medical educators need to address some pressing issues, with particular emphasis on systems-based practice and inter-professional teamwork [1]. As part of the new "collaboratively effective system", a physician needs to embrace a markedly expanded range of competencies [1]. The new basic sciences now incorporate the behavioral sciences, which include psychology, especially its critical domain, cognitive psychology. Others have already advocated for basic training in cognitive psychology as a preparation for medicine [40,41]. But, most importantly, there needs to be specific training in *rational decision making*. The dimensions of rationality have been described in detail by psychologists and are reviewed in Chapter 8 of this text. Among present-day cognitive psychologists, rationality is considered superordinate to both critical thinking and intelligence [42]. With the recent advent of tests to evaluate rational thinking, the Comprehensive Assessment of Rational Thinking (CART) is now available [43]. CART aims to be to rationality what IQ tests have been to intelligence. It may well serve as a useful adjunct in the screening process of applicants to medical school, as well as a test to determine what aspects of medical decision making might need remediation in those already in training and practice.

For medical educators, there is an inescapable logic here. If we accept that the ultimate attribute of a well-calibrated thinker is rationality, and if a significant part of diagnostic failure is due to flawed clinical reasoning, then we need to strongly promote rationality in clinical reasoning and decision making. Basically, we are arguing that the goals of medical education need to evolve. In addition to producing clinicians who are certifiably competent in the diagnostic process, we also want clinicians who are rationally well-calibrated, reflective, and mindful.

SUMMARY POINTS

- There are two types of medical knowledge: Knowing content (declarative) and knowing how to think about content (procedural). Historically, medical education has done a much better job with the former than the latter.

- It is important to recognize that multiple factors contribute to the complex process of clinical reasoning and decision making, and clinical epidemiology and biostatistics are only part of the training required for optimal decision making.

- There is good evidence in cognitive science literature that medical decision makers do not reach adequate normative standards of performance. In particular, there is a need for greater awareness of the impact of bias on the decision-making process.

- New models of the clinical reasoning process include a variety of sub-processes that deserve attention.
- Rationality has recently emerged as the most important characteristics in the cognitive make-up of the clinical decision maker and is now measurable using CART. Medical educators should consider using CART as both an adjunct for medical school admission testing, as well as a potential tool for identifying areas of clinical reasoning and decision making in need of remediation.

References

1. Lucey CR. Medical education: Part of the problem and part of the solution. *JAMA Intern Med*. 2013;173(17):1639–43.
2. Rao G. *Rational Decision Making: A Case-Based Approach*. New York: McGraw-Hill Medical; 2007.
3. Brush JE. *The Science of the Art of Medicine: A Guide to Medical Reasoning*. Richmond, VA: Dementi Milestone Publishing; 2015.
4. Trowbridge RL, Rencic JJ, Durning SJ, editors. *Teaching Clinical Reasoning: ACP Teaching Medicine Series*. Philadelphia, PA: American College of Physicians; 2015.
5. Cooper N, Frain J, editors. *ABC of Clinical Reasoning*. Indianapolis, IN: Wiley; 2017.
6. Hamm RM. Theory about heuristic strategies based on verbal protocol analysis: The emperor needs a shave. *Med Decis Making*. 2004;24(6):681–86.
7. Stanovich KE. On the distinction between rationality and intelligence: Implications for understanding individual differences in reasoning. In: Holyoak KJ, Morrison RG, editors. *The Oxford Handbook of Thinking and Reasoning*. Oxford: Oxford University Press; 2012. pp. 343–64.
8. Riegelman RK. *Minimizing Medical Mistakes: The Art of Medical Decision Making*. Boston, MA: Little, Brown; 1991.
9. Kassirer JP, Kopelman RI. *Learning Clinical Reasoning*. Baltimore, MD: Williams & Wilkins; 1991.
10. Kassirer J, Wong J, Kopelman RI. *Learning Clinical Reasoning*. 2nd ed. Baltimore, MD: Lippincott, Williams & Wilkins; 2010.
11. Detmer DE, Fryback DG, Gassner K. Heuristics and biases in medical decision-making. *J Med Educ*. 1978;53(8):682–83.
12. Stiegler MP, Neelankavil JP, Canales C, Dhillon A. Cognitive errors detected in anaesthesiology: A literature review and pilot study. *Br J Anaesth*. 2012;108(2):229–35.
13. David CV, Chira S, Eells SJ, Ladrigan M, Papier A, Miller LG, Craft N. Diagnostic accuracy in patients admitted to hospitals with cellulitis. *Dermatol Online J*. 2011;17(3):1.
14. Croskerry P. Achieving quality in clinical decision making: Cognitive strategies and detection of bias. *Acad Emerg Med*. 2002;9(11):1184–204.
15. Croskerry P. The importance of cognitive errors in diagnosis and strategies to prevent them. *Acad Med*. 2003;78:1–6.

16. Redelmeier DA. Improving patient care: The cognitive psychology of missed diagnoses. *Ann Intern Med.* 2005 Jan;142(2):115–20.
17. Vickrey BG, Samuels MA, Ropper AH. How neurologists think: A cognitive psychology perspective on missed diagnoses. *Ann Neurol.* 2010;67(4):425–33.
18. Fargen KM, Friedman WA. The science of medical decision making: Neurosurgery, errors, and personal cognitive strategies for improving quality of care. *World Neurosurg.* 2014;82(1–2):e21–9.
19. Dunphy BC, Cantwell R, Bourke S, Fleming M, Smith B, Joseph KS, Dunphy SL. Cognitive elements in clinical decision-making: Toward a cognitive model for medical education and understanding clinical reasoning. *Adv Health Sci Educ Theory Pract.* 2010;15(2):229–50.
20. Margo CE. A pilot study in ophthalmology of inter-rater reliability in classifying diagnostic errors: An underinvestigated area of medical error. *Qual Saf Health Care.* 2003;12(6):416–20.
21. Foucar E. Error in anatomic pathology. *Am J Clin Pathol.* 2001;116(Suppl): S34–46.
22. Crowley RS, Legowski E, Medvedeva O, Reitmeyer K, Tseytlin E, Castine M, Jukic D, Mello-Thomas C. Automated detection of heuristics and biases among pathologists in a computer-based system. *Adv Health Sci Educ Theory Pract.* 2013;18(3):343–63.
23. Singh H, Thomas EJ, Wilson L, Kelly PA, Pietz K, Elkeeb D, Singhal G. Errors of diagnosis in pediatric practice: A multisite survey. *Pediatrics.* 2010;126(1):70–79.
24. Jenkins MM, Youngstrom EA. A randomized controlled trial of cognitive debiasing improves assessment and treatment selection for pediatric bipolar disorder. *J Consult Clin Psychol.* 2016 Apr;84(4):323–33.
25. Crumlish N, Kelly BD. How psychiatrists think. *Adv Psychiat Treat.* 2009;15(1):72–79.
26. Sabih DE, Sabih A, Sabih Q, Khan AN. Image perception and interpretation of abnormalities; can we believe our eyes? Can we do something about it? *Insights Imaging.* 2011;2(1):47–55.
27. Bruno MA, Walker EA, Abujudeh HH. Understanding and confronting our mistakes: The epidemiology of error in radiology and strategies for error reduction. *Radiographics.* 2015;35(6):1668–76.
28 Shiralkar U. *Smart Surgeons, Sharp Decisions: Cognitive Skills to Avoid Errors and Achieve Results.* Shropshire, UK: TFM Publishing; 2010.
29 Boyle DJ. Cognitive errors. In: Stahel PF, Mauffrey C, editors. *Patient Safety in Surgery.* London: Springer; 2014. pp. 19–32.
30. Hershberger PJ, Markert RJ, Part HM, Cohen SM, Finger WW. Understanding and addressing cognitive bias in medical education. *Adv Health Sci Educ Theory Pract.* 1996;1(3):221–26.
31. Gillon SA, Radford ST. Zebra in the intensive care unit: A metacognitive reflection on misdiagnosis. *Crit Care Resusc.* 2012;14(3):216–20.
32. Dror IE, Thompson WC, Meissner CA, Kornfield I, Krane D, Saks M, Risinger M. Context management toolbox: A linear sequential unmasking (LSU) approach for minimizing cognitive bias in forensic decision making [letter]. *J Forensic Sci.* 2015; 60(4):1111–12.
33. Brereton M, De La Salle B, Ardern J, Hyde K, Burthem J. Do we know why we make errors in morphological diagnosis? An analysis of approach and decision-making in haematological morphology. *EBioMedicine.* 2015;2(9):1224–34.

34. Hicks EP, Kluemper GT. Heuristic reasoning and cognitive biases: Are they hindrances to judgments and decision making in orthodontics? *Am J Orthod Dentofacial Orthop.* 2011;139(3):297–304.
35. CReME. http://www.creme.org.uk/index.html. Accessed July 9, 2016.
36. Sibinga EMS, Wu AW. Clinician mindfulness and patient safety. *JAMA.* 2010; 304(22):2532–33.
37. Stanovich, KE. *What Intelligence Tests Miss: The Psychology of Rational Thought.* New Haven, CT/London: Yale University Press; 2009.
38. Nisbett RE. *Mindware: Tools for Smart Thinking.* New York: Farrar, Straus and Giroux; 2015.
39. Berner ES, Graber ML. Overconfidence as a cause of diagnostic error in medicine. *Am J Med.* 2008;121(5 Suppl):S2–23.
40. Redelmeier DA, Ferris LE, Tu JV, Schull MJ. Problems for clinical judgement: Introducing cognitive psychology as one more basic science. *CMAJ.* 2001;164(3):358–60.
41. Elstein AS. Thinking about diagnostic thinking: A 30-year perspective. *Adv Health Sci Educ Theory Pract.* 2009;14(Suppl 1):7–18.
42. Stanovich KE, West RF, Toplak ME. Intelligence and rationality. In: Sternberg R, Kaufman SB, editors. *Cambridge Handbook of Intelligence.* 3rd ed. Cambridge, UK: Cambridge University Press; 2011. pp. 784–826.
43. Stanovich KE, West RF, Toplak ME. *The Rationality Quotient: Toward a Test of Rational Thinking.* Cambridge, MA: MIT Press; 2016.

15

Cognitive Bias Mitigation: Becoming Better Diagnosticians

Pat Croskerry

CONTENTS

Introduction

Many human behaviors are extremely resistant to change. Virtually all behavior is preceded by cognition on some level; therefore, any discussion about behavioral change is really about changing cognition. Considerable challenges are experienced in a wide variety of domains; therapists working in the field of addiction behaviors face formidable difficulties in undoing acquired, harmful behaviors; the judicial system constantly seeks to change behaviors unacceptable to society; doctors face the ongoing task of motivating their patients to change unhealthy behaviors; coaches need to continuously refine and shape the behavior of their athletes; teachers need to instill new behaviors in their learners but also extinguish old ones; knowledge uptake and implementation in many fields, especially science, is a major challenge, and throughout there are abundant biases in human decision making that continuously undermine rationality. Secular and nonsecular thoughts and belief systems are often characterized by severely biased thinking and an ideological extremism that has accounted for human suffering on an extraordinary scale. Creating unbiased, balanced, rational thinkers is perhaps the greatest challenge that societies face. For us all, it is a lifelong journey for which cognitive debiasing is a vital and necessary tool. Given that cognitive factors appear to underlie the majority of diagnostic failures, cognitive debiasing strategies for clinical reasoning are a critically important issue.

Broadly speaking, clinical decisions about patient diagnoses are made in one of two modes: either *rational* or *intuitive*. The former is fairly reliable, safe, and effective, but slow and resource-intensive (Chapters 3 and 7). The latter is faster, more commonly used, and usually effective, but more commonly associated with failure. The intuitive mode of decision making is characterized by heuristics, such as short-cuts, abbreviated ways of thinking, and maxims like "I've seen this many times before". It is a rule of thumb among cognitive psychologists that we spend about 95% of our time in the intuitive mode [1]. We perform many of our daily activities through serial associations—one thing automatically triggering the next, with few episodes of conscious, deliberate, focused, analytical thinking. Thus, we have a prevailing disposition to use heuristics, and while they work well most of the time, they are intrinsically vulnerable to error. Our systematic errors are termed biases [2], of which there are many, including over one hundred cognitive biases [3] and probably about a dozen or so affective biases (ways in which our feelings influence our judgment) [4].

Pervasiveness of Bias

Bias is inherent in human judgment and decision making [5]. It is the principal factor underlying erroneous decision making (Chapter 8: The Rational Diagnostician). Its importance has been recognized beyond the individual at an organizational level in healthcare (Figure 15.1) [6,7] and by the broader scientific community [8]. Seshia et al. [6] use the term *cognitive biases plus* to describe the collective influence of cognitive biases, logical fallacies, conflicts of

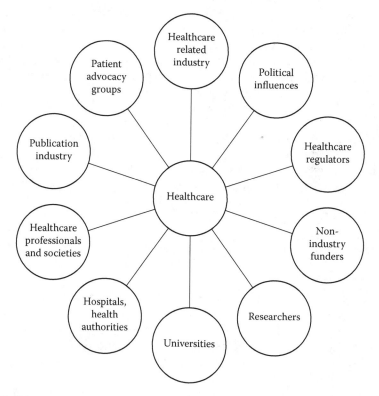

FIGURE 15.1
Ten major organizations in healthcare that are vulnerable to the influence of cognitive biases, fallacies, conflicts of interest, and ethical violations. (Reprinted with permission from Seshia, S.S. et al., *J Eval Clin Pract.*, 20(6), 735, 2014 [6,7].)

interest, and ethical violations on individuals and organizations in healthcare. All four of them lead to distorted reasoning and decision making. Biases have been described as "predictable deviations from rationality" [9]. Many biases that diagnosticians hold are often recognized and corrected by themselves. Essentially, this is the process that underlies learning and the refinement of clinical behavior. We may learn an inappropriate response to a particular situation that leads to a maladaptive habit, but then some insight or revelation occurs and we change our ways to achieve a more successful outcome. However, the persistence of particular biases that appear resistant to change has attracted the interest of research studies and is the focus of cognitive debiasing [10–15], perhaps more accurately known as cognitive bias mitigation (CBM) as it is more likely that we can reduce as opposed to eliminate bias. The basic argument is that if we can effectively reduce bias in our thinking, we will become better thinkers and improve our clinical reasoning skills.

Besides the general vulnerability of the human mind toward biases in decision making, there is clear evidence that the quality of decision making is also influenced by ambient factors, or prevailing conditions in the immediate environment in which decisions are being made, including context, team factors, patient factors, resource limitations, physical plant design, ergonomic factors (Chapters 11 and 17) and individual homeostatic factors such as affective state, general fatigue, cognitive loading, decision fatigue, interruptions and distractions, sleep deprivation, and sleep-debt (Chapter 9). Thus, the tendency toward biased decisions may be exacerbated by ambient conditions. Individual factors such as personality, intelligence, rationality, gender, and other variables are also known to impact decision making (Chapter 9). Psychopathology, nonsecular beliefs, postmodernism, deconstructuralism, and magical thinking are generally not considered in these discussions, although all clearly have the potential to exert powerful influences on rationality.

Cognitive Bias Mitigation

There are two questions: Firstly, can we improve our decisions by CBM? This means appropriately alerting the analytic mode to situations in which a bias might arise in the intuitive mode so that it can be detected and a CBM intervention applied. As Burton notes [16], there remains some polarization on this issue. Daniel Kahneman, who wrote *Thinking, Fast and Slow* [2] appears to be generally pessimistic about whether we can change our cognitive failings to improve decision making (although in his book he does offer a number of CBM suggestions). In contrast, another prominent cognitive scientist, Steven Pinker, points to a significant body of evidence showing that we have been able to change a variety of our behaviors for the better over time [17].

Recent developments would appear to support the more optimistic view [18]. Secondly, can we mitigate the impact of adverse ambient conditions by improving conditions in the decision-making environment? Various strategies described below do suggest that we might be equally optimistic about extra-cognitive interventions, that is, changes to the workplace.

Principle Strategies for Cognitive Debiasing

Getting Clinicians past the Precontemplative Stage

Diagnosticians first need to accept that their thinking is often biased and that change may be necessary. Many clinicians are simply unaware and uninformed about cognitive biases and their effect on thinking—after all, this has not been part of traditional medical education and the concepts and terms will be alien to many [16]. There likely also prevails a not invented here (NIH) bias [18,19], in that many clinicians would not be willing to incorporate developments in cognitive science, as they are outside the discipline of medicine. Those that are aware of bias may be disinclined to believe that their thinking is biased due to blind spot bias [18,20]; that is, they do not believe they are biased and therefore are not in need of remediation. Another possibility is that they might accept their judgments to be biased at times but believe they do not result in serious consequences.

We need to appreciate that cognitive change rarely comes about through a discrete, single event but instead involves moving through a succession of stages—from a state of lack of awareness and disinterest (precontemplation stage), to considering a change (contemplation stage), to deciding to change, to initiating strategies to accomplish change, and finally maintaining the change. These are the key steps outlined in the Transtheoretical Model of Change (Figure 15.2) [21]. A significant number of clinicians are presently at the *precontemplative* level; they are most likely unaware of their cognitive biases and hence see no reason to take any action to change their thinking.

Rounds, seminars, workshops, journal articles and other forms of communication serve to introduce these ideas and raise awareness. Sometimes a sentinel event can catalyze the uptake of an important idea, such as the publication of Groopman's book *How Doctors Think* [22]. Another important way in which a single event or experience can change thinking is if it is emotionally charged. For example, if a physician misdiagnoses a headache as benign and the patient subsequently dies from a subarachnoid hemorrhage, the impact of this experience might forever change the physician's approach toward patients presenting with headaches. Affect and arousal are especially effective motivators in reasoning, in the formation of beliefs, and in change.

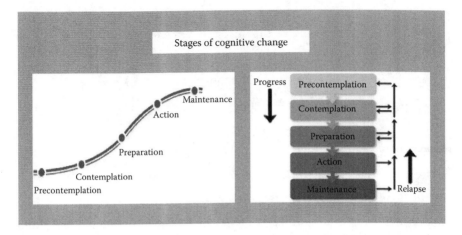

FIGURE 15.2
Transtheoretical Model of Change, as described by Prochaska, J.O. et al., *Am Psychol.*, 47(9), 1102–1104, 1992 [21].

A surprisingly wide variety of CBM strategies have been proposed [10,11,13,14]. Medicine itself has developed some intrinsic operating characteristics that historically have served a bias-mitigating function for some of the more common biases (Table 15.1). Some of these emerged prior to the heuristics and biases literature that began in earnest in the 1970s [23]. Others have come from domains outside of medicine and many are recently developed.

Once clinicians are past the precontemplative stage and open to more ideas, a variety of approaches may be useful for CBM; they are described in more detail in the following sections.

Dual Process Theory Training (DPTT)

Promoting awareness and understanding of the rationale for CBM to occur requires insight into the basic processes by which decisions are made. As a foundation, knowledge of dual process theory (DPT) as it applies to diagnostic reasoning [24], the properties of the dual process model (Chapter 3), the nature and extent of heuristics, and the range of cognitive and affective biases are all important. This is the strategy of providing training around the overall schema of decision making, including its strengths and its fallibilities.

Bias Inoculation (BI)

Specifically, teaching about particular cognitive and affective biases and giving multiple clinical examples in different contexts is important, as well as proposing particular debiasing strategies [25]. For example, for search satisficing, educators should illustrate the bias in orthopedics, toxicology, and

TABLE 15.1

Existing Strategies in Medicine That May Mitigate Cognitive and Affective Bias

Strategy	Purpose	Examples of Potential Biases Addressed
History and physical exam	Deliberate and systematic gathering of data	Unpacking principle Ascertainment bias
Differential diagnosis	Forces consideration of diagnostic possibilities other than the obvious or the most likely	Anchoring and adjustment Search satisficing Premature diagnostic closure Availability Representativeness Confirmation bias
Use of "Not Yet Diagnosed" (NYD)	Keeps diagnostic possibilities open	Premature closure Diagnosis momentum Confirmation bias
Clinical prediction rules	Force a scientific, statistical assessment of patient's signs and symptoms, and other data to develop numerical probabilities of the presence/absence of a disease or an outcome	Base rate fallacy Errors or reasoning Errors in estimating probabilities
Evidence-based medicine	Establishes imperative for objective scientific data to support analytic decision making	Many biases
Checklists	Ensure that important issues have been considered and completed, especially under conditions of complexity, stress and fatigue, but also when routine processes are being followed.	Anchoring and adjustment Availability Memory failures
Mnemonics	Protect against memory failures and ensure a full range of possibilities is considered in the differential diagnosis	Availability Anchoring and adjustment Premature closure
Pitfalls	Alert inexperienced clinicians to predictable failures commonly encountered in a particular discipline	Many biases
Rule out worst-case scenario (ROWS)	Ensures that the most serious condition in a particular clinical setting is not missed	Anchoring and adjustment Premature diagnostic closure
Until proven otherwise (UPO)	Ensures that a particular diagnosis cannot be made unless other specific diagnoses have been excluded	Anchoring Confirmation bias Diagnosis momentum Premature closure
Caveats	Offer discipline-specific warnings to ensure important rules are followed to avoid missing significant conditions	Many biases
Red flags	Specific signs and symptoms to look out for, often in the context of commonly presenting conditions, to avoid missing serious conditions.	Anchoring Confirmation bias Diagnosis momentum Premature closure

soft tissue foreign bodies. The general rule is when you see one abnormality, be careful to search for others. Demonstrate how the bias works in a particular clinical context, and how transfer of the concept applies. If possible, re-test and reinforce after a short time interval, and re-test/reinforce again and again. This is a form of cognitive vaccination with booster shots – it is also known as *cognitive engineering*.

Specific Educational Interventions (SEI)

Teaching specific skills may mitigate particular biases by providing the learner with foundational knowledge that allows greater insight into problems. For example, people who have taken courses in probability, statistical reasoning, and experimental research are less likely to commit base rate errors and will be more sensitive to flaws in non-evidence-based claims. Thus, giving medical students specific training in critical thinking (Chapters 2 and 9), argumentation, and basic research design might make them less likely to accept unwarranted assumptions from various sources (patients, colleagues, pharmaceutical representatives, media).

Interactive Serious Computer Games (ISCG)

Several studies have now demonstrated the significant benefit from training with interactive serious computer games (ISCG) targeted at debiasing specific cognitive biases [26–30]. Repetitive training with the game significantly improves retention. These initiatives originated from the Sirius Project at the Intelligence Advanced Research Projects Activity (IARPA) of the United States Intelligence Agency, and appear to be particularly promising CBM interventions.

Cognitive Tutoring Systems (CTS)

Another recent innovation is the development of software that can monitor decision making in clinical cases and detect cognitive biases according to pre-set criteria. Thus, a learner's profile of decision making across a series of clinical cases in pathology has been constructed to provide feedback on specific biases [31]. Providing such feedback to the learner, along with specific strategies to avoid certain biases, might be a powerful CBM tool.

Get More Information (I)

Heuristics and biases are often driven by insufficient information. System 1 (where most heuristics and biases occur) is typically activated under conditions of limited information, and a response will be initiated, especially when the information appears to be a good story. Kahneman refers to this as "what you see is all there is" (WYSIATI), that is, making the assumption that what you see is all there is to see and sufficient to make a decision [2]. It is also referred to as shallow or narrow thinking [32]. It is easier to construct a good story under conditions of limited information because, paradoxically, more coherence can be given to the story if there are fewer pieces to integrate together. We have, Kahneman says, a strong tendency to ignore what we don't know [2]. Thus, getting more information opens up more options and may force a consideration of alternative interpretations.

Structured Data Acquisition (SDA)

Diagnostic error may arise when clinicians focus on salient, prototypical features in a patient's presentation. System 1 is engaged and premature closure of diagnostic options may occur through various biases such as anchoring and search satisficing. This typically happens with "corridor consultations" and "drive-by" diagnoses (Chapter 4). Instead, forcing a deliberate structured acquisition of data can avoid this trap by ensuring that less obvious areas are covered; for example, for decision making in psychiatry, routine training in an explicit structured approach toward the clinical interview for DSM disorders (SCID: Structured Clinical Interview for DSM disorders) improves diagnostic performance by nullifying biases [33]. Combining standardized diagnostic interviews (SDIs) with expert opinion and information from the medical record, exemplified in the LEAD (Longitudinal, Expert and All Data) approach [34], may also mitigate bias.

Being More Skeptical (S)

The prevailing tendency in human thinking is to believe rather than disbelieve [35]. When we engage System 1 to make sense of something, we have a strong tendency to view something as more "tidy, predictable, and coherent than it really is" [2]. In medicine, we are surrounded by uncertainty and, therefore, should aim to disbelieve what is put in front of us. Overall, we should be more skeptical and willing to challenge the apparent veracity and coherence of data.

Affective Debiasing (AD)

Virtually all decision making involves some degree of affective input, nevertheless, a broad distinction is often made between cognitive and affective biases. Many affective biases are hardwired, although some will be implicitly or explicitly learned due to affective associations with prior experience. As noted, affect and the arousal that often goes with it provide strong motivation for the formation of strong beliefs. But affective bias mitigation is particularly challenging because clinicians are often unaware of intrusions of affective influences on their decision making. Overviews are available of the influence of affect on decision making [36] and a preliminary taxonomy of affective biases has been proposed [37]. In the context of a strong emotional bias counteracting reason, Gigerenzer recommends using a conflicting and stronger emotion [38].

Forcing Functions (FF)

There are a variety of forcing functions that can be built into clinical behavior around decision making. These are probably our most important tools for debiasing. They require an interface at which the forcing function can be applied. They do not all need to be explicit; sometimes it is possible to gently nudge people in a particular direction in order to get better outcomes, a practice Thaler and Sunstein have termed paternal libertarianism [39]. Examples of forcing functions are as follows:

1. *Generate alternatives:* To be able to generate alternatives is a good overall strategy, especially for dealing with narrow thinking [40]. It may mitigate a number of biases, such as anchoring, confirmation bias, diagnostic momentum, premature closure, and others. The generation of a differential diagnosis in medicine is an example of an intrinsic forcing function that has stood the test of time. It forces a consideration of relevant, competing alternatives. Generating alternatives appears more effective when each alternative is considered separately [41].

2. *Decision support systems (DSS):* Technical systems (e.g., DxPlain, Isabel) automatically provide a differential diagnosis once the patient's demographics, symptoms and signs have been entered. The differential diagnosis list forces consideration of a diagnosis that may not have been considered. These are often effective prompts.

3. *Cognitive forcing strategies (CFS):* These are a special case of forcing functions but require the clinician to internalize the forcing function and apply it deliberately in a context-specific way. They represent a systematic change in clinical practice. CFSs may range from universal to generic to specific [42].

4. *Disconfirming strategy (DS):* Confirmation bias (see Appendix I) is the tendency to seek information that supports or confirms an already favored hypothesis, and is generally considered pervasive in reasoning. While confirming strategies often strengthen the belief in an existing hypothesis, the most powerful falsifiability strategy is a disconfirming strategy. Falsifiability of a hypothesis allows the ultimate demarcation into scientific or not scientific, that is, true or not true.

5. *Data blinding:* Various forms of data blinding strategies exist that essentially protect the decision maker from being influenced by the thinking of others, or ambient influences. It is a critical part of the randomized double-blind clinical trial. But it can also be used on

an individual basis; for example, some emergency physicians prefer not to hear the opinions of others, and choose not to read the triage nurse's notes in order to avoid "cognitive contamination" and maintain their cognitive independence. In forensic science, a variant of this is termed *linear sequential unmasking* (LSU), in which the decision maker is protected from potentially biasing task-irrelevant information in the course of developing an opinion and making a decision [43].

6. *Standing rules*: These may be used in certain clinical settings (e.g., in an emergency department) which require that a given diagnosis not be made unless other must-not-miss diagnoses have been ruled out; for example, a diagnosis of acute myocardial infarction cannot be made until a chest x-ray has been done and blood pressure measured in both arms (to rule out thoracic aorta dissection).

7. *Prospective hindsight*: This addresses the pervasive disposition toward overconfidence [44] associated with the tendency to overestimate the chances of success in dealing with a new problem with optimism bias toward one's own decision making. It involves imagining a point in the future when the results of your decision can be seen to have failed [45]. For example, an emergency physician might imagine that the patient he has just diagnosed and discharged appears at the emergency department the next day with what was clearly a wrong diagnosis. The process of contemplating an imagined past failure may generate diagnostic possibilities that were not apparent in foresight.

8. *General diagnostic rules in clinical practice*: These are general rules that have evolved to avoid predictable pitfalls. For example, no diagnosis of a patient who has neurological symptoms can be considered until the blood sugar is measured; pulmonary embolus should always be considered in patients with any type of chest pain; every diabetic patient with systemic symptoms automatically gets a 12 lead electrocardiogram; anxiety disorder, somatization disorder, and conversion reaction cannot be diagnosed except by exclusion.

9. *Specific forcing strategies*: Clinicians may follow specific strategies, such as "Rule Out Worst-Case Scenario" (ROWS) to avoid missing certain diagnoses. For example, any patient with presenting symptoms of wrist sprain automatically has a scaphoid exam; a diagnosis of musculoskeletal chest pain cannot be made unless acute coronary syndrome, pneumothorax, aortic dissection, pneumonia, and pulmonary embolus have been considered; patients with back pain always force an exclusion of cauda equina syndrome. A similar strategy is "Until Proved Otherwise" (UPO), where the physician is obliged to rule out other specific possibilities before accepting a diagnosis. For example, any athlete with an on-field head injury has

a neck injury UPO; any new onset neurological condition is hypo-glycemia UPO; the agitated, belligerent patient is hypoxic UPO.

10. *Checklists* are a basic form of forcing function. They are a recognized standard in areas such as aviation, and have now been incorporated into medicine, in intensive care units [46], surgery [47], and in the diagnostic process [48]. They are a simple way of saying "what else might this be?", part of the strategy that has been termed "strong inference" [49].

11. *Structured report templates:* In some settings, semi-structured report templates may be used, which remind the decision maker to take a second look at specific aspects, areas of the problem, and features of the data [50]. This strategy essentially forces the decision maker to ensure that key areas are covered, and take a second look themselves.

12. *Stopping rules (SR):* Most of our problem solving and decision-making behavior depends very much on how we search for informa-tion [51]. The information search determines the number and quality of options that will be considered, as well as the ultimate choice [52]. Stopping rules are an important type of forcing function—the rule specifies the point at which enough information has been gathered so that a considered and optimal decision can be made. Typically, they specify some criterion that has to be reached before a diagnosis can be safely made. For example, when a fracture is found on an x-ray, the search does not stop until a second fracture or significant soft tissue injury has been excluded; one troponin does not exclude an acute coronary syndrome; the examination of an injured joint does not stop until the joint above and below have been examined.

Metacognition, Mindfulness, and Reflection (MMR)

The process of metacognition or thinking about thinking and reflection are represented in the dual process model as the System 2 check on System 1 (executive control). Mamede et al. have shown the benefits of reflective strat-egies on decision making [53,54]. A physician's diagnostic accuracy suffers when diagnoses are made early in the assessment process and improves when an effort is made to slow down [55]. *Mindfulness* is defined as non-judgmental awareness of the present moment. It has considerable overlap with metacognition and involves reflection. A variety of mindfulness quali-ties have been described that may be used to reduce specific biases [56–58].

Slowing Down (SD) Strategies

The adoption of "slowing down" is a deliberate strategy that facilitates transition from the intuitive mode (Type 1 processing) to the analytic mode (Type 2) and provides an opportunity to reflect on the situation [59]. These intermissions may be determined by the situation at hand, or proactively planned, for example, a planned time out in the operating room [60].

Re-Biasing (RB)

This involves using one bias to offset another. In a sense, it is a forcing strategy in that when the bias is detected, the decision maker automatically substitutes it with another bias, forcing a compensation of the original bias. For example, there are biases associated with the diagnosis and management of psychiatric patients. Their medical problems are often minimized and they suffer more adverse events in hospitals than non-psychiatric patients. Being aware of this bias, a clinician can re-bias themselves by being more attentive than usual to psychiatric patients with medical complaints, fully examining them, and conducting whatever investigations might be necessary.

Group Decision Strategy (GDS)

Sometimes the wisdom of the crowd exceeds that of an individual decision-maker [61]. Group rationality tends to exceed individual rationality [62]. Although it is time-consuming and not always practical, in complex situations, it may be worth having a case conference to reach an optimal solution, like tumor boards for example. At a minimum, it is sometimes worth bouncing one's decision making off colleagues to run a check on one's own thinking.

Public Policy Decision Making

While the majority of the interventions proposed here are aimed at an individual level, it follows that decisions made by organizations which are vulnerable to bias may also be in need of bias mitigation. Thus, individuals within organizations could be protected from the effects of bias by policies that have built-in CBM that would result in ideological reform. The World Bank, for example, in their World Development Report (2015) looked at how individual staff made decisions [63,64]. Not surprisingly, evidence was found of several common biases at work. Once identified, policies and practices can be implemented to mitigate them at an organizational level.

Personal Accountability (PA)

When people know their decisions will be scrutinized and they will be held accountable for them, the amount of effort increases and people generally perform better. Strategies that improve personal accountability generally lead to better decisions [65].

Educating Intuition

This is more about improving the overall quality of decision making than bias mitigation *per se*. The overall approach is aimed at improving the environment in which decisions are made, that is, creating less "wicked" and more supportive environments [66]. Although particular workplaces may present challenges, clinicians should try to make the environment friendlier and more supportive. They should avoid taking on too much and cognitively overloading themselves. They should deal with as few as possible problems at any one time, and arrange regular breaks to avoid decision fatigue. Protocols, decision rules, clinical guidelines, and patient care pathways should be readily available. Rules should be in place about when and how team members may interrupt each other. Roles and responsibilities should be clarified. Hogarth has further recommendations to make the environment more supportive of good decision making [66].

Sparklines

Sparklines are information mini-graphics embedded in context in clinical data. For example, a simple graphic showing the trends in prevalence of several pediatric respiratory viruses by month can give an immediate and accurate estimate of respective base rates and trends. These graphics provide a powerful visual augmentation of data, and can immediately mitigate specific biases [67–69].

Cultural Training

Given that some biases are learned, implicitly or explicitly, it follows that cultural pressures to avoid certain biases may occur. In Eastern cultures, for example, the tendency toward dialectical reasoning may lead to different inferences and assumptions being made about the world that make it less likely that certain biases (fundamental attribution error, confirmation bias, susceptibility to the interview illusion) will be expressed; there is also less vulnerability to contextual influences [70].

Are All Biases Created Equal?

A tacit assumption prevails in the medical literature that all biases are created equal, that all are equally difficult to overcome, and that some common CBM strategies might be effective. However, as Larrick points out, many biases have multiple determinants, and it is unlikely that there is a "one-to-one mapping of causes to bias, or of bias to cure" [71]. Neither is it likely that one-shot debiasing interventions will usually be effective [72]. From DPT and other work of cognitive psychologists, we know that most biases are associated with heuristics and typically are Type 1 (intuitive) processes. Further, Stanovich [73] has categorized these "autonomous" Type 1 processes according to their origins; there are four main groups as follows:

1. *Processes that are hardwired.* These were naturally selected (in the Darwinian sense) in our evolutionary past for their adaptation value. Those who had them passed their genes onto the next generation and they have been genetically preserved in modern brains, although they may no longer be adaptive in certain settings. This is the argument made by evolutionary psychologists (Chapter 6). Although there is no acid test for defining an evolutionary bias, some likely examples are: meta-heuristics (anchoring and adjustment, representativeness, availability), search satisficing, overconfidence, and others.

2. *Processes that are regulated by our emotions.* These too may be evolved adaptations (hardwired) and can be classified into six major categories: happiness, sadness, fear, surprise, anger, and disgust. Fear of snakes, for example, is universally present in all cultures. They may also be socially constructed (acquired, learned), or combinations of the two—hardwired processes modified by learning, for example, visceral reactions against particular types of patients [74].

3. *Processes that become firmly embedded in our cognitive and behavioral repertoires through over-learning.* These might include explicit cultural and social mores, but also those associated with specific knowledge domains. Intubation is a good example—through many repetitions of psychomotor, visual, and haptic (sense of touch) responses, anesthetists and others become very competent and comfortable in intubating patients smoothly and effortlessly. An example of a bias acquired through repetitive exposures might be a "frequent flyer" in a family doctor's office or in the emergency department where the bias might be the expectation that no significant problem will be found.

4. *Processes that have developed through implicit learning.* It is now well recognized that we learn in two fundamental ways. Firstly, we learn through deliberate explicit learning, such as that which occurs in school and in formal training, and secondly, we learn through implicit learning, which is without intent or conscious awareness. Implicit learning plays an important role in the development of skills, and in our perceptions, attitudes and overall behavior. It allows us to detect and appreciate incidental covariance and complex relationships between things in the environment without necessarily being able to articulate that understanding. Thus, some biases may be acquired unconsciously. Medical students and residents might subtly detect and acquire particular biases by simply spending time in environments where others have these biases, even though the bias is never deliberately articulated or overtly expressed to them. Examples might be the acquisition of such biases as ageism, socioeconomic status, gender, race, psychiatric patients, and others [75].

[Although Type 1 processes appear the most vulnerable to bias and suboptimal decision making, they are not the sole repository of impaired judgment. Arkes points out that error due to biases also occur with Type 2 processes, which are included here under category 5] [76].

5. *Errors that arise through biases that have become established through inferior strategies or imperfect decision rules,* that is, even though the decision-maker may be deliberately and analytically applying accepted deliberate strategies or rules, they may be flawed. Thus, there may have been a problem in the initial selection of a strategy, which may underestimate or overestimate a diagnosis. Of the two, it would seem preferable to always overestimate (e.g., ROWS) so that important things do not get missed, but this can lead to a waste of resources. Generally, suboptimal strategies get selected when the stakes are not high.

Prescriptive Decision Making

Are There Specific Cognitive Pills for Cognitive Ills?

It is evident that biases are not easily eliminated from our decision making. However, some degree of CBM can be achieved if we adopt optimal approaches. Given the differing etiologies of bias, might we expect that some are more robust and therefore more resistant to change than others? Should there then be different approaches to mitigation?

We might expect that the hardwired "evolutionary" biases would be the most resistant to change, and we may need several different CBM strategies as well as multiple interventions. Cultural, sociocentric and other biases that have been established through learning may be a little easier to change, although these biases should ideally not be allowed to form in the first place. Good modeling, good teaching programs, and optimal learning environments will minimize them. Locally acquired biases might be the least intransigent and the most amenable to change. Strong affective biases may need fundamentally different strategies from general cognitive biases.

Recent literature is becoming more specific about biases and their defining characteristics. Various taxonomic strategies have been proposed, such as those by Arkes [76], Campbell et al. [77], and Arnott [9]. We may find that we can start predicting which strategies will work for biases of particular properties, and that certain types of strategies might work for certain taxonomic classes of biases, as Arkes has proposed [76]. Table 15.2 illustrates the taxonomy that was initially developed by Arnott [9]. He notes that considerable overlap is likely between categories in terms of their definitions and their effects. For each of his categories, possible generic CBM strategies may be effective.

Recently, several issues have become increasingly clear. Anyone who is involved in clinical decision making in the care of patients should be grounded in basic decision theory, the dual process model and its basic operating characteristics, and the origin of biases. They should also have some awareness of cognitive and affective biases, how pervasive they are, and the need for CBM. In the meantime, we need some idea of what CBM strategies have been tried. Table 15.3 lists the main types of bias, with some of their determining characteristics, as well as some potential strategies for debiasing, although these are speculative at this stage. Interestingly, there is a surprising correspondence between the bias categories described by Stanovich [73] and the psychological barriers to clear reasoning originally described by Francis Bacon in his book *Novum Organum* published in 1620 [78]. Bacon identified four types of reasoning problems, or fixations (idols) of the mind (cave/cavern). *Idols of the tribe* were archetypal, hardwired beliefs; *Idols of the cave* were individual characteristics based on personality, education, habits and environmental influences; *idols of the marketplace* arose from language and semantics—the ways in which words might be used to substitute for ideas and to mislead others; and *idols of the theatre* were beliefs that had been developed and incorporated as dogma and status quo, and subsequently went unchallenged.

Anyone who is involved in clinical decision making around patient care should receive general training in basic decision making, understand DPT and its major operating characteristics, and be aware of where in the process bias originates. All CBM initiatives should initially receive a presentation on the general properties of heuristics and biases.

Are There Specific Situations in Which Biases Are More Frequent?

Many physicians can think of situations where they appear to be particularly vulnerable to bias. Some will set the physician up for exposure to particular biases whereas others will produce exposure to a wide range of biases. Some common situations are described in Table 15.4 [79].

How Does CBM Actually Work?

Some degree of debiasing is part of everyday living. We learn the consequences of certain actions and take steps to avoid falling into the same traps. Often we can do this using forcing strategies or deliberately suppressing impulsivity in certain situations. We can't find our car keys, usually at a time when we are in a hurry, so we (some of us at least) learn the forcing strategy of always putting them in a specific place as soon as we arrive home [15]. In other domains, we have learned that it is a good idea to suppress belief and be skeptical when we are offered deals that are too good to be true, such as the Nigerian email looking for our financial support. Interestingly, higher intelligence does not necessarily protect against such follies [73].

TABLE 15.2

Taxonomy of Biases by Mechanism

Bias Category	Mechanism	Examples	Possible Generic CBM Strategies												
			I	S	DPTT	BI	TS	DS	RF	SEI	MMR	IE	U	CTS	ISCG
Memory	Interfering with storage and recall of information	Hindsight, availability, recall, search, representativeness, testimony	+	+	+	+	+				+			+	+
Statistical	Information processing that violates the normative principles of probability theory	Chance, conjunction, correlation, disjunction, sample, subset	+	+	+					+	+			+	+
Confidence	Leads the decision maker to a distorted view of their ability, often resulting in search satisficing and premature closure	Completeness, control confirmation, desire, overconfidence, redundancy, selectivity, success, test	+	+	+	+		+			+			+	+
Adjustment	Undue emphasis on salient features and failure to establish and adjust to more objective reference points	Anchoring and adjustment, conservatism, reference, regression	+	+	+						+	+	+	+	+
Presentation	The particular way the data is presented distorts the way in which information is perceived	Framing, linear, mode, order, scale	+	+	+				+		+	+	+	+	+
Situational	Determined by the particular situation or context in which the decision is made	Attenuation, complexity, escalation, habit, inconsistency, rule	+	+							+	+	+	+	+

Note: **BI**, Bias inoculation; **CBM**, Cognitive bias mitigation; **CTS**, Cognitive tutoring systems; **DPTT**, Dual process theory training; **DS**, Disconfirming strategy; **I**, Get more information; **IE**, Improve environment; **ISCG**, Interactive serious computer games; **MMR**, Mindfulness, metacognition, reflection; **RF**, Reframing; **S**, Skepticism; **SEI**, Specific educational intervention; **TS**, Technology strategy (DSS); **U**, Unfreezing.

TABLE 15.3

Taxonomy of Biases by Origin

Type of Bias — Determinants and Characteristics of Bias — Examples	Potential CBM Strategies																		
	DPTT	ISCG	BI	MAS	TO	SEI	RB	AD	GDS	CFS	DS	CTO	S	M&R	SD	DRT	F	IE	CTS
1. Evolutionary																			
a) Hardwired Directed at modular activities Adaptive in ancient environments																			
• Search satisficing	+	+	+			+				+			+		+				
• Representativeness	+	+	+				+									+	+		+
• Anchoring	+	+	+				+					+			+	+	+		+
• Availability	+	+	+				+					+		+		+	+		+
• Overconfidence	+	+	+				+					+		+		+	+		+
• Premature closure	+	+	+				+					+		+	+		+		+
• Base rate neglect	+	+	+				+							+		+	+		+
• Confirmation bias	+	+					+				+	+		+		+	+		+
2. Emotionally-driven																			
a) Those with an evolutionary origin are hardwired																			
• Fear, anxiety, anger	+	+		+	+				±						+				+
b) Acquired emotional dispositions associated with particular individual life-history experiences																			
• Counter transference reaction against borderline patient	+	+		+	+				±										+
• Aversion to drug seekers	+	+		+	+														+

(Continued)

TABLE 15.3 (CONTINUED)

Taxonomy of Biases by Origin

↓ Type of Bias	↓ Determinants and Characteristics of Bias	↓ Examples	Potential CBM Strategies																			
			DPTT	ISCG	BI	MAS	TO	SEI	RB	AD	GDS	CFS	DS	CTO	S	M&R	SD	DRT	F	IE	CTS
3. Over-learned	a) Acquired cultural/racial/ social/ professional mores and biases	• Racial stereotyping	+	+				+		+											+
		• Profiling classes of patients	+	+				+				+				+					+
	b) Specific biases acquired through over-learning of particular response	• Ageism	+	+	+			+		+		+									+
		• Obesity	+	+	+			+		+		+									+
		• Patients with psychiatric comorbidity	+	+	+			+		+		+									+
		• Status quo bias	+	+				+										+	+			+
4. Implicitly-learned	a) Acquired unconsciously through observation and experience in specific environments Taking cues from others	• "Frequent flyer" bias	+	+	+			+		+		+							+	+	+
		• Psychiatric comorbidity bias	+	+	+			+		+		+							+	+	+
		• Status quo bias	+	+				+										+				+

(Continued)

TABLE 15.3 (CONTINUED)

Taxonomy of Biases by Origin

↓ Type of Bias ↓ Determinants and Characteristics of Bias ↓ ↓ Examples	Potential CBM Strategies																		
	DPTT	ISCG	BI	MAS	TO	SEI	RB	AD	GDS	CFS	DS	CTO	S	M&R	SD	DRT	F	IE	CTS
5. Strategy-based error																			
a) Type 2 based selection of particular strategy that deals with an issue that is not perceived to be of high consequence.																			
• Taking an abbreviated history	+	+				+	±												+
• Doing a limited exam	+	+				+	+												+

Note: **AD**, Affective debiasing; **BI**, Bias inoculation; **CFS**, Cognitive forcing strategy; **CTO**, Consider the opposite; **CTS**, Cognitive tutoring systems; **DPTT**, Dual process theory training; **DRT**, Decision rule training; **DS**, Disconfirmation strategy; **F**, Feedback; **GDS**, Group decision strategy; **IE**, Improve environment; **ISGG**, Interactive serious computer games; **MAS**, Meta-affect strategy; **M&R**, Mindfulness, metacognition, reflection; **RB**, Re-biasing; **S**, Skepticism; **SD**, Slow down; **SEI**, Specific educational intervention; **TO**, Timeout.

TABLE 15.4

Bias Exposure in Risk Situations

High Risk Situations	Potential Biases
Was this patient handed off to me from a previous shift?	Diagnosis momentum, framing
Was the diagnosis suggested to me by the patient, nurse or another MD?	Premature closure, framing bias
Did I just accept the first diagnosis that came to mind?	Anchoring, availability, search satisficing, premature closure
Did I consider other organ systems besides the obvious one?	Anchoring, search satisficing, premature closure
Is this a patient I don't like, or like too much, for some reason?	Affective bias
Have I been interrupted/distracted excessively while evaluating this patient?	All biases
Did I sleep badly last night/Am I feeling fatigued right now?	All biases
Am I cognitively overloaded or over-extended right now?	All biases
Am I stereotyping this patient?	Representative bias, affective bias, anchoring, fundamental attribution error, psych-out error
Have I effectively ruled out must-not-miss diagnoses?	Overconfidence, anchoring, confirmation bias

Source: Adapted from Graber, M.L. et al., *Diagnosis (Berl).*, 1(3), 223–31, 2014.

Medical training is often sprinkled with precautionary caveats and at its completion we are probably at our most cautious because of heightened uncertainty. Experience accumulates, but does not necessarily bring expertise with it. Inevitably, we develop our own debiasing strategies to avoid the predictable pitfalls that we have learned at our own expense or secondhand through the experience of others. Morbidity and mortality rounds are a good opportunity for such vicarious learning. However, with experience there is an increased likelihood of biases of the 3a type (see Table 15.3), the Type 1 judgments that have been established through repetitive exposure in Type 2. Other changes that are noted with experience are a progressive loss of empathy as training progresses, which may impact the emotion biases. We might also expect that certain classes of patient might engender less tolerance after repeated exposure.

Overall, we are faced with the continuous challenge of debiasing our judgments throughout our careers, and various ideas have been proposed for how this should work. In order to examine them in detail, we need to review the dual process model and its major properties (Figure 15.3).

The intuitive system is schematized as Type 1 processes. It contains the four channels depicting the first four classes in Table 15.3. The analytic system is the Type 2 processes. There are eight major features of the model:

- Type 1 processing is fast, autonomous, and where we spend most of our time—this is where most heuristics and biases occur.

- Type 2 processing is slower, deliberate and generally more accurate.
- Most errors occur in Type 1 (intuitive) processes—the predictable deviations from rationality.
- Repetitive processing in Type 2 (analytic) processes may lead to processing in Type 1.
- Type 2 processes can override Type 1 (executive override function)—this is crucial to debiasing.
- Type 1 processes can override Type 2 (dysrationalia override function)—this works against debiasing.
- The decision maker can toggle (T) back and forth between the two systems—shown as a broken line.
- The brain generally tries to default to Type 1 processing whenever it can (Cognitive Miser Function).

Figure 15.4 is a modified version of the dual process model of diagnosis with some parts omitted to better visualize the expansion of Type 1 Processes.

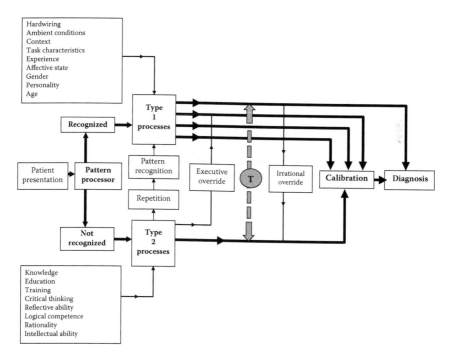

FIGURE 15.3
Dual Process Model for decision making. (Adapted with permission from Croskerry, P., Singhal, G., Mamede, S., *BMJ Qual Saf.*, 22(Suppl 2), ii58–ii64, 2013; with permission of BMJ Publishing Group Ltd. [12].)

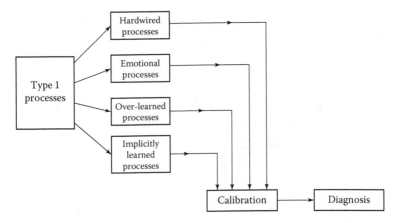

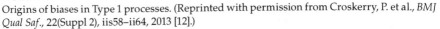

FIGURE 15.4
Origins of biases in Type 1 processes. (Reprinted with permission from Croskerry, P. et al., *BMJ Qual Saf.*, 22(Suppl 2), ii58–ii64, 2013 [12].)

Bazerman suggests the key to debiasing is *unfreezing* [52]. The three steps in the process are unfreezing, moving, and refreezing. Firstly, some disequilibrium of the decision maker needs to occur such that the individual wants to unfreeze from a previously established response and change. This could come about by the individual simply being informed of a potential bias, acknowledging that their past judgment has shown the influence of bias, or developing insight into the adverse consequences of bias. This critical step may be more than simply becoming aware of the existence of biases and their causes; sometimes a vivid, major revelation needs to occur—a cognitive intervention. The next step, *moving*, involves learning how the change will occur and what alternate strategies need to be learned—the purpose of this chapter. The final step, *refreezing*, occurs when the new approach is incorporated into the cognitive make-up of the decision maker and (with maintenance) becomes part of their regular thinking behavior. Referring back to the dual process model, when the decision maker achieves enlightenment and understands there is a problem with decision making, there is a need to de-couple cognition from the intuitive system; this corresponds to Bazerman's unfreezing step.

In Stanovich's view [73], Type 2 (analytical) processing occurs on two levels (Figure 15.5). The first is the *algorithmic* mind, which is associated with fluid intelligence, known as Gf [80]. It is that feature of general intelligence that provides us with the capacity to think logically and solve problems in novel situations, without necessarily having experienced specialized learning about the topic. It includes both inductive reasoning (the logic of experience) and deductive reasoning, and is especially applicable to scientific and technical reasoning. A critical feature of such thinking is the ability to suppress automatic responses in the intuitive mode by decoupling from it. This is depicted in the model as the executive override function that goes from the

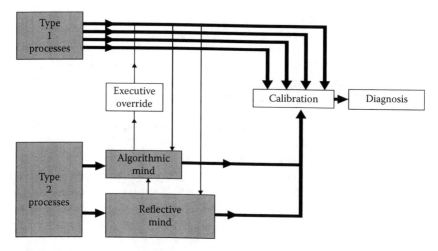

FIGURE 15.5
Cognitive Bias Mitigation (CBM): Analytic decoupling through the algorithmic mind from the reflective mind.

analytic mode to the intuitive mode (Figure 15.3). However, overall monitoring of the need to decouple resides at a second level in the *reflective* mind, associated with crystallized intelligence (Gc) [80]. Crystallized intelligence is the other part of overall intelligence that is measured in IQ tests. It is the intelligence that we gain throughout life, reflecting the depth and breadth of our knowledge, beliefs, skills, goals and experience. So, the true override function and the measure of our rationality reside in the reflective mind. If we are to unfreeze ourselves from a bias, we must initiate the action at this level. These interrelationships are schematized in Figure 15.5, which is based on Stanovich's tripartite model [73]. For simplicity, parts of Figure 15.3 have been omitted to show the expanded workings of the analytic mode in Figure 15.5.

Conclusions

The problem of cognitive and affective debiasing needs serious attention. If cognitive and affective biases are the major cause of cognitive failure leading to diagnostic error, then CBM becomes a major imperative. It is, arguably, the most important thing to do in clinical decision making, and it is doable. Again, it should be emphasized that CBM is already an integral part of everyday human cognition, and of medicine. We need to be constantly vigilant about our hardwired biases, ensuring they do not lead us into difficulties. We need to de-couple from biases that have been acquired through the over-learning of suboptimal decision making, as well as detect and un-bias ourselves from decision making acquired through implicit learning. Ideally, we should direct some effort at designing environments that minimize the acquisition of poor implicitly learned behaviors, as it is far better to acquire good habits from the beginning. There are several directions forward:

Firstly, we can continue to depend on many decision makers recognizing at least some of their biases and putting measures in place to control them. However, the interface between patient and doctor is a unique, dynamic place that challenges the best minds. There are many contextual influences that are difficult to control: the patient's characteristics and personality, the demographics and presentation (both typical and atypical) of the disease process itself, the knowledge, experience, expertise, personality and other characteristics of the physician (see Chapter 9), as well as the ambient conditions under which the decision will be made. Cultural and other individual differences will also play a role in the effectiveness of CBM initiatives. Overall, even the most motivated clinicians will deserve continuing support and encouragement.

Secondly, we should be aware that simplistic approaches toward debiasing are unlikely to be effective. Except, perhaps, in cases of significant affective arousal, we cannot expect that one-shot interventions will usually work. Nor can we expect that one particular type of intervention will be sufficient. It seems certain that CBM will inevitably require repeated training using a variety of strategies. As with any cognitive skill, further maintenance will also be necessary for retention.

Thirdly, there is ground to be made in Hogarth's direction—educating intuition. We can create better environments in which trainees acquire their skills—making the scientific method intuitive, and providing better mentoring, better feedback, and fewer homeostasis insults (more rest, sleep, and minimal cognitive overloading).

Fourthly, there is the strategy of *nudging*—steering healthcare providers toward better choices through "choice architecture"– which gently maneuvers people to do the right thing. For example, the bias toward inaction often leads to default options. If we make the available default option the safest, we are more likely to minimize diagnostic error. An example is the increased use of "Not Yet Diagnosed" (NYD) at discharge from emergency departments [81]. This tactic minimizes diagnosis momentum and premature closure and is a safer default option than the physician making a premature guess at the patient's diagnosis when there is insufficient evidence.

We must have defenses in place against some of the omnipresent intractable biases: blind spot bias, overconfidence, confirmation bias, premature closure, search satisficing, anchoring, representativeness, and others. We need to accept that CBM is not easily done but, at the same time, be optimistic that with the improved insights into decision making developed over the last decade, together with the increased armamentarium of CBM strategies that are now available, the prospect is less daunting than Fischoff first saw it 30 years ago [82]. Finally, the *Maintenance* stage of the Transtheoretical Model of Change [21] requires that clinicians recognize the need for constant vigilance and surveillance of their thinking. There is an ongoing imperative to self-monitor for bias, and work to prevent relapses into inappropriate System 1 decisions. This is a lifelong journey.

SUMMARY POINTS

- Cognitive biases are extremely common and pervasive. They may have a significant impact on clinical reasoning and rationality.
- Cognitive debiasing results in the temporary or permanent removal of a bias whereas CBM is aimed at an overall reduction in the impact of biases.
- Many biases have multiple determinants, and it is unlikely that there is a one-to-one mapping of causes to bias, or of bias to cure. Neither is it likely that one-shot debiasing interventions will usually be effective.
- Certain biases are more likely in particular ambient conditions.
- Although pessimism has prevailed toward CBM, a broad range of CBM strategies have been developed from both medical and non-medical sources that demonstrably reduce bias.
- Historically, medicine has developed a number of CBM strategies that appear to be effective.
- A critical feature of CBM is the ability to suppress automatic responses in the intuitive mode by decoupling from it and exerting executive override from the analytic system.

References

1. Lakoff G, Johnson M. *Philosophy in the Flesh: The Embodied Mind and Its Challenge to Western Thought*. New York: Basic Books; 1999.
2. Kahneman D. *Thinking, Fast and Slow*. Canada: Doubleday; 2011.
3. Jenicek M. *Medical Error and Harm: Understanding, Prevention, and Control*. New York: Productivity Press, Taylor and Francis Group; 2011.
4. Croskerry P, Abbass AA, Wu AW. How doctors feel: Affective issues in patients' safety. *Lancet*. 2008; 372(9645):1205–1206.
5. Croskerry P. Bias: A normal operating characteristic of the diagnosing brain. *Diagnosis*. 2014; 1(1): 23–27.
6. Seshia SS, Makhinson M, Phillips DF, Young GB. Evidence-informed person-centered healthcare Part I: Do "cognitive biases plus" at organizational levels influence quality of evidence? *J Eval Clin Pract*. 2014; 20(6): 734–47.
7. Seshia SS, Makhinson M, Young GB. Evidence-informed person-centred health care (Part II): Are "cognitive biases plus" underlying the EBM paradigm responsible for undermining the quality of evidence? *J Eval Clin Pract*. 2014; 20(6): 748–58.
8. Editorial. Let's think about cognitive bias. *Nature*. 2015; 526(7572): 163.

9. Arnott D. Cognitive biases and decision support systems development: A design science approach. *Info Systems J*. 2006;16(1): 55–78.

10. Croskerry P. The importance of cognitive errors in diagnosis and strategies to minimize them. *Acad Med*. 2003;78(8): 775–80.

11. Graber ML, Kissam S, Payne VL, Meyer AN, Sorensen A, Lenfestey N, Tant E, Henriksen K, Labresh K, Singh H. Cognitive interventions to reduce diagnostic error: A narrative review. *BMJ Qual Saf*. 2012;21(7): 535–57.

12. Croskerry P, Singhal G, Mamede S. Cognitive debiasing 1: Origins of bias and theory of debiasing. *BMJ Qual Saf*. 2013;22(Suppl 2): ii58–ii64.

13. Croskerry P, Singhal G, Mamede S. Cognitive debiasing 2: Impediments to and strategies for change. *BMJ Qual Saf*. 2013;22(Suppl 2): ii65–ii72.

14. Lambe KA, O'Reilly G, Kelly BD, Curristan S. Dual-process cognitive interventions to enhance diagnostic reasoning: A systematic review. *BMJ Qual Saf*. 2016; (10):808-2015.

15. Croskerry P. When I say … cognitive debiasing. *Med Educ*. 2015;49(7): 656–57.

16. Burton A. "Black box thinking" and "Failure: Why science is so successful." *New York Times*. 2015 Dec 29. Available at: http://www.nytimes.com/2016/01/03/books/review/black-box-thinking-and-failure-why-science-is-so-successful.html?_r=0. Accessed January 8, 2016.

17. Pinker S. *The Better Angels of Our Nature: Why Violence Has Declined*. New York: Penguin Books; 2011.

18. Croskerry P. Our better angels and black boxes. *Emerg Med J*. 2016;33(14): 242–44.

19. Antons D, Piller FT. Opening the black box of "not invented here": Attitudes, decision biases, and behavioral consequences. *Acad Manage Perspect*. 2015;29(2):193–217.

20. Pronin E, Gilovich T, Ross L. Objectivity in the eye of the beholder: Divergent perceptions of bias in self versus others. *Psychol Rev*. 2004;111(3):781–99.

21. Prochaska JO, DiClemente CC, Norcross JC. In search of how people change: Applications to addictive behaviors. *Am Psychol*. 1992;47(9):1102–1104.

22. Groopman J. *How Doctors Think*. New York: Houghton Mifflin Co; 2007.

23. Croskerry P. Medical decision making. In: V Thompson and L Ball, editors. *International Handbook of Thinking and Reasoning*. Florence, Kentucky: Psychology Press. (Forthcoming 2017).

24. Croskerry P. A universal model for diagnostic reasoning. *Acad Med*. 2009; 84(8):1022–28.

25. Jenkins MM, Youngstrom EA. A randomized controlled trial of cognitive debiasing improves assessment and treatment selection for pediatric bipolar disorder. *J Consult Clin Psychol*. 2016; 84(4): 323–33.

26. Mullinix G, Gray O, Colado J, et al. *Heuristica: Designing a serious game for improving decision making*. Paper presented at the Interactive Games Innovation Conference (IGIC) IEEE International, 2013.

27. Clegg BA, Martey RM, Stromer-Galley J, Strzalkowski T. Game-based training to mitigate three forms of cognitive bias. 2014; Interservice/Industry Training, Simulation, and Education Conference (I/ITSEC) Paper No. 14180. 2014; 1–12.

28. Dunbar NE, Miller CH, Adame BJ, Elizondo J, Wilson SN, Lane BL Kauffman AA, Bessarabova E, Jensen ML, Straub SK, Lee Y-H, Burgoon JK, Valacich JJ, Jenkins J, Zhang J. Implicit and explicit training in the mitigation of cognitive bias through the use of a serious game. *Computers in Human Behavior*. 2014; 37:307–18.

29. Wheaton KJ. Game based learning interventions for bias identification and mitigation. Workshop at the Society for Medical Decision Making. 15th Biennial European SMDM Meeting. Antwerp, Belgium. June 8–10, 2014.

30. Morewedge CK, Yoon H, Scopelliti I, Symborski C, Korris J, Kassam KS. Debiasing decisions: Improved decision making with a single training intervention. *Policy Insights from the Behavioral and Brain Sciences*. 2015;2(1): 129–40.

31. Crowley RS, Legowski E, Medvedeva O, Tseytlin E, Roh E, Jukic D. Evaluation of an intelligent tutoring system in pathology: Effects of external representation on performance gains, metacognition, and acceptance. *J Am Med Inform Assoc*. 2007;14(2): 182–90.

32. Bond SD, Carlson KA, Keeney RL. Improving the generation of decision objectives. *Decision Analysis*. 2010;7:238–55.

33. Whaley AL, Geller PA. Toward a cognitive process model of ethnic/racial biases in clinical judgment. *Review of General Psychology*. 2007; 11(1):75–96.

34. Jensen-Doss A, Youngstrom EA, Youngstrom JK, Feeny NC, Findling RL. Predictors and moderators of agreement between clinical and research diagnoses for children and adolescents. *J Consult Clin Psychol*. 2014;82(6): 1151–62.

35. Shermer M. *The Believing Brain: From Ghosts and Gods to Politics and Conspiracies: How We Construct Beliefs and Reinforce Them as Truths*. New York: Times Books, Henry Holt and Company; 2011.

36. Vohs KD, Baumeister RF, Loewenstein G, editors. *Do Emotions Help or Hurt Decision Making? A Hedgefoxian Perspective*. New York: Russell Sage Foundation; 2007.

37. Croskerry P, Abbass A, Wu A. Emotional issues in patient safety. *J Patient Safety*. 2010; 6:1–7.

38. Gigerenzer G. *Risk Savvy. How to Make Good Decisions*. London: Penguin Books; 2014.

39. Thaler RH, Sunstein CR. *Nudge: Improving Decisions about Health, Wealth, and Happiness*. New York: Penguin Books; 2009.

40. Soll JB, Milkman KL, Payne JW. A user's guide to debiasing. In: Keren G, Wu G, editors. *The Wiley-Blackwell Handbook of Judgment and Decision Making*. Hoboken, NJ: Wiley-Blackwell; 2016.

41. Keeney RL. Value-focused brainstorming. *Decision Analysis*. 2012; 9(4): 303–13.

42. Croskerry P. Cognitive forcing strategies in clinical decisionmaking. *Ann Emerg Med*. 2003;41(1): 110–20.

43. Dror IE, Thompson WC, Meissner CA, Kornfield I, Krane D, Saks M, Risinger M. Context management toolbox: A linear sequential unmasking (LSU) approach for minimizing cognitive bias in forensic decision making. *J Forensic Sci*. 2015; 60(4): 1111–12.

44. Berner ES, Graber ML. Overconfidence as a cause of diagnostic error in medicine. *Am J Med*. 2008; 121(5 Suppl):S2–23.

45. Russo JE, Schoemaker PJH. Overconfidence. In: Augier M, Teece DJ, editors. *The Palgrave Encyclopedia of Strategic Management*. London: Palgrave Macmillan UK; 2016.

46. Pronovost P, Needham D, Berenholtz S, Sinopoli D, Chu H, Cosgrove S, Sexton B, Hyzy R, Welsh R, Roth G, Bander J, Kepros J, Goeschel C. An intervention to decrease catheter-related bloodstream infections in the ICU. *N Engl J Med*. 2006; 355(26):2725–32.

47. Haynes AB, Weiser TG, Berry WR, Lipsitz SR, Breizat AH, Dellinger EP, Herbosa T, Joseph S, Kibatala PL, Lapitan MC, Merry AF, Moorthy K, Reznick RK, Taylor B, Gawande AA. Safe Surgery Saves Lives Study Group. A surgical safety checklist to reduce morbidity and mortality in a global population. *N Engl J Med*. 2009; 360(5): 491–99.

48. Ely JW, Graber ML, Croskerry P. Checklists to reduce diagnostic errors. *Acad Med*. 2011; 86(3):307–13.

49. Platt JR. Strong inference: Certain systematic methods of scientific thinking may produce much more rapid progress than others. *Science*. 1964; 146(3642): 347–53.

50. Bruno MA, Walker EA, Abujudeh HH. Understanding and confronting our mistakes: The epidemiology of error in radiology and strategies for error reduction. *Radiographics*. 2015;35(6):1668–76.

51. Simon HA. *The Sciences of the Artificial*. Cambridge, MA: MIT Press; 1981.

52. Bazerman MH. *Judgment in Managerial Decision Making*. 5th ed. New York: Wiley; 2002.

53. Mamede S, Schmidt HG, Rikers R. Diagnostic errors and reflective practice in medicine. *J Eval Clin Pract*. 2007;13(1):138–45.

54. Mamede S, van Gog T, van den Berge K, Rikers RM, van Saase JL, van Guldener C, Schmidt HG. Effect of availability bias and reflective reasoning on diagnostic accuracy among internal medicine residents. *JAMA*. 2010; 304(11):1198–203.

55. Elstein AS, Shulman LS, Sprafka SA. *Medical Problem Solving: An Analysis of Clinical Reasoning*. Cambridge, MA: Harvard University Press; 1978.

56. Sibinga EM, Wu AW. Clinician mindfulness and patient safety. *JAMA*. 2010;304(22): 2532–33.

57. Hafenbrack AC, Kinias Z, Barsade SG. Debiasing the mind through meditation: Mindfulness and the sunk-cost bias. *Psychol Sci*. 2014;25(2):369–76.

58. Lueke A, Gibson B. Mindfulness meditation reduces implicit age and race bias: The role of reduced automaticity of responding. *Soc Psychol Pers Sci*. 2015; 6: 284–91.

59. Moulton CA, Regehr G, Mylopoulos M, MacRae HM. Slowing down when you should: A new model of expert judgment. *Acad Med*. 2007; 82(10 Suppl): S109–16.

60. Moulton CA, Regehr G, Lingard L, Merritt C, MacRae H. Slowing down to stay out of trouble in the operating room: Remaining attentive in automaticity. *Acad Med*. 2010;85(10):1571–77.

61. Surowiecki J. *The Wisdom of Crowds*. New York: Anchor Books; 2005.

62. Bornstein G, Yaniv I. Individual and group behavior in the Ultimatum Game: Are groups more "rational" players? *Experimental Economics*. 1998;1:101–108.

63. World Bank Group. *World Development Report 2015: Mind, Society, and Behavior*. Washington, DC: World Bank; 2014.

64. McKee M, Stuckler D. Reflective practice: How the World Bank explored its own biases? *Int J Health Policy Manag*. 2016; 5(2): 79–82.

65. Tetlock PE, Kim JI. Accountability and judgment processes in a personality prediction task. *J Pers Soc Psychol*. 1987; 52(4): 700–709.

66. Hogarth RM. *Educating Intuition*. Chicago, IL: University of Chicago Press; 2001.

67. Radecki RP, Medow MA. Cognitive debiasing through sparklines in clinical data displays. *AMIA Annu Symp Proc*. 2007: 1085.

68. Bauer DT, Guerlain S, Brown PJ. The design and evaluation of a graphical display for laboratory data. *J Am Med Inform Assoc*. 2010; 17(4): 416–24.

69. Dye GA. Graphic organizers to the rescue! Helping students link and remember information. *Teach Excep Children*. 2000;32(3):72–76.
70. Nisbett RE. *Mindware Tools for Smart Thinking*. New York: Farrar, Straus, and Giroux; 2015. p. 48.
71. Larrick RP. Debiasing. In: Koehler DJ, Harvey N, editors. *The Blackwell Handbook of Judgment and Decision Making*. Oxford: Blackwell Publishing; 2004. pp. 316 –37.
72. Lilienfeld SO, Ammirati R, Landfield K. Giving debiasing away: Can psychological research on correcting cognitive errors promote human welfare? *Perspect Psychol Sci*. 2009; 4(4): 390–98.
73. Stanovich KE. *Rationality and the Reflective Mind*. New York: Oxford University Press; 2011. p. 19.
74. Groves JE. Taking care of the hateful patient. *N Eng J Med*. 1978;298(16): 883–87.
75. Croskerry P, Nimmo GR. Better clinical decision making and reducing diagnostic error. *J R Coll Physicians Edin*. 2011; 41:155–62.
76. Arkes HR. Costs and benefits of judgment errors: Implications for de-biasing. *Psychol Bull*. 1991; 110(3): 486–98.
77. Campbell SG, Croskerry P, Bond WF. Profiles in patient safety: A "perfect storm" in the emergency department. *Acad Emerg Med*. 2007; 14(8): 743–49.
78. Francis Bacon, Novum Organum (1620). In: Burtt EA, editor. *The English Philosophers from Bacon to Mill*. New York: Random House; 1939.
79. Graber ML, Sorensen AV, Biswas J, Modi V, Wackett A, Johnson S, Lenfestey N, Meyer AND, Singh H. Developing checklists to prevent diagnostic error in emergency room settings. *Diagnosis (Berl)*. 2014;1(3): 223–31.
80. Cattell RB. *Abilities: Their Structure, Growth, and Action*. New York: Houghton Mifflin; 1971.
81. Campbell SG. Advances in emergency medicine: A 10 year perspective –16 milestones –10 years. *Canadian Journal of Diagnosis*. 2003:115–18.
82. Fischhoff B. Debiasing. In: Kahneman D, Slovic P, Tversky A, editors. *Judgment under Uncertainty; Heuristics and Biases*. Cambridge, England: Cambridge University Press; 1982. pp. 422–44.

16

Diagnostic Support from Information Technology

Karen Cosby

CONTENTS

Introduction

Throughout this book, we have described diagnostic work as a series of complex tasks, beginning with the cognitive thought process of clinicians (individually or in teams); dependent upon communication between providers, patients, and services; and reliant on numerous interlinked processes and systems of care. Unfortunately, we have also demonstrated that the diagnostic process is often flawed. One cannot help but wonder how it can be that healthcare is still plagued by poor design for even mundane processes. The challenge is made even more difficult by a healthcare system that is made up of loosely connected entities, an amalgam of private offices, ambulatory care centers, clinics, and hospitals. The system is a behemoth that is hard to tame. But there are solutions to managing such beasts, and many believe that information technology (IT) may provide answers.

Healthcare Systems Lag Behind Modern Technology

We have entered a time in history aptly described as "The Information Age," brought about by a "Digital Revolution." The ability to process massive amounts of data and disseminate information instantaneously has changed how we perform even simple daily tasks. We shop using the Internet. Most of us bank, manage our finances, and file our taxes electronically. Our smartphones navigate for us; Siri guides us to local restaurants for our favorite cuisine. We can push, pull, and spread documents through the Internet cloud to remote areas around the globe. We hold virtual meetings and even view and interact with people from different places remote from one another, aided by computer applications. None of these remarkable realities is viewed as exceptional; even two-year-olds "read" from electronic tablets. And yet, we still misidentify patients and have imperfect processes to save, store, and share personal medical information or reliably perform basic tests and communicate results. Our systems of care suffer from an *ad hoc* infrastructure plagued by gaps and flaws for even routine tasks. Our focus on customized care for individual patients prioritizes flexibility, adaptability, and variability at the expense of organized structure for the general case, the routine and the ordinary. Whether or not this is viewed as a philosophical matter or a structural one, failure to develop an adequate infrastructure to meet the general needs of the predictable normal processes of care creates flaws that impact care individually and collectively. Improvements offered by health information technology (HIT) may begin to lay some of the basic skeletal structure to undergird many of our routine processes.

Hope and Change for Healthcare and Health Information Technology

In 2009, the American Recovery and Reinvestment Act provided federal incentives for the development and adoption of HIT and electronic health records (EHRs) [1]; Congress approved a budget of $27 billion to incentivize providers and healthcare organizations to adopt EHRs and finally replace paper records

[2]. At last, it seemed as though the time was right for healthcare organizations to invest in the design of much-needed infrastructure. These developments provided impetus for the explosion of IT solutions within healthcare.

The potential for HIT to redesign our workspace and improve our work product seems virtually unlimited. Examples of IT applications abound. EHRs ensure that legible and (seemingly) accurate information is accessible to multiple providers at distant sites at all hours. No longer do we need to search for and hoard paper medical records; no longer do we need to copy and fax reams of paper or reproduce "hard copies" of poor-quality radiographic images. Even better, with electronic resources, we acquire new functionality, complete with bells and whistles to poke, prod, and nudge providers to comply with evidence-based guidelines and treatment protocols. EHRs now provide notes that auto-populate with information from past visits. Intelligent systems can even be designed to prompt the clinician to ask questions based on the risk profile of the patient and current symptoms (along the lines of "Did you consider the possibility of...?"). Triggers can be developed to recognize patients who might be septic and alert the provider to look for a source of infection, check a lactate, or order blood cultures. Computers can be augmented with info buttons to provide immediate access to clinical references such as *UpToDate*®. As the clinician considers his plan, differential diagnosis generators can help to broaden the range of conditions considered [3–6]. Checklists can be incorporated in care sets and bundles to remind providers to consider diagnoses, or include actions that might otherwise come late or be easily forgotten altogether (e.g., "Did you remember to order the dysphagia screen for your patient with a stroke?") [7]. Clinical prediction rules can be supplied for clinicians to use, or even integrated in such a way that the provider can be alerted to important predictors of outcome. Alerts can warn clinicians that their patient has a high illness severity score and urge them to consider an intensive care unit (ICU) consult.

Electronic orders for tests can include advice on the most appropriate choice, a functionality that is especially helpful in light of the growing menu of sophisticated tests. Orders for radiology can require the ordering clinician to justify the radiation exposure based on evidence-based guidelines, then prompt him to supply clinical context to help the radiologist answer the question at hand. Electronic requests can be designed to capture data from the medical record and populate with age, weight, allergies, renal function, and pregnancy status. In turn, decision support features can provide guidance on the best type of image and optimal technique to ensure that the right image is obtained to best address the clinical question.

Much of routine primary care can be automated and improved with electronic reminders [8]. Some office-based IT systems can detect abnormal measurements of weight entered during routine visits, add them to the problem list, and even send an electronic suggestion to the provider to discuss diet and lifestyle [9,10]. Routine prompts can be set to remind providers and patients of due dates for vaccinations and cancer screening [11]. A social history noting cigarette smoking can trigger an alert to the provider to discuss smoking cessation, or automatically supply information to the patient about the health risks of smoking [12]. Providers can be reminded to do routine assessments for dementia in the elderly [13] and mental health screens for populations at risk [14]. Emergency department (ED) visits can set off a message to the patient's primary care provider to inform them that their patient has had an unexpected visit and probably needs a follow-up phone call or office visit.

Electronic orders also offer new opportunities for improving communication between healthcare professionals and patients, a common source for diagnostic delays and failures. Critical laboratory results (noting potentially dangerous problems that need immediate attention) can be automatically sent to pagers or other wireless devices to alert the ordering physician, a process shown to improve time to intervention [15]. Electronic test result managers have been designed to help providers view and respond to test data, providing electronic means to send notes to patients and order new consultations and additional laboratory tests or medications as needed [16]. Some automated systems not only file alerts in response to amended reports or abnormalities but also require the alert to be acknowledged; if a result goes unnoticed or is not acknowledged, the result is forwarded to a backup system that tracks the patient until the loop is closed [17–19]. Systems can be devised to place mandatory consults for conditions that are high risk and need additional evaluation: a positive hemoccult can trigger a consult for endoscopy; a positive surgical biopsy can populate a database of newly diagnosed cancer patients and order an oncology consultation [20]. Patients can be given direct access to their medical records through patient portals, where they can view their medical information, test results, and appointment schedules [21]. Some practices provide access to their providers via e mail, so that patients and providers can share updates and concerns outside their regularly scheduled office visits [21–24]. These are a few of the innovative ways that HIT can support and improve the diagnostic process; in fact, they address many of the system problems we identified in Chapter 11 that contribute to diagnostic failures. While these examples describe IT support for diagnosis, a similar and even longer list of electronic applications exists for treatments and other therapeutic interventions.

Does Our Reality Fulfill the Promise of HIT?

The implementation of HIT inevitably changes the workplace, the work product, and the people in the workplace. Moreover, we are now realizing that HIT inevitably changes medical practice in ways we did not anticipate. In order to successfully design and implement HIT, the nature of the work needs to be well understood in the natural environment in which it takes place, and implementation needs to consider sociotechnical factors [25]. A number of prophetic voices warned IT enthusiasts that the path to implementation would not be easy. Patel argued that "decision technology does not merely facilitate or augment decision-making, rather it reorganizes decision-making practices" and can "produce unintended consequences" [26]. Others predicted that the EHR would increase complexity and create new sources of error [27]. They were right. Many of these innovative developments fall disappointingly short of their expectations in real life. The work is harder than many expected.

There are four criteria by which we can assess the effects of HIT on diagnosis:

- Can HIT make us smarter?
- Does HIT make us (and our patients) happy?
- Can HIT improve communication?
- Does HIT improve the accuracy and timeliness of diagnosis?

It is simply too early to know some of these answers. Our hope and expectation is for improvement, but changes to unwieldy systems can be unpredictable.

Can HIT Make Us Smarter?

Experienced clinicians have reacted to the EHR with horror and frustration. The standardized, template-based EHR has changed the nature of their most basic and essential clinical task, the "history and physical exam (H&P)," beginning with how information is gathered and organized [26,28,29]. The medical record used to be an artifact of work, the work product of deliberate cognitive effort that helped gather and summarize relevant data, create a narrative and time line, generate diagnostic hypotheses, explore ideas, and lead to a differential diagnosis and an action plan. The process of interviewing and examining a patient, then developing and recording the H&P, allowed the author to wind her way through the facts to generate a story, the "narrative," exploring possible cause and effects, generating a time line of events (or symptoms), and connecting the facts to a storyline [30]. The work of writing the narrative account was a sort of synthesis necessary to construct meaning from the patient's story [31]. In contrast, an overly mechanistic boilerplate requiring "yes" and "no" responses to standardized checklists, which is common to many EHRs, constrains the storyteller, impedes the synthesis, and preserves facts at the expense of the whole. The resulting record is rich in detail but lacking in synthesis [32,33].

If the EHR impedes the cognitive process essential to diagnosis, it likely interferes with clinical reasoning and may degrade the work. The problem is not necessarily one of *usability*, as some suggest, but rather *usefulness*. In contrast to the narrative that clinicians use, computer records tend to chunk discrete items of information without connecting their relatedness. Users complain that they have facts, but do not have a sense of the whole picture [32].

Educators are left wondering how this might influence young trainees who enter the profession totally unaware of the distinction between a narrative and the forced organization of a template-based interview [34,35]. Verghese describes the change as trainees seem to tend to the needs of the EHR at the expense of the actual patient [36]. One can observe the pattern of physicians who digest all the medical history from the electronic record before even meeting the patient. Does the pre-screening review improve their history, or diminish it? Does it reduce their powers of observation, or help them gather the most relevant concerns in the context of the past medical history? Might it frame their thinking and set them up for the cognitive and affective biases we discussed in Chapter 7, even before they encounter the patient? Admitting medical teams have been observed to initiate their H&P two floors away, complete with History of Present Illness and Review of Systems, without having met or even greeted the subject they write about, almost as if they were assessing an "e-patient."

Strange things happen as clinicians seek to be thorough but efficient. Daily notes are filled with "cut-and-paste" features [37] that fill the note with redundant and irrelevant facts that are no longer current, contributing to lengthy "note bloat" [38]. Irrelevant and inaccurate information is passed along from day to day, and even between admissions. Fitzgerald describes an encounter with a student who reported that her patient had undergone bilateral below-the-knee amputations (BKAs) [39]. Upon examining the patient, Fitzgerald asked the trainee how she accounted for the patient's two apparently healthy legs. When confronted with the real patient, the student sheepishly admitted that she had accepted the information in the medical record at face value. The error had apparently crept into the record when "DKA times 2" (meaning diabetic ketoacidosis) was mistaken for "BKA times 2," a fact thereafter "enshrined" in the medical record—not just for this student, but for several teams of clinicians who preceded her [39]. Perhaps this is an example of *automation bias*, the tendency to accept something as true, or fail to question it, if the information presents itself in an electronic version that is perceived to be authoritative [40].

Depending on the design, HIT has the potential to improve understanding by enhancing the visual display of data. However, the design should be purposeful and guided by the people and processes it supports. Many clinical units have developed paper flowcharts that capture data and display it in a manner that brings meaning to those who use them. If these are replaced without fully understanding how they are used, the ability to track trends and patterns can be obfuscated. When manual flowsheets for one ICU were replaced by an electronic version, clinicians found it difficult to know what was going on with their patients [32]. Even when impressive graphical displays of data are available, overreliance on them may still pose risks. Dare we ask if our trainees (or clinicians) will ever spend enough time at the bedside to detect the early warning signs of a patient's deteriorating condition—a light bead of sweat on the forehead, the subtle conversational dyspnea of early decompensating heart failure, the look of mild anxiety in a febrile septic patient—or will they wait to see the downward trend on a graphic in the electronic health record before they recognize the urgency for action?

Assuming that clinicians escape the snare posed by the EHR and actually connect to patients and their stories, is there a role for decision support in augmenting diagnostic work? Most existing applications for clinical decision support focus on chronic disease management, medication optimization, and preventive care [41]. Only a few address diagnosis. Limited neural computational aids have been developed for the detection of myocardial infarctions [42,43] and estimations of risk for coronary artery disease [44]. Clinical guidelines and prediction rules are simpler tools that are available for the diagnosis of pulmonary emboli, back pain, dementia, appendicitis, mood disorders, and ischemic strangulation in small bowel obstructions [43,45]. However, these rules do not actually replace clinical assessment by a clinician—they still rely on someone to trigger their use and provide clinical details.

While interest in neural networks and clinical decision support has stalled, there has been some progress in the development of differential diagnosis generators [3–6]. Two of the best performing systems, Isabel and DxPlain, have modest accuracy; however, they produce a long list of up to 30 conditions and the user is left to tediously work through all the possibilities. These systems have not yet been fully integrated into existing EHRs and thus require data entry separate from the medical record. As they are not practical for use in every case, the clinician must decide when and if they are likely to help. Since clinicians are not likely to recognize that they have missed a diagnosis, these tools may help in particularly challenging cases, but are unlikely to reduce diagnostic errors. And while they might contain the correct diagnosis, users may be just as likely to abandon a correct diagnosis for an incorrect one [4].

Alternatively, checklists may provide a low-tech method for expanding the differential diagnosis. They have generated interest as a practical tool to help explore etiologies for common complaints [7]. The checklists can be adapted to include "most common," "most serious," and "can't-afford-to-miss" conditions. This simple, easy version of their computer counterpart may help prevent overreliance on memory and allow a physician and patient to explore diagnostic possibilities together [7,46,47].

Another potential use for HIT is surveillance and detection of possible missed or delayed diagnoses. A number of healthcare systems have developed trigger tools for the surveillance of events that are surrogate markers for diagnosis errors. For example, unexpected revisits or admissions within 10–14 days of a primary care visit were found to have a missed or delayed diagnosis in 16% of cases [48,49]. Methods set in motion to detect these markers can be used to monitor the system and rescue patients from potential flaws and gaps in care. Other triggers can be used to detect patients who may have fallen through the cracks simply because they were lost to follow-up. The surveillance of patients with iron-deficiency anemia, elevated prostate-specific antigen, positive fecal occult blood tests, or hematochezia can be undertaken to make sure that they complete appropriate screening for cancer [50].

IT can help organize work, record details, and even provide decision support, but will not in itself necessarily make clinicians reason more rationally. By providing the right information at the right moment, HIT can serve as a source of useful mindware—helpful bits of knowledge for optimal decision making, such as a Bayesian calculator or clinical guideline. A well-developed IT system may protect against cognitive miserliness by providing information that we might otherwise not have the time or the inclination to go looking for. However, it can also lure the overly trusting and unsuspecting clinician into a trap if the information recorded is wrong, or if too much trust is placed in an electronic medium via *automation bias*. And even the best IT system cannot substitute for sound, rational, expert cognition.

Does Our HIT Make Us Happy?

That seems like a silly question, but worker satisfaction relates well with how easy it is to get tasks done and how fulfilling the work is. Happiness is a property of an effective system. A highly functioning system that is accurate, efficient, effective, and safe provides satisfaction, even happiness, in the workplace. Discontent may be a sign of a poor fit of technology to task [25]. Certainly, if HIT achieved its promise, we should all be content and satisfied. Unfortunately, the development and implementation of HIT has unsettled many. A RAND Corporation survey noted that healthcare workers have decreased satisfaction with their professional work, attributed in part to the implementation of the EHR [51,52]. The intense reaction to the EHR has to do with usability that fails to match clinical work flow. Clinicians complain of overly time-consuming data entry, an overwhelming number of prompts and alerts, work that detracts from face-to-face patient care, and a growing distrust of the accuracy of the medical record [53,54]. A recent web survey of Veterans Administration (VA) primary care providers found that the majority felt oppressed by excessive prompts and alerts, and most felt that they could not adequately address them all [55]. Internal medicine residents now spend a paltry 12% of their days in direct patient care, but 40% of their time on their computers [56]. Documentation duties on average take up more than four hours of a resident's workday [57]. A typical shift in one community ED practice requires 4000 mouse clicks to accomplish tasks, including eight clicks to obtain a chest x-ray and six to order an aspirin [58]. And that is assuming that the computer does not "go down" or require a fresh login. Even ED practices that typically require intense patient contact note that 43%

of clinical time is spent with the EHR, leaving only 28% for direct time with patients [58]. In the middle of a night shift in the ED, the sound of clicking keyboards has replaced the drone of heart monitors as patients sleep.

Can HIT Improve Communication?

Electronic media offer new mechanisms to enhance communication [59]. Electronic prompts, automated reminders, and alerts make it easy to push information between providers and patients. Patient portals help make information visible and available for those with most at stake. However, this deluge of electronic data has unintended consequences.

IT has changed the way that patients and providers communicate and interact. There is a stranger in the exam room: the computer [36]. For some, the presence of the computer seems to change the dynamic of the medical encounter. Wachter argues that the computer interferes with the sacred bond between doctor and patient [60]. The distancing of the doctor from the patient is poignantly illustrated in a crayon drawing by a pediatric patient who showed herself sitting all alone on the exam table while her doctor faced the wall, typing away at his computer [61]. Even when the computer is placed near the patient, Shinsky describes the trend of "texting while doctoring" as the clinician taps away at data entry, only partially listening to the patient [62]. This electronic stranger is not totally disquieting to everyone, however. One family practice clinic gradually implemented their EHR and found that many patients liked knowing that their records were stored safely and securely [63]. Patients appreciated the opportunity to participate in checking the facts in their record, and sometimes searched for Internet-based education resources during their visits with their provider [63,64]. Perhaps satisfaction with the electronic record may depend on how it is implemented and how well clinicians integrate it into their work habits. Does it augment their work, or is it considered an extra task that adds to the burden of work?

In hospital systems with HIT, all work requires the computer. All documentation and all orders for tests, imaging, and medications are filtered through it. Thus, nothing can be done without it. The result is a "siloing" of clinicians, who talk to each other through an electronic intermediary [65]. Consultants come and go without so much as a wave, eager to leave their impressions and recommendations in a note authored somewhere at a work station removed from the patient and other clinicians, appearing anxious to be efficient and reluctant to slow down for a conversation. Doctors and nurses meet briefly at the bedside of an unstable patient, only to scurry away to their respective corners to check the comments and orders they leave for one another. Even the time-honored trek following morning rounds by internal medicine residents to the radiology suite has changed—once a place to examine images and expand on possible diagnoses, it is now reduced to dark corners occupied with computer workstations; radiologists who used to delight in mingling with medical teams seeking their expertise now seem to view them as interlopers interrupting their work [60]. "Just read my note!" is the new mantra. All these changes produce reports that are well preserved and accessible to all. But collegial communication—the active exchange of information and opinion, conversation, and the use of discussion and debate to enlighten and expand clinical reasoning—has suffered.

Does HIT Improve Accuracy and Timeliness of Diagnosis?

Frankly, we do not know and cannot really assess how much has changed with the implementation of EHRs. Most of what we did before is unknown, unmeasured, and unmeasurable. The electronic management of diagnostic processes ought to improve diagnosis, although it may be difficult to prove since we have few reliable measures of our pre-intervention state.

We quickly acclimatize to change. Most of us cannot imagine a time when we did not have access to prior electrocardiograms (EKGs), even though it was not that long ago when we often had to assume that all EKG abnormalities were new ones. Many clinical workstations have Picture Archiving and Communication System (PACS) images available at our fingertips, with the ability to view and compare new and old imaging studies side by side. Ready access to official historical reports of surgical pathology, imaging, and laboratory tests is a recent development not widely available before the integrated EHR. The new system is not perfect but it has made meaningful and substantive change by making clinical information easily accessible.

Cost of HIT and Added Risk

The United States has allocated $27 billion for initial efforts to incentivize the adoption of the EHR. The actual cost is tough to estimate [2]. Kaiser Permanente reports that it cost $4 billion to initiate an EHR across its system of providers [66]. Wake Forest Hospital System spent $13.3 million on its initial investment with *Epic*, another $8 million to implement it, then incurred $26.6 million in lost revenue during the go-live phase [67]. Novant Medical Group estimated that the cost to convert to its EHR was $600 million over five years [67].

However, the actual cost goes well beyond the financial investment. There is an unsettling awareness that the implementation of new technology brings with it new risks and new types of error. In 2007, Wiener coined the term *e-iatrogenesis* to describe patient harm from incidents related to HIT [68] and we now have a growing collection of events from which to learn [69]. A National Patient Safety webinar audience cited safety of the EHR as the top patient safety hazard of 2013 [70]. At the request of the Office of the National Coordinator for Health Information Technology, the Institute of Medicine issued a report in 2012 to address concerns about HIT safety [71] and ultimately called for an independent federal entity for monitoring patient safety related to HIT [72]. The Joint Commission has joined in the efforts to investigate, analyze, and disseminate information about HIT-related deaths, serious injury, and harm; it now includes assessment of HIT as part of its regulatory activities [73,74].

We have experienced improvements, but even with our new-fangled technology the fact remains that nothing is perfect. Automated notifications of abnormal laboratory results still fail in up to 10% of cases [75]. The VA alert system, designed to track radiology reports, notes that 40% of its alerts go unacknowledged, and even with aggressive tracking and referral, 4% of abnormal reports are still lost to follow-up at four weeks [17]. Although not related to diagnosis *per se*, the use of computerized provider order entry has proved to have its own inherent risks, which tend to be difficult to detect and correct [76]. Some of the problems now traced to HIT are old ones known to plague healthcare even in our pre-EHR era, namely, patient identification, medication errors, wrong site surgery, and delays in treatment [73,74]. Some safety issues are new to HIT and are yet to be elucidated.

Clinicians may not fully understand the inherent limitations of HIT and the type of risks it adds; indeed, some of the risks are actually invisible to the frontline users [77,78]. The design and implementation of HIT need to be undertaken with a primary focus on safety, and with support systems in place to detect, track, and rectify flaws [79–82].

Beyond HIT: Technological Innovations That May Improve Diagnosis

The EHR is not the only change that we have witnessed. A number of other innovations offer new methods for diagnostic work; some are already in use, while others are on the near horizon. The development of PACS and digitized imaging has made radiology services accessible to providers at the point of care and added improved functionality. Telemedicine (teleradiology, telepathology, teledermatology, teleophthalmology), or the use of remote consultants to review digitized images, provides a means for image, slide, or photo review by consultants distant from the patient [83–87]. This can improve access to services for patients in resource-limited settings. They can also be an added resource for expertise needed during off-hours and holidays, and they can even provide a convenient source of second reviews in difficult cases. In addition, newer methods of computer-aided detection can supplement interpretive services to improve the accuracy of mammography and cervical cytology, as well as increase the sensitivity for the detection of pulmonary nodules [88–90].

New biosensor technology has brought diagnostic tools and tests directly to the patient's home and workplace. Fitness enthusiasts are already familiar with wearable technology capable of tracking heart rate and blood pressure. Increasingly, biosensor technology is being developed for home monitoring for a variety of medical applications; some can easily be incorporated into clothing items and even undergarments. Wearable biosensors can be used to monitor cardiac rhythm, heart rate, and oxygenation [91]. Implantable chip devices can store medical identifiers linked to a protected web site, where first responders and medical staff can find personal medical information in case of an emergency [92]. New devices can be attached to iPhones to record single-lead EKGs [93,94]. "Smart" socks that detect low oxygen levels or localized pressure in feet can alert patients to the risk of diabetic and ischemic foot ulcers [95,96]. Google contact lenses are now available that continuously monitor blood glucose and blood alcohol levels, so diabetics may be able to skip the finger stick, and people can know when it is best to call a cab [97].

While our initial experience with EHRs in medicine has been humbling and even disappointing, a number of burgeoning advancements in IT may yet revolutionize healthcare. IBM has developed a supercomputer on a technology platform that uses natural-language processing and machine learning. The world was first introduced to "Watson," the supercomputer, in 2011, when it challenged and soundly defeated two all-time champions of the game show *Jeopardy!*. After digesting the entire content of Medline and PubMed and thousands of pages of patient case material, it was renamed "Dr. Watson" and has since been used to facilitate treatment decisions for complex cases at Memorial Sloan Kettering and MD Anderson Cancer Centers [98]. Eventually, supercomputers like Dr. Watson may simplify clinical decisions and augment the work that clinicians provide at the point of care.

Dr. Watson promises to revolutionize patient care on an individual level, but other developments in computer technology may revolutionize scientific research. The increased processing speed of ever-faster computers has enabled a rapidly expanding industry of "Big Data" analytics that exploits large data warehouses and real-time data to detect patterns and even predict events [99,100]. Big Data technology has already been used to detect trends and make forecasts for business, economics, politics, and sporting events. The Hospital for Sick Children in Toronto has used Big Data analytics to detect babies with neonatal sepsis as much as a day earlier, before they develop noticeable clinical signs of infection [101]. Computer scientists from Johns Hopkins used data from Twitter to track influenza and outperformed traditional disease tracking conducted by the Centers for Disease Control and Prevention [102]. Big Data technology may lead to *personalized medicine* for individuals (also referred to as *precision medicine*) based on their genomic profile, environmental exposures, and lifestyle, to anticipate likely diseases before they even develop. Precision medicine can also characterize disease better and help physicians design specific treatment based on improved understanding of disease [103]. The U.S. government has announced initiatives to "unlock data insights" for better understanding of factors that contribute to chronic illnesses [104].

If we have learned anything from our experience with HIT, we know that our optimism should be cautious and tempered. Big Data analytics detects trends and finds correlations, but huge data sets are likely to find *spurious correlations* that are meaningless, and even harmful [105,106]. Examples of spurious correlations illustrate the absurdity of some results: for example, the divorce rate in Maine apparently correlates well with the per capita consumption of margarine; the number of people who die by falling in a pool each year correlates with the number of films in which Nicholas Cage appears [107]. Overzealous use of Big Data may suffer from the *cum hoc* fallacy, that is, extracting meaning from random correlations, or confusing *correlation* with *causality* [105]. Big Data technology brings new challenges and new questions: How do we ensure that the conclusions from Big Data are valid and the patterns seen are meaningful? How do we protect patient privacy? Who owns healthcare data? Do we focus on population-based care or individuals? How will the information be used and for whom [108]?

Conclusion

We are only just now beginning to see how difficult and challenging change in healthcare can be, and how difficult diagnosis is [109]. While technological innovations are amazing achievements, successful integration into practice is often harder and slower than we imagine. IT not only provides important structure and support for diagnosis, but also adds complexity. Problems with the implementation of the EHR are an example of the need for caution. New ways of doing things rely on a foundational understanding of our work, and some suggest that we will make better progress in future design if we add expertise from human factors engineers, cognitive psychologists, and medical sociologists [101]. Our first generation of HIT may be clunky, but with improved design we can expect that it will live up to more of its potential. Clinicians have stubbornly resisted change that does not seem to support the work as they envision it, but they will likely contribute actively to the effort for redesign if they see that it has value and impacts on their ability to provide care. Our success depends in part on how much we contribute to these efforts [110]. Diagnosis has much to gain from the resources that technology offers, but it will take time to achieve all the benefit we seek.

SUMMARY POINTS

- Healthcare systems have suffered from lack of sufficient infrastructure to support diagnostic work.
- IT may provide many of the solutions we need to improve diagnostic processes.
- Recent gains have been made in the development and implementation of HIT, and electronic medical records are now widely disseminated in most settings.
- HIT offers many features that ought to streamline processes, facilitate communication, and improve diagnostic accuracy.
- The current design of existing HIT fails to meet expectations. New problems are introduced by HIT that add risk. Success in achieving all the potential benefits of IT will require ongoing work to customize and design products that support diagnostic work.

References

1. Blumenthal D. Launching HITECH. *N Engl J Med*. 2010 Feb 4;362(5):382–85.
2. Schilling B. The Federal Government Has Put Billions into Promoting Electronic Health Record Use: How Is It Going? Quality Matters: Innovations in Healthcare Quality Improvement. June/July 2011. Available at: http://www.commonwealthfund.org/publications/newsletters/quality-matters/2011/june-july-2011/in-focus. Accessed February 18, 2016.
3. Bond WF, Schwartz LM, Weaver KR, Levick D, Giuliano M, Graber ML. Differential diagnosis generators: An evaluation of currently available computer programs. *J Gen Intern Med*. 2012 Feb;27(2):213–19.
4. Berner ES, Webster GD, Shugerman AA, Jackson JR, Algina J, Baker AL, Ball EV, Cobbs CG, Dennis VW, Frenkel EP, Hudson LD, Mancall EL, Rackley CE, Taunton OD. Performance of four computer-based diagnostic systems. *N Engl J Med*. 1994 June 23;330(25):1792–96.
5. Graber ML, Mathew A. Performance of a web-based clinical diagnosis support system for internists. *J Gen Intern Med*. 2008 Jan;23 Suppl 1:37–40.
6. Friedman CP, Elstein AS, Wolf FM, Murphy GC, Franz TM, Heckerling PS, Fine PL, Miller TM, Abraham V. Enhancement of clinicians' diagnostic reasoning by computer-based consultation: A multisite study of 2 systems. *JAMA*. 1999 Nov 17;282(19):1851–56.
7. Ely JW, Graber ML, Croskerry P. Checklists to reduce diagnostic errors. *Acad Med*. 2011 Mar;86(3):307–13.
8. Frank O, Litt J, Beilby J. Opportunistic electronic reminders: Improving performance of preventive care in general practice. *Aust Fam Physician*. 2004 Jan–Feb;33(1–2):87–90.

9. Lee NJ, Chen ES, Currie LM, Donovan M, Hall EK, Jia H, John RM, Bakken S. The effect of a mobile clinical decision support system on the diagnosis of obesity and overweight in acute and primary care encounters. *ANS Adv Nurs Sci.* 2009 Jul–Sep;32(3):211–21.

10. Schriefer SP, Landis SE, Turbow DJ, Patch SC. Effect of a computerized body mass index prompt on diagnosis and treatment of adult obesity. *Fam Med.* 2009 Jul–Aug;41(7):502–507.

11. Bright TJ, Wong A, Dhurjati R, Bristow E, Bastian L, Coeytaux RR, Samsa G, Hasselblad V, Williams JW, Musty MD, Wing L, Kendrick AS, Sanders GD, Lobach D. Effect of clinical decision-support systems: A systematic review. *Ann Intern Med.* 2012 Jul 3;157(1):29–43.

12. Unrod M, Smith M, Spring B, DePue J, Redd W, Winkel G. Randomized controlled trial of a computer-based, tailored intervention to increase smoking cessation counseling by primary care physicians. *J Gen Intern Med.* 2007 Apr;22(4):478–84.

13. Downs M, Turner S, Bryans M, Wilcock J, Keady J, Levin E, O'Carroll R, Howie K, Iliffe S. Effectiveness of educational interventions in improving detection and management of dementia in primary care: Cluster randomised controlled study. *BMJ.* 2006 Mar 25;332(7543):692–96.

14. Cannon DS, Allen SN. A comparison of the effects of computer and manual reminders on compliance with a mental health clinical practice guideline. *J Am Med Inform Assoc.* 2000 Mar–Apr;7(2):196–203.

15. Kuperman GJ, Teich JM, Tanasijevic MJ, Ma'Luf N, Rittenberg E, Jha A, Fiskio J, Winkelman J, Bates DW. Improving response to critical laboratory results with automation: Results of a randomized controlled trial. *J Am Med Inform Assoc.* 1999 Nov–Dec;6(6):512–22.

16. Poon EG, Wang SJ, Gandhi TK, Bates DW, Kuperman GJ. Design and implementation of a comprehensive outpatient Results Manager. *J Biomed Inform.* 2003 Feb–Apr;36(1–2):80–91.

17. Singh H, Arora HS, Vij MS, Rao R, Khan MM, Petersen LA. Communication outcomes of critical imaging results in a computerized notification system. *J Am Med Inform Assoc.* 2007 Jul–Aug;14(4):459–66.

18. Brenner RJ. To err is human, to correct divine: The emergence of technology-based communication systems. *J Am Coll Radiol.* 2006 May;3(5):340–45.

19. Brantley SD, Brantley RD. Reporting significant unexpected findings: The emergence of information technology solutions. *J Am Coll Radiol.* 2005 Apr;2(4):304–307.

20. Humphrey LL, Shannon J, Partin MR, O'Malley J, Chen Z, Helfand M. Improving the follow-up of positive hemoccult screening tests: An electronic intervention. *J Gen Intern Med.* 2011 Jul;26(7):691–97.

21. Hassol A, Walker JM, Kidder D, Rokita K, Young D, Pierdon S, Deitz D, Kuck S, Ortiz E. Patient experiences and attitudes about access to a patient electronic health care record and linked web messaging. *J Am Med Inform Assoc.* 2004 Nov–Dec; 11(6):505–13.

22. Leong SL, Gingrich D, Lewis PR, Mauger DT, George JH. Enhancing doctor-patient communication using email: A pilot study. *J Am Board Fam Pract.* 2005 May–Jun;18(3):180–88.

23. Slack WV. A 67-year-old man who e-mails his physician. *JAMA.* 2004 Nov 10;292(18):2255–61.

24. Slack WV. Patient-computer dialogue: A hope for the future. *Mayo Clin Proc.* 2010 Aug;85(8):701–703.
25. Berg M. Patient care information systems and health care work: A sociotechnical approach. *Int J Med Inform.* 1999 Aug;55(2):87–101.
26. Patel VL, Kaufman DR, Arocha JF. Emerging paradigms of cognition in medical decision-making. *J Biomed Inform.* 2002 Feb;35(1):52–75.
27. Ash JS, Berg M, Coiera E. Some unintended consequences of information technology in health care: The nature of patient care information system-related errors. *J Am Med Inform Assoc.* 2004 Mar–Apr;11(2):104–12.
28. Patel VL, Kushniruk AW, Yang S, Yale JF. Impact of a computer-based patient record system on data collection, knowledge organization, and reasoning. *J Am Med Inform Assoc.* 2000 Nov–Dec;7(6):569–85.
29. Noteboom C, Qureshi S. How Can Physician's Knowledge be Activated to Provide Better Healthcare? Explaining Electronic Health Record Adaptation by Physicians. Presented at: 2013 46th Hawaii International Conference on System Sciences (HICSS); pp. 812–21; January 7–10, 2013; Wailea, HI. Available at: https://www.computer.org/csdl/proceedings/hicss/2013/4892/00/4892a812.pdf. Accessed February 26, 2016.
30. Kay S, Purves IN. Medical records and other stories: A narratological framework. *Methods Inf Med.* 1996 Jun;35(2):72–87.
31. Hartzband P, Groopman J. Off the record: Avoiding the pitfalls of going electronic. *N Engl J Med.* 2008 Apr 17;358(16):1656–58.
32. Varpio L, Day K, Elliot-Miller P, King JW, Kuziemsky C, Parush A, Roffey T, Rashotte J. The impact of adopting EHRs: How losing connectivity affects clinical reasoning. *Med Educ.* 2015 May;49(5):476–86.
33. Varpio L, Rashotte J, Day K, King J, Kuziemsky C, Parush A. The EHR and building the patient's story: A qualitative investigation of how EHR use obstructs a vital clinical activity. *Int J Med Inform.* 2015 Dec;84(12):1019–28.
34. Ober KP, Applegate WB. The electronic health record: Are we the tools of our tools? *Pharos Alpha Omega Alpha Honor Med Soc.* 2015 Winter;78(1):8–14.
35. Pageler NM, Friedman CP, Longhurst CA. Refocusing medical education in the EMR era. *JAMA.* 2013 Dec 4;310(21):2249–50.
36. Verghese A. Culture shock: Patient as icon, icon as patient. *New Engl J Med.* 2008 Dec 25;359(26):2748–51.
37. Hirschtick RE. A piece of my mind. Copy-and-paste. *JAMA.* 2006 May 24;295(20):2335–36.
38. Shoolin J, Ozeran L, Hamann C, Bria W 2nd. Association of Medical Directors of Information Systems consensus on inpatient electronic health record documentation. *Appl Clin Inform.* 2013 Jun 26;4(2):293–303.
39. Fitzgerald FT. Curiosity. *Ann Intern Med.* 1999 Jan 5;130(1):70–72.
40. Goddard K, Roudsari A, Wyatt JC. Automation bias: A systematic review of frequency, effect mediators, and mitigators. *J Am Med Inform Assoc.* 2012 Jan–Feb;19(1):121–27.
41. Bates DW, Gawande AA. Improving safety with information technology. *N Engl J Med.* 2003 Jun 19;348(25):2526–34.
42. Baxt WG, Shofer FS, Sites FD, Hollander JE. A neural computational aid to the diagnosis of acute myocardial infarction. *Ann Emerg Med.* 2002 Apr;39(4):366–73.
43. Baxt WG. Application of artificial neural networks to clinical medicine. *Lancet.* 1995 Oct 28;346(8983):1135–38.

44. Kline JA, Zeitouni RA, Hernandez-Nino J, Jones AE. Randomized trial of computerized quantitative pretest probability in low-risk chest pain patients: Effect on safety and resource use. *Ann Emerg Med.* 2009 Jun;53(6):727–35.
45. Garg AX, Adhikari NK, McDonald H, Rosas-Arellano MP, Devereaux PJ, Beyene J, Sam J, Haynes RB. Effects of computerized clinical decision support systems on practitioner performance and patient outcomes: A systematic review. *JAMA.* 2005 Mar 9;293(10):1223–38.
46. Sibbald M, de Bruin AB, Yu E, van Merrienboer JJ. Why verifying diagnostic decisions with a checklist can help: Insights from eye tracking. *Adv Health Sci Educ Theory Pract.* 2015 Oct;20(4):1053–60.
47. Sibbald M, de Bruin AB, van Merrienboer JJ. Checklists improve experts' diagnostic decisions. *Med Educ.* 2013 Mar;47(3):301–308.
48. Singh H, Giardina TD, Forjuoh SN, Reis MD, Kosmach S, Khan MM, Thomas EJ. Electronic health record-based surveillance of diagnostic errors in primary care. *BMJ Qual Saf.* 2012 Feb;21(2):93–100.
49. Singh H, Thomas EJ, Khan MM, Petersen LA. Identifying diagnostic errors in primary care using an electronic screening algorithm. *Arch Intern Med.* 2007 Feb 12;167(3):302–308.
50. Murphy DR, Laxmisan A, Reis BA, Thomas EJ, Esquivel A, Forjuoh SN, Parikh R, Khan MM, Singh H. Electronic health record-based triggers to detect potential delays in cancer diagnosis. *BMJ Qual Saf.* 2014 Jan;23(1):8–16.
51. Friedberg MW, Chen PG, Van Busum KR, Aunon F, Pham C, Caloyeras J, Mattke S, Pitchforth E, Quigley DD, Brook RH, Crosson FJ, Tutty M. *Factors Affecting Physician Professional Satisfaction and Their Implications for Patient Care, Health Systems, and Health Policy.* Santa Monica, CA: Rand Corporation, 2013.
52. Sinsky CA, Beasley JW. Texting while doctoring. *Ann Intern Med.* 2014 Apr 15;160(8):584.
53. Friedberg MW, Crosson FJ, Tutty M. Physicians' Concerns About Electronic Health Records: Implications and Steps Toward Solutions. The RAND blog. March 12, 2014. Available at: www.rand.org/blog/2014/03/physicians-concerns-about-electronic-health-records.html. Accessed February 23, 2016.
54. Payne TH, Corley S, Cullen TA, Gandhi TK, Harrington L, Kuperman GJ, Mattison JE, McCallie DP, McDonald CJ, Tang PC, Tierney WM, Weaver C, Weir CR, Zaroukian MH. Report of the AMIA EHR-2020 Task Force on the status and future direction of EHRs. *J Am Med Inform Assoc.* 2015 Sep;22(5):1102–10.
55. Singh H, Spitzmueller C, Petersen NJ, Sawhney MK, Sittig DF. Information overload and missed test results in electronic health record-based settings. *JAMA Intern Med.* 2013 Apr 22;173(8):702–704.
56. Block L, Habicht R, Wu AW, Desai SV, Wang K, Silva KN, Niessen T, Oliver N, Feldman L. In the wake of the 2003 and 2011 duty hours regulations, how do internal medicine interns spend their time? *J Gen Intern Med.* 2013 Aug;28(8):1042–47.
57. Oxentenko AS, West CP, Popkave C, Weinberger SE, Kolars JC. Time spent on clinical documentation: A survey of internal medicine residents and program directors. *Arch Intern Med.* 2010 Feb 22;170(4):377–80.
58. Hill RG Jr, Sears LM, Melanson SW. 4000 clicks: A productivity analysis of electronic medical records in a community hospital ED. *Am J Emerg Med.* 2013 Nov;31(11):1591–94.
59. Singh H, Naik AD, Rao R, Petersen LA. Reducing diagnostic errors through effective communication: Harnessing the power of information technology. *J Gen Intern Med.* 2008 Apr;23(4):489–94.

60. Wachter R. *The Digital Doctor: Hope, Hype, and Harm at the Dawn of Medicine's Computer Age*. New York; McGraw-Hill; 2015.

61. Toll E. A piece of my mind: The cost of technology. *JAMA*. 2012 Jun 20;307(23):2497–98.

62. Sinsky CA, Beasley JW. Texting while doctoring: A patient safety hazard. *Ann Intern Med*. 2013 Dec 3;159(11):782–83.

63. Shield RR, Goldman RE, Anthony DA, Wang N, Doyle RJ, Borkan J. Gradual electronic health record implementation: New insights on physician and patient adaptation. *Ann Fam Med*. 2010 Jul–Aug;8(4):316–26.

64. Doyle RJ, Wang N, Anthony D, Borkan J, Shield RR, Goldman RE. Computers in the examination room and the electronic health record: Physicians' perceived impact on clinical encounters before and after full installation and implementation. *Fam Pract*. 2012 Oct;29(5):601–608.

65. Stoller JK. Electronic siloing: An unintended consequence of the electronic health record. *Cleve Clin J Med*. 2013 Jul;80(7):406–409.

66. Rudin RS, Fuglesten J. The Little Exchange That Could … Transform the U.S. Health Care System. The RAND blog. January 28, 2015. Available at: http://www.rand.org/blog/2015/01/the-little-exchange-that-could-transform-the-us-health.html. Accessed February 23, 2016.

67. Craver R. Wake Forest Baptist has cost overruns, revenue loss with electronic records system. *Winston-Salem Journal*. April 26, 2014. Available at: http://www.journalnow.com/business/business_news/local/wake-forest-baptist-has-cost-overruns-revenue-loss-with-electronic/article_c2801866-9e0c-11e2-bf84-0019bb30f31a.html. Accessed February 23, 2016.

68. Weiner JP, Kfuri T, Chan K, Fowles JB. "e-Iatrogenesis": The most critical unintended consequence of CPOE and other HIT. *J Am Med Inform Assoc*. 2007 May–Jun;14(3);387–88.

69. Harrington L, Kennerly D, Johnson C. Safety issues related to the electronic medical record (EMR): Synthesis of the literature from the last decade, 2000–2009. *J Healthc Manage*. 2011 Jan–Feb;56(1):31–43.

70. Denham CR, Classen DC, Swenson SJ, Henderson MJ, Zeltner T, Bates DW. Safe use of electronic health records and health information technology systems: Trust but verify. *J Patient Saf*. 2013 Dec;9(4):177–89.

71. IOM (Institute of Medicine). *Health IT and Patient Safety: Building Safer Systems for Better Care*. Washington, DC: The National Academies Press, 2012a.

72. Sittig DF, Classen DC, Singh H. Patient safety goals for the proposed Federal Health Information Technology Safety Center. *J Am Med Inform Assoc*. 2015 Mar;22(2):472–78.

73. The Joint Commission: Investigations of Health IT-related Deaths, Serious Injuries or Unsafe Conditions. March 30, 2015. Available at: https://www.healthit.gov/sites/default/files/safer/pdfs/Investigations_HealthIT_related_SE_Report_033015.pdf. Accessed February 23, 2016.

74. The Joint Commission: Safe use of Health Information Technology, Sentinel Event Alert #54, March 31, 2015. Available at: http://www.jointcommission.org/assets/1/18/SEA_54.pdf. Accessed February 23, 2016.

75. Singh H, Thomas EJ, Sittig DF, Wilson L, Espadas D, Khan MM, Petersen LA. Notification of abnormal lab test results in an electronic medical record: Do any safety concerns remain? *Am J Med*. 2010 Mar;123(3):238–44.

76. Koppel R, Metlay JP, Cohen A, Abaluck B, Localio AR, Kimmel SE, Strom BL. Role of computerized physician order entry systems in facilitating medication errors. *JAMA*. 2005 Mar 9;293(10):1197–203.

77. Karsh BT, Weinger MB, Abbot PA, Wears RL. Health information technology: Fallacies and sober realities. *J Am Med Inform Assoc*. 2010 Nov–Dec;17(6):617–23.

78. Vest JR, Gamm LD. Health information exchange: Persistent challenges and new strategies. *J Am Med Inform*. 2010 May–Jun;17(3):288–94.

79. Walker JM, Carayon P, Leveson N, Paulus RA, Tooker J, Chin H, Bothe A Jr, Stewart WF. EHR safety: The way forward to safe and effective systems. *J Am Med Inform Assoc*. 2008 May–Jun;15(3):272–77.

80. Wears RL. Can we make health IT safe enough for patients? *Work*. 2012;41 Suppl 1:4484–89.

81. Sittig DF, Ash JS, Singh H. The SAFER guides: Empowering organizations to improve the safety and effectiveness of electronic health records. *Am J Manag Care*. 2014 May;20(5):418–23.

82. Wears RL, Leveson NG. "Safeware": Safety-critical computing and health care information technology. In: Henriksen K, Battles JB, Keyes MA, Grady ML, eds. *Advances in Patient Safety: New Directions and Alternative Approaches*. Vol. 4. Technology and Medication Safety. AHRQ Publication No. 08-0034-4. Rockville, MD: Agency for Healthcare Research and Quality; August 2008.

83. Wilson LS, Maeder AJ. Recent directions in telemedicine: Reviews of trends in research and practice. *Healthc Inform Res*. 2015 Oct;21(4):213–22.

84. Kalyanpur A. The role of teleradiology in emergency radiology provision. *Radiol Manage*. 2014 May–Jun;36(3):46–49.

85. Graham AR, Bhattacharyya AK, Scott KM, Lian F, Grasso LL, Richter LC, Carpenter JB, Chiang S, Henderson JT, Lopez AM, Barker GP, Weinstein RS. Virtual slide telepathology for an academic teaching hospital surgical pathology quality assurance program. *Hum Pathol*. 2009 Aug;40(8):1129–36.

86. Whited JD. Teledermatology. *Med Clin North Am*. 2015 Nov;99(6):1365–79.

87. Surendran TS, Raman R. Teleophthalmology in diabetic retinopathy. *J Diabetes Sci Technol*. 2014 Mar 17;8(2):262–66.

88. Shiraishi J, Li Q, Applebaum D, Doi K. Computer-aided diagnosis and artificial intelligence in clinical imaging. *Semin Nucl Med*. 2011 Nov;41(6):449–62.

89. Rao VM, Levin DC, Parker L, Cavanaugh B, Frangos AJ, Sunshine JH. How widely is computer-aided detection used in screening and diagnostic mammography? *J Am Coll Radiol*. 2010 Oct;7(10):802–805.

90. Kok MR, van Der Schouw YT, Boon ME, Grobbee DE, Kok LP, Schreiner-Kok PG, van der Graaf Y, Doornewaard H, van den Tweel JG. Neural network-based screening (NNS) in cervical cytology: No need for the light microscope? *Diagn Cytopathol*. 2001 Jun;24(6):426–34.

91. Yang YL, Chuang MC, Lou SL, Wang J. Thick-film textile-based amperometric sensors and biosensors. *Analyst*. 2010 Jun;135(6):1230–34.

92. Halamka J. Straight from the shoulder. *N Engl J Med*. 2005 Jul 28;353(4):331–33.

93. Saxon LE. Ubiquitous wireless EKG recording: A powerful tool physicians should embrace. *J Cardiovas Electrophysiol*. 2013 Apr;24(4):480–83.

94. Haberman ZC, Jahn RT, Bose R, Tun H, Shinbane JS, Doshi RN, Chang PM, Saxon LA. Wireless smartphone ECG enables large-scale screening in diverse populations. *J Cardiovasc Electrophysiol*. 2015 May;26(5):520–26.

95. Mertz L. Sending out an SOS ... and more: Next-generation textiles and EEG headsets transport vital biomed information. *IEEE Pulse*. 2015 March–Apr;6(2):30–36.

96. Coxworth B. Smart socks keep watch over diabetics' feet. Gizmag [serial online]. January 27, 2016. Available at: www.gizmag.com/sensego-diabetic-socks-41529. Accessed February 26, 2016.

97. Kosir S. Wearables in Healthcare. Wearable Technologies. April 15, 2015. Available at: www.wearable-technologies.com/2015/04/wearables-in-health-care/. Accessed February 18, 2016.

98. Lee H. Paging Dr. Watson: IBM's Watson supercomputer now being used in healthcare. *J AHIMA*. 2014 May;85(5):44–47.

99. Raghupathi W, Raghupathi V. Big data analytics in healthcare: Promise and potential. *Health Inf Sci Syst*. 2014 Feb 7;2(3). doi:10.1186/2047-2501-2-3.

100. Fernandes L, O'Connor M, Weaver V. Big data, bigger outcomes: Healthcare is embracing the big data movement, hoping to revolutionize HIM by distilling vast collection of data for specific analysis. *J AHIMA*. 2012 Oct;8(10):38–43.

101. Horowitz BT. IBM InfoSphere, Big Data Help Toronto Hospital Monitor Premature Infants. eWeek [serial online]. September 26, 2013. Available at: www.eweek.com/enterprise-apps/ibm-infosphere-big-data-help-toronto-hos-pital-monitor-premature-infants.html. Accessed February 23, 2016.

102. Paul MJ, Dredze M, Broniatowski D. Twitter improves influenza forecasting. *PLoS Curr*. 2014 Oct 28;6. doi:10.1371/currents.outbreaks.90b9ed0f59bae4ccaa6 83a39865d9117.

103. Collins FS, Varmus H. A new initiative on precision medicine. *New Engl J Med*. 2015 Feb 26;372(9):793–95.

104. The White House, Office of the Press Secretary. FACT SHEET: New Patient-Focused Commitments to Advance the President's Precision Medicine Initiative. July 8, 2015. Available at: https://www.whitehouse.gov/the-press-office/2015/07/08/fact-sheet-new-patient-focused-commitments-advance-pres-ident%E2%80%99s-precision. Accessed February 23, 2016.

105. Calude CS, Long G. The Deluge of Spurious Correlations in Big Data. Centre for Discrete Mathematics and Theoretical Computer Science Report 488 (CDMTCS-488), Revision 3. Available at: https://www.cs.aukland.ac.nz/research/groups/CDMTCS/researchreports/index.php?download&paper_file=606. Accessed February 24, 2016.

106. Flockhart D, Bies RR, Gastonguay MR, Schwartz SL. Big data: Challenges and opportunities for clinical pharmacology. *Br J Clin Pharmacol*. 2016 Feb 4. doi:10.1111/bcp.12896.

107. Vigen T. *Spurious Correlations: Correlation Does Not Equal Causation*. New York: Hachette Books; 2015.

108. Wears RL, Williams DJ. Big questions for "big data." *Ann Emerg Med*. 2016 Feb;67(2):237–39.

109. Wears RL. What makes diagnosis hard? *Adv Health Sci Educ Theory Pract*. 2009 Sep;14 Suppl 1:19–25.

110. Bailey JE. Does health information technology dehumanize health care? *Virtual Mentor*. 2011 Mar 1;13(3):181–85.

17

What Is the Patient's Role in Diagnosis?

Karen Cosby

CONTENTS

Introduction

The process of diagnosis is modeled on the scientific method of inquiry, which is largely viewed as an analytical process using logic and objective reasoning, and is therefore expected to be reproducible and highly accurate. Diseases are sorted and classified on the basis of their underlying causes and are identified by their typical physical manifestations. Doctors gather evidence and apply diagnostic tests to come to a final conclusion or

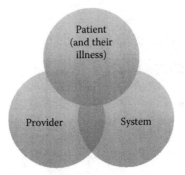

FIGURE 17.1
Overlapping factors that influence diagnosis. Factors from each component may contribute to
success or diagnostic delays and failures.

answer—the diagnosis. This model would lead one to think that clinical
problems are simply an interaction of a disease state with provider(s) and
the healthcare system. If so, most clinical problems could be solved on
paper, with little need to even meet the patient, except perhaps to gather
specimens for analysis. This model is obviously flawed. Patients are not
passive receptacles of illness or disease states. Each patient has his or her
own unique combination of conditions, manifestations of illness, and
personal response to them: the ability to recognize the need for a medical evaluation, seek and find appropriate care, and engage the healthcare
system may largely dictate the efficiency and accuracy of the diagnostic
process. In turn, the ability of clinicians to apply their knowledge is determined not just by their clinical acumen and reasoning skills, but also by
their ability to interact with patients: to gain their trust, hear their story,
correctly interpret their signs and symptoms, and engage the patient in a
strategy for testing and follow-up care. We don't care for diseases *per se*,
but rather for the patients who have diseases. Science provides a framework to understand disease, but effective diagnostic work involves understanding the complex psychosocial aspects of illness, skill in interacting
with patients who have the disease, an understanding of how the illness
impacts their life, and how the patient's personal values and preferences
determine acceptable options for his or her care. Clinicians can be bril-
liant scientists, yet fail to apply their skills effectively if they lack the ability to engage patients in the diagnostic process. Much of education—and
this book for that matter—focuses on scientific clinical reasoning and the
system of care necessary for diagnostic evaluation, but a more complete
picture of diagnosis emerges when we include the patient as both a factor
in disease and a necessary partner in diagnosis, a relationship illustrated
in Figure 17.1.

Patients as Contributing Factors to Disease and Diagnostic Failure

Some patient characteristics contribute to the risk of disease, as well as to difficulties in participating in timely and accurate diagnosis. Vulnerable populations include the homeless, the disabled, the very young and the very old, and those with dementia, cognitive dysfunction, or psychiatric illness. In fact, most of us will at one time or another, either through illness, injury, or untoward circumstances, find ourselves in positions where we experience need, even if only temporary problems impair our mobility, our ability to reason, or our access to care. Some risk factors are voluntary lifestyle choices, including smoking, substance abuse, and risky behaviors such as unprotected sex, adventure/thrill-seeking, or gang affiliations, to name just a few. Some patients, through no act or will of their own, are too acutely ill to even participate in their care. Others have increased risk because of body habitus (difficult airways and poor venous access) or even inheritable characteristics. Some of these factors are amenable to intervention and some are not. These features may pose increased risk of illness, or complex issues in diagnosis and management that are particularly prone to failure. Patient factors are cited as common contributing factors to morbidity in emergency department patients [1,2].

Patients as Partners in Diagnosis and Their Role in Diagnostic Delays and Failures

Patients by necessity make the first move to engage the healthcare system. They must first be aware of the need for care, and be willing and able to seek it. Unfortunately, as many as 30% of patients delay seeking care for symptoms related to cancer [3]. In one series of newly reported lung cancer, half of patients reported experiencing symptoms for more than a year before seeking care, some reporting delays greater than 2 years [4]. Once they are initially evaluated, 44% of patients with suspected lung cancer have further delays in completing their diagnostic evaluation due to missed appointments and nonadherence with their management plan [5]. Women of lower socioeconomic status and minorities have worse outcomes with breast cancer, attributable to delays in diagnosis and treatment [6]. Reasons identified include fear, mistrust, failure to appreciate the significance of their symptoms, financial considerations, transportation, access to care, difficulty with scheduling appointments, childcare, and competing priorities [6]. Other factors identified in diagnostic delays include patient denial, attributing their symptoms to other underlying problems, feeling unworthy of care, guilt,

TABLE 17.1

Difficulties Patients Encounter during Diagnosis

How patients feel	• They are fearful and afraid to complain • Feel "powerless," sick, and scared • Sometimes don't recognize when they are sick or need care • Unsure how to access and navigate the healthcare system
How healthcare professionals act	• Sometimes dismiss patients' complaints • Fail to listen or address new concerns • Label patients' symptoms as psychiatric • Attribute patient problems to substance abuse • Are hurried and appear too busy to be bothered
How the healthcare system performs	• Care lacks coordination • Poor communication between providers, consultants, and services • Fail to inform patients about important results • Lose test results, fail to review or follow up • Fail to disclose errors or take accountability

Source: Adapted from McDonald, K.M., et al., *BMJ Qual Saf.*, 22 Suppl 2, ii33–ii39, 2013; Frosch, D.L., et al., *Health Aff (Milwood).*, 31(5), 1030–38, 2012.

uncertainty, and anxiety [3]. Some people may try to avoid the expense and formality of the traditional medical system and seek help first from alternative sources, a topic we discuss in Section II. Healthcare illiteracy, cognitive decline, poor functional capacity, and language barriers may also contribute to an inability to advocate for themselves or navigate the healthcare system [7]. Even well-educated and fully engaged patients may find numerous challenges in the process of completing their diagnostic evaluations, as summarized in Table 17.1 [8,9]. Patients may be unfamiliar with the diagnostic process, lack insight into when the diagnosis is uncertain or evolving, or not recognize when the process may be off-track (as in misdiagnosis) [8]. Patients may not ask questions for fear of being viewed as a "difficult" or "problem patient." Too many do not feel respected by their physician [9].

Difficult Patients, Difficult Doctors

Most would say that effective doctors in general like their patients and enjoy helping them. But some patients evoke strong negative emotional reactions in their providers. Even attempts to maintain a professional attitude cannot always disguise the frustration, despair, and impatience experienced when providers meet "problem patients" [10]. Groves was one of the first to acknowledge and write publicly of "hateful patients," including the "dependent clingers," the "entitled demanders," the "manipulative help-rejecters," and the "self-destructive deniers" [11]; in most cases, these were patients who defied all efforts to help them and exhausted his efforts. Others describe "heartsinks"—patients who may exasperate, defeat, and overwhelm their providers. These patients are hard to care for, and are at risk of being abandoned by the very systems designed to help them [12]. Sometimes, provider reactions are clearly recognized and identified, but some are not. A negative emotional response to a subset of patients may increase the risk of missed diagnoses (affective bias) [13]. Mortality and morbidity (M&M) conferences for emergency medicine are filled with examples of this; some are common enough to become part of the lore of the specialty and are ingrained in the training. For example, trainees are often warned to always do a neurological examination on the chronic inebriant (to avoid missing a subdural hematoma, as they all fall and all have periods of semi-consciousness while intoxicated), and to beware of the frequent visitor who is judged to be a malingerer (for even malingerers can have new diagnoses). Clinicians are warned to monitor their "affective temperature" lest they allow their emotions to negatively impact clinical reasoning [14]. Difficult patients use up provider's energy, leaving fewer resources available for actual reasoning. Mamede demonstrated that difficult patients tend to deplete physicians' cognitive resources; physicians made more mistakes and had poorer recall in patient vignettes describing disruptive patient attitudes and behaviors [15,16]. Affective bias can also harm patients when the provider identifies positively with the patient, such that he doesn't want to consider the possibility of a bad diagnosis in someone for whom he has a particular fondness (the chagrin factor, or outcome bias) [17–19].

The designation of "difficult patient" may not always reflect the patient's behavior as much as the doctor's mood. Physicians who report a high degree of frustration with patients are more likely to be young, work longer hours (more than 55 hours/week), and have higher levels of stress and anxiety [20]. Although doctors are typically high performers, they are people too and are subject to human frailties, including depression, mental illness, substance abuse, and personality disorders—human limitations that are likely to be reflected in their interaction with patients. We would be remiss if we failed to recognize that many patients who suffer diagnostic errors find their doctors to be "difficult" too and often describe them as arrogant, callous, or indifferent to their concerns. This may be in part attributable to the culture of medical narcissism, created by the need to be confident and assertive, all the while facing irreducible uncertainty in clinical decisions [21,22]. Every clinical encounter, every challenge to a clinician's thinking, is a reminder of how uncertain and precarious clinical decisions are—a fact that can be most unsettling. Doctors may develop defense mechanisms that avoid frank and honest disclosure of uncertainty, reduce their ability to incorporate new information to revise their diagnoses, and contribute to diagnostic errors [21]. We can only hope that these problems are the exception to an otherwise open and caring environment for patients.

Overall, most patients are neither difficult nor disruptive, and problems in their interaction with providers are not due to any fault of their own. Similarly, most doctors (or so we hope) are not uncaring, or intentionally disrespectful. Conflict may arise, in part, out of the stressful situations that patients and doctors find themselves in—patients in crisis and fearful, doctors stressed from unhealthy lifestyles with unrealistic demands placed upon them, or situations for which there are no easy answers or adequate resources [23]. Patients can benefit from understanding what to expect during their medical encounters and how to optimize their visit. Doctors too must learn the need to achieve a balanced and healthy lifestyle if they hope to have a productive and rewarding career. They can be coached to better see illness from a patient's perspective and communicate more effectively with them. A good match of a fully engaged patient with a compassionate and knowledgeable physician forms a foundation to optimize diagnosis. The respective needs of doctor and patient are illustrated well in the YouTube video produced by MedStar Health, "Please See Me" (see Figure 17.2) [24].

Patient's Role in the Diagnostic Process: The Medical Interview

Patients and clinicians often approach the medical interview from very different perspectives. Doctors are taught to quickly and efficiently characterize symptoms into categories to generate possible diagnoses. A basic complaint of chest pain will likely elicit a series of questions meant to categorize the symptom as cardiac, vascular, pulmonary, gastrointestinal, or musculoskeletal, and then proceed with confirming questions to wrap up the encounter with a concise, efficient, and clean diagnosis. An emergency physician will likely decide "dangerous or not," "admit or not" within the first minute of his or her encounter with the patient. In fact, the specialty boasts of developing intuitive gestalt that is necessary for rapid recognition of life-threatening conditions and the efficient management of an emergency department. In contrast, a sick patient, likely not at his best, may not know how to describe his illness; to the physician, the complaints may sometimes be vague and not easily characterized. A harried physician, short on time, might simply dismiss vagueness for "atypical" and send the patient off with false reassurance; the patient has not had a chance to tell his story and may leave completely dissatisfied with the encounter. Worse, he may experience a dangerous diagnostic error.

On the other hand, patients often come with concerns but are not sure how to describe them. Patients may not understand their doctor's brusque manner, or be familiar with medical phrases. It takes patience and skill to interview patients in such a way as to elicit what is troubling them, determine exactly what symptoms they have experienced, how their symptoms have impacted their level of functioning, and to what they ascribe the likely cause. In simulated cases with standardized patients, physicians missed almost 50% of the relevant and important medical information ultimately necessary for diagnosis [25,26]. This is caused by a physician-centric interview style that is interruptive and uses closed-ended questions meant to specify features of the complaint and narrow diagnostic possibilities. The physician knows what he wants to get from the interview and thus controls content and pace, often without recognizing the need to allow the patient to better explain all his concerns. Once interrupted, patients seldom return to the free flow of information they came to share [27]. When patients were asked to list three items of concern they hoped to address with their physician, most identified the last

Please see me

Both:	We are here, in a hospital.
Patient:	In your office.
Doctor:	In my office.
Patient:	Where I am your patient.
Doctor:	Where I have many patients.
Patient:	I came in today for a check-up to check in with the hope that you can help me feel better.
Doctor:	I come here every day to make people feel better, for check-ups, to check in with the nurses and patients I care for.
Patient:	I came here for answers about my illness and pain. I'm afraid of what you might find, afraid you may not find.
Both:	An answer.
Doctor:	It is my job to find answers.
	I went to school for much of my life in order to care for yours. It is my job.
Both:	We are here, in a hospital.
Patient:	In your office.
Doctor:	In my office.
Patient:	Where I am your patient.
Doctor:	Where I have many patients.
Patient:	I am more than my cells, more than my chart can tell! I have a wife and children, four grandchildren.
Both:	Please see me!
Doctor:	I have a family too. People I care for, but I am here to care for you.
Patient:	The hospital is large, full of people waiting for their sickness to be named. Do you remember my name?
Doctor:	Yes I remember you, although often I am tired. I leave for work too early and get home too late. I have a long list of patients to see and not enough hours in the day. I am not a god.
Both:	Please see me.
Patient:	Everyone hurries here and I'm afraid to ask a question, or I forget my question and then you are gone. I need to know what might be wrong.
Both:	Please talk to me.
Doctor:	I want to have the answers. I studied for years for this. I learned the names of every bone and muscle, of diseases and how to treat them.
Patient:	I am not a doctor. The names of pills and procedures sound foreign to me but this is my body. What will this medicine do to me? How much do I owe per pill, per minute of your time?
Both:	Please see me.
Doctor:	I read your chart, your family history, but I need your help to understand what paper cannot tell.
Both:	Please help me.
Patient:	I want to be healthy, but I need to know that I can trust you.
Doctor:	Please don't be afraid—talk honestly to me.
Both:	I am here.
Patient:	I am your patient.
Doctor:	You are my patient.
Patient:	Please listen to my story carefully, carefully.
Doctor:	I want to hear your story, to care for you.
Both:	Help me understand. Give me the tools I need.
Patient:	To be healthy.
Doctor:	To help you stay healthy.
Patient:	To be a partner in my care.
Doctor:	To be a partner in your care.
Both:	Please see me.
Patient:	I am your patient.
Doctor:	You are my patient.
Both:	I am here with you.
Patient:	A Patient.
Doctor:	A Doctor.
Both:	A Person.
Patient:	In your office.
Doctor:	In the hospital.
Both:	We are working.
Patient:	To get better.
Doctor:	To be better.
Both:	We are working, Patient and Doctor together. Please see me!

FIGURE 17.2

Transcript of lyrics from MedStar Health. "Please See Me," YouTube video [24]. (Reproduced with permission.)

one mentioned as most important, the one that provided the context for why they were concerned about the first two items on their list [27]. For example, a patient may come with a cough: perhaps he is just curious to understand what might cause it, or it might be keeping him from getting a good night's sleep. He may even secretly be afraid that he has lung cancer. The most appropriate actions for various scenarios differ, from a review of possible causes and reassurance, versus a prescription for a cough suppressant, versus a chest x-ray and discussion about the need for smoking cessation. Once interrupted, 85% of patients never get to the point of divulging their most significant concern during their visit [28,29]. Failure of the medical interview to elicit the most relevant cause of concern is evidenced by the patient who discloses, just as he is ready for discharge, a new concern, a sort of "hidden agenda" missed in the initial encounter that itself may be the most relevant problem, or the one most in need of addressing [30]. One example is the patient who comes "Just to check my blood pressure" but casually mentions on his way out, "By the way doc, you don't think this chest pain is anything to worry about?" An exasperated physician may feel blindsided by this unexpected revelation, and likely has no clue why the patient did not mention it before.

The ubiquitous occurrence of patients who leave against medical advice (AMA) is further evidence of a mismatch between patient expectations and physician concerns. What may be lacking in some "AMAs" is a trusting relationship and a meaningful conversation [31]. How do we explain the differences in concerns and communication styles between physician and patient?

Communication: Provider–Patient, Patient–Provider

Physicians are taught a structured interview process: elicit the chief complaint, characterize that complaint and associated symptoms, then briefly review the past medical history. Physicians on average interrupt the patient 18 seconds after their first greeting, then take charge of the remainder of the interview to complete their standard assessments [27]. Niceties aside, physicians do not see this as rude or interruptive; it is simply the nature of acquiring all the facts they need to make a diagnosis. However, this "physician-centric" style is foreign to most patients, and stifles an open discourse with the patient.

In contrast, a "patient-centric" approach to the medical interview allows the patient to complete his thought, invites him to expand on what he means, and pauses to allow time him to elaborate and "tell his story." Although many physicians imagine that this type of open-structure interviewing will take longer and fail to answer all their structured questions, others have shown that it typically leads to a better exchange of information that is both effective and efficient. When patients are allowed to complete, without interruption, their response to why they sought care, most completed their thought in less than 60 seconds [27]. Interviews that asked, "What concerns do you have?" and "Anything else?" until all concerns were revealed added only 6 seconds to the encounter time [30]. A patient-centric interview provides a more complete view of the patient's illness, more accurately and quickly identifies the patient's pre-visit concerns, improves treatment adherence, improves both patient and provider satisfaction, and adds to the sense of empathy and respect experienced by patients [26,32–35]. Two of the most common cognitive biases reported in diagnostic errors—premature closure and anchoring—may in fact be due in part to truncated interview styles that stop the patient from contributing information that is important and vital to the diagnosis; information that later becomes obvious in hindsight.

Strategies for Improvement

What Doctors Need to Know

When patients describe their medical encounters, they reveal that they come seeking help about very personal, and possibly very private, concerns [8]. They are sometimes sick, often worried, and almost always vulnerable. They enter the clinical world largely unfamiliar with medical terminology and procedures. They do not always know what information is most relevant to their diagnosis, what details are most helpful, or what will be seen as trivial or disinteresting. They need space and time to explain their concerns and describe their symptoms in their own words.

Helen Haskell, founder of Mothers Against Medical Error, explains her experience and the concerns of patients who have experienced diagnostic errors:

> *"The vast majority of diagnostic errors and delays reported by patients concern not issues of process, but problems in the patient-provider relationship. More specifically, diagnosticians' attitudes and preconceptions toward individual patients can and do result in failure to listen, really listen, to what the patient is saying."[36]*

Although communication seems like such a basic concept, even an innate human skill, it may not come naturally to clinicians, who are trained to be primarily academic thinkers. Without specific mentoring and guidance in how to approach people who are dependent upon them but inexperienced in medical settings, they may flounder when they face patients who are overwhelmed with illness and the stress that accompanies it. Communication skills are now included as an important part of professional training in medical schools. Efforts to include communication skills in training have been modestly successful, and include the following recommendations [26,33]:

- Encourage open-ended questions
- Make specific efforts to elicit patient concerns

- Attempt to understand the patient's illness from his or her point of view
- Explore the impact of illness on the patient's life
- Encourage patient questions
- Express empathy
- Verify patient understanding

Another simple recommendation is offered by Barrier. While exploring the chief complaint (the main reason for the visit), keep asking "What else?" until the patient reveals all his or her immediate concerns, and then set an agenda for what can reasonably be accomplished in a single visit [37].

When a group of expert clinicians were interviewed about how doctors could be more effective communicators, Dhaliwal gave this advice [38]. Before completing the patient interview, ask the patient three questions:

1. What is your idea about what is going on?
2. What are you most worried about?
3. What are you expecting that I can do?

These three simple patient-centric questions identify the patient's main concern, address his fears, and reveal his expectations for the visit.

Additionally, clinicians need to be aware of how their own emotional state, and their reaction to patients, may influence their diagnostic skills and intuitive decision making. Specific training geared toward acknowledging, recognizing, and addressing affective bias is necessary to help providers avoid the negative influence of affect on their diagnostic accuracy [13]. When encountering patients known to elicit affective bias, clinicians can be equipped with techniques and strategies to mitigate the impact of bias [18,19,39–41].

In some practices, repeat visitors who are judged as "difficult" can be better managed with a case management plan, especially when the patient is involved with and agrees with the plan. A well-thought-out plan allows clinicians to set acceptable and consistent management plans in advance of a visit [12]. In some difficult patients, finding a consistent approach empowers physicians to attend to the patient with less discomfort, and likely provide better care [10,23].

What Patients Need to Know

Unlike doctors, there is no training for being a patient. Yet, patients have a central role in their diagnosis and the most to gain or lose in the process. They can be prepared to help along the way if given appropriate aids and resources; a few are shown in Figure 17.3.

Resources for patients

Patient toolkit, the society to improve diagnosis in medicine. Available at: http://www.improvediagnosis.org/page/PatientToolkit (44)

Smart partners guide to your health. Smart checklist. Smart script. A product for the national Patient safety foundation. Available at:

https://c.ymcdn.com/sites/www.npsf.org/resource/collection/930A0426-5BAC-4827-AF94-1CE1624CBE67/SMART-Partners-Guide1.pdf (45)

Speakup. The joint commission. Available at: https://www.jointcommissionorg/speakup.aspx (47)

Ask me 3: Good questions for your good health. National patient safety foundation. Available at: http://www.npsf.org/?page=askme3 (46)

I wish i had asked that! The informed medical decision foundation. Available at: http://cdn-www.informedmedicaldecisions.org/imdfdocs/Patient Visit Guide.pdf (48)

FIGURE 17.3
Resources for patients.

Patients can be encouraged to become a "more knowledgeable consumer" [8]. While some sources of information may be unreliable, increasingly, resources are available that provide excellent information about diseases and conditions. In particular, online symptom checkers can help patients explore possible causes for their symptoms and determine the urgency with which they should seek care, such as the one offered by Isabel Healthcare [42]. Patients will be better able to have a full conversation and optimize the time with their doctor if they prepare for their visit. When they are seeking help for a new problem or symptom, the type of information they give is especially likely to influence the direction and pace of their workup. A patient engagement committee with the Society to Improve Diagnosis in Medicine (SIDM) advises patients to "tell their story well," and describe the exact date of onset of symptoms and the sequence of events with their illness [8]. Patients should be encouraged to prepare summaries of past medical events and current medications in advance of their visit. They should write down the three most important questions or concerns they wish to address during their visit and take these with them [43]. Patients should use online patient resources to help summarize relevant medical history, such as the SIDM patient toolkit [44], or the Kaiser Permanente Smart Partners Guide [45]. They could also

---

Questions to guide my doctor's visit

Before each visit decide: What is the issue that I want help with today?

Questions to ask during my appointment:

1. What is my diagnosis?

2. What are my treatment options?

3. For each option, what are the side effects? Costs?

4. How can I get more information?

FIGURE 17.4
Questions patients should ask during their doctor visits. (Adapted from The Informed Medical Decision Foundation. I wish I had asked that! Available at: http://cdn-www.informedmedical-decisions.org/imdfdocs/Patient_Visit_Guide.pdf [48].)

bring an advocate with them for their visit to help keep track of important information. Before concluding the visit, the National Patient Safety Foundation recommends that the patient follows their guide to "Ask Me 3"—three questions to help get the right diagnosis: "What could be causing my problem? What else could it be? When will I get my test results and what should I do to follow up?" [46]. Patients should not be afraid to ask their doctors how confident they are in their diagnosis [8]. The Joint Commission encourages patients to "Speak Up" and provides useful tips online to guide patients during their medical encounters [47]. The Informed Medical Decision Foundation provides a worksheet with questions to complete during a doctor's visit to ensure that patients get all the facts needed to make sound decisions; some are shown in Figure 17.4 [48]. Engaged and empowered patients should know their test results, keep copies of their records, and ask questions. If something unexpected happens, they should notify their doctor. Unexpected problems may be the first sign that the diagnosis is wrong. When possible, patients need to avoid doctor-shopping or switching systems of care; when they do, the second provider often begins again at the beginning and may fail to advance the workup. Patients need to understand which practice settings are best suited to further their diagnosis. Granted, emergency departments and urgent care centers are conveniently open all hours, but they cannot replace an established relationship with a consistent provider, and while they are excellent for detecting dangerous conditions that are immediately threatening to life or limb, they will not provide answers for many other complex medical conditions.

Resources to Improve Diagnosis

There are many approaches to improving the patient's role in diagnosis. From a public health perspective, efforts to inform and educate the public can promote awareness of early warning signs of disease. One example is stroke awareness campaigns by the American Stroke Association. Education can be provided in schools, in advertisements, and even in days or months devoted to activities that raise public consciousness.

Diagnostic services can be provided within local neighborhoods; mobile mammography vans and "clinics on wheels" are examples of efforts to take diagnosis directly to the patient. Some organizations provide doctor home visits for those who are unable to travel. For those with language barriers, interpreter services by phone are widely available and accessible to facilitate medical interviews. Many healthcare organizations now utilize patient portals for ready access to relevant healthcare information. Some physicians now communicate directly with their patients via e mail, exchanging messages and updating their progress in between visits. Healthcare navigators can help identify and sometimes address barriers to care, although they are not widely available for all patients [6]. Patients can find additional online resources from organizations and patient advocacy groups.

Despite all these resources, the diagnostic process is still an uncertain one, and often has many unexpected twists and turns. Ultimately, an engaged patient may be the single most important factor for diagnosis; one who avails of resources and takes charge of his or her course can compensate for many other flaws and imperfections in the diagnostic process. Healthy and respectful relationships between providers and patients can go a long way to improving diagnosis.

SUMMARY POINTS

- Medical evaluations involve a relationship between patients and their illness, their provider, and the healthcare system.
- A patient's ability to access the healthcare system and successfully engage in his or her care is an important aspect of diagnostic work. Efforts to improve diagnosis should include support for patient engagement.
- Patients and providers may not understand one another; both providers and patients can benefit from tools to help them communicate more effectively.
- Resources are available for patients to help them navigate their diagnostic evaluations; a few are referenced here.

References

1. Cosby KS. A framework for classifying factors that contribute to error in the emergency department. *Ann Emerg Med*. 2003;42(6):815–23.
2. Cosby KS, Roberts R, Palivos L, Ross C, Schaider J, Sherman S, Nasr I, Couture E, Lee M, Schabowski S, Ahmad I, Scott RD 2nd. Characteristics of patient care management problems identified in emergency department morbidity and mortality investigations during 15 years. *Ann Emerg Med*. 2008;51(3):251–61.
3. Corner J, Hopkinson J, Roffe L. Experience of health changes and reasons for delay in seeking care: A UK study of the months prior to diagnosis of lung cancer. *Soc Sci Med*. 2006;62(6):1381–91.
4. Corner J, Hopkinson J, Fitzsimmons D, Barclay S, Muers M. Is late diagnosis of lung cancer inevitable? Interview study of patients' recollections of symptoms before diagnosis. *Thorax*. 2005;60(4):314–19.
5. Singh H, Hirani K, Kadiyala H, Rudomiotov O, Davis T, Khan MM, Wahls TL. Characteristics and predictors of missed opportunities in lung cancer diagnosis: An electronic health record-based study. *J Clin Oncol*. 2010;28(20):3307–15.
6. Battaglia TA, Roloff K, Posner MA, Freund KM. Improving follow-up to abnormal breast cancer screening in an urban population: A patient navigation intervention. *Cancer*. 2007;109 Suppl 2:359–67.
7. Carman KL, Dardess P, Maurer M, Sofaer S, Adams K, Bechtel C, Sweeney J. Patient and family engagement: A framework for understanding the elements and developing interventions and policies. *Health Aff (Millwood)*. 2013;32(2):223–31.
8. McDonald KM, Bryce CL, Graber ML. The patient is in: Patient involvement strategies for diagnostic error mitigation. *BMJ Qual Saf*. 2013;22 Suppl 2:ii33–ii39.
9. Frosch DL, May SG, Rendle KA, Tietbohl C, Elwyn G. Authoritarian physicians and patients' fear of being labeled "difficult" among key obstacles to shared decision making. *Health Aff (Millwood)*. 2012;31(5):1030–38.
10. Gerrard TJ, Riddell JD. Difficult patients: Black holes and secrets. *BMJ*. 1988;297(6647):530–32.
11. Groves JE. Taking care of the hateful patient. *N Engl J Med*. 1978;298(16):883–87.
12. O'Dowd TC. Five years of heartsink patients in general practice. *BMJ*. 1988;297(6647):528–30.
13. Croskerry P, Abbass A, Wu AW. Emotional influences in patient safety. *J Patient Saf*. 2010;6(4):199–205.
14. Park DB, Berkwitt AK, Tuuri RE, Russel WS. The hateful physician: The role of affect bias in the care of the psychiatric patient in the ED. *Am J Emerg Med*. 2014;32(5):483–85.
15. Mamede S, van Gog T, Ce Schuit S, Van den Berge K, LA Van Daele P, Bueving H, Van der Zee T, W Van den Broek W, Lcm Van Saase J, Schmidt HG. Why patients' disruptive behaviours impair diagnostic reasoning: A randomised experiment. *BMJ Qual Saf*. 2016. pii: bmjqs-2015-005065. doi:10.1136/bmjqs-2015-005065. [Epub ahead of print]
16. Schmidt HG, van Gog T, Ce Schuit S, Van den Berge K, LA Van Daele P, Bueving H, Van der Zee T, W Van den Broek W, Lcm Van Saase J, Mamede S. Do patients' disruptive behaviours influence the accuracy of a doctor's diagnosis? A randomised experiment. *BMJ Qual Saf*. 2016. pii: bmjqs-2015-004109. doi:10.1136/bmjqs-2015-004109. [Epub ahead of print]

17. Groopman, J. *How Doctors Think*. New York: Houghton Mifflin; 2007.
18. Croskerry P. The importance of cognitive errors in diagnosis and strategies to minimize them. *Acad Med*. 2003;78(8):775–80.
19. Croskerry P. Critical thinking and reasoning in emergency medicine, in *Patient Safety in Emergency Medicine*, eds Croskerry P, Cosby K, Schenkel SM, Wears RL, 213–218. Philadelphia, PA: Walters Kluwer/Lippincott Williams & Wilkins. 2009.
20. Krebs EE, Garrett JM, Konrad TR. The difficult doctor? Characteristics of physicians who report frustration with patients: An analysis of survey data. *BMC Health Serv Res*. 2006;6:128.
21. Banja, J. *Medical Errors and Medical Narcissism*. Sudbury, MA: Jones and Barlett; 2005.
22. Alexander GC, Humensky J, Guerrero C, Park H, Loewenstein G. Brief report: Physician narcissism, ego threats, and confidence in the face of uncertainty. *J App Soc Psychol*. 2010;40(4):947–55.
23. Serour M, Al Othman H, Al Khalifah G. Difficult patients or difficult doctors: An analysis of problematic consultations. *Eur J Gen Med*. 2009;6(2):87–93.
24. MedStar Health. Please See Me [video]. YouTube. https://www.youtube.com/watch?v=380MiMDoddI. Published April 3, 2015. Accessed May 24, 2016.
25. Roter DL, Hall JA. Physician's interviewing styles and medical information obtained from patients. *J Gen Intern Med*. 1987;2(5):325–29.
26. Maguire P, Pitceathly C. Key communication skills and how to acquire them. *BMJ*. 2002;325(7366):697–700.
27. Beckman HB, Frankel RM. The effect of physician behavior on the collection of data. *Ann Intern Med*. 1984;101(5):692–96.
28. Rost K, Frankel R. The introduction of the older patient's problem in the medical visit. *J Aging Health*. 1993;5(3):387–401.
29. Frankel RM. Clinical care and conversational contingencies: The role of patients' self-diagnosis in medical encounters. *Interdiscip J Stud Discourse*. 2001;21(1–2):83–111.
30. Marvel, MK, Epstein RM, Flowers K, Beckman HB. Soliciting the patient's agenda: Have we improved? *JAMA*. 1999;281(3):283–87.
31. Alfandre DJ. "I'm going home": Discharges against medical advice. *Mayo Clin Proc*. 2009;84(3):255–60.
32. Smith RC, Hoppe RB. The patient's story: Integrating the patient- and physician-centered approaches to interviewing. *Ann Intern Med*. 1991;115(6):470–77.
33. Rao JK, Anderson LA, Inui TS, Frankel RM. Communication interventions make a difference in conversations between physicians and patients: A systematic review of the evidence. *Med Care*. 2007;45(4):340–49.
34. Bhadula R. A piece of my mind: The Good Physician. *JAMA*. 2013 Sep 4;310(9):909.
35. Chipidza FE, Wallwork RS, Stern TA. Impact of the doctor-patient relationship. Prim Care Companion *CNS Disord*. 2015;17(5). doi:10.4088/PCC.15f01840. eCollection 2015.
36. Haskell HW. What's in a story? Lessons from patients who have suffered diagnostic failure. *Diagnosis*. 2014;1(1):53–54.
37. Barrier PA, Li JT, Jensen NM. Two words to improve physician-patient communication: What Else? *Mayo Clin Proc*. 2003;78(2):211–14.
38. Dhaliwal, G. Ask patients These three simple questions. In: The Experts: How to Improve Doctor-Patient Communication. *The Wall Street Journal*. April 12, 2013. Available at: http://www.wsj.com/articles/SB10001424127887324050304578411251805908228. Accessed May 24, 2016.

39. Croskerry P. Cognitive forcing strategies in clinical decisionmaking. *Ann Emerg Med*. 2003;41(1):110–20.

40. Croskerry P, Singhal G, Mamede S. Cognitive debiasing 1: Origins of bias and theory of debiasing. *BMJ Qual Saf*. 2013;22 Suppl 2:ii58–ii64.

41. Croskerry P, Singhal G, Mamede S. Cognitive debiasing 2: Impediments to and strategies for change. *BMJ Qual Saf*. 2013;22 Suppl 2:ii65–ii72.

42. Isabel Symptom Checker. Available at: http://symptomchecker.isabelhealthcare.com. Accessed May 24, 2016.

43. Post DM, Cegala DJ, Miser WF. The other half of the whole: Teaching patients to communicate with physicians. *Fam Med*. 2002;34(5):344–52.

44. The Society to Improve Diagnosis in Medicine. The Patient Toolkit. Available at: http://www.improvediagnosis.org/page/PatientToolkit. Accessed May 24, 2016.

45. Kaiser Permanente Smart Partners Guide. Available at: https://c.ymcdn.com/sites/www.npsf.org/resource/collection/930A0426-5BAC-4827-AF94-1CE1624CBE67/SMART-Partners-Guide1.pdf. Accessed May 24, 2016.

46. National Patient Safety Foundation. Ask Me 3. Available at: http://www.npsf.org/?page=askme3. Accessed May 24, 2016.

47. Joint Commission. Speak Up. Available at: https://www.jointcommission.org/facts_about_speak_up/. Accessed May 24, 2016.

48. The Informed Medical Decision Foundation. I wish I had asked that! Available at: http://cdn-www.informedmedicaldecisions.org/imdfdocs/Patient_Visit_Guide.pdf. Accessed May 25, 2016.

Afterword

Medicine is a noble profession. Those who devote their life's energy to the practice of medicine provide service to humanity, and their efforts are honorable. But we suffer from an outdated view of medicine based on false beliefs, and ultimately those beliefs have led us astray. Without recognizing it, we have created a model of medicine that suggests that we can achieve (even expect, or demand) perfection. We can (and must) be certain, that is, if we are minimally competent. And if we fail to be anything less than perfect or certain, then we must simply have failed to apply our scientific advances well and need only study more, or try harder to achieve success.

It's not that we haven't made progress. The origins of early medicine were based largely on superstition; diseases were sometimes considered punishment for sins, or attributed to an imbalance of humors. In the last century, developments in science have led to an understanding of many diseases at the genetic, molecular, and cellular levels. Our progress in understanding the basis of disease is staggering! But we have also adopted some equally false beliefs based on our success.

We are a bit infatuated with our model of medical knowledge, somewhat like the Greek mythological character Pygmalion was infatuated with his creation, the statue of the loveliest, most beautiful woman. Like Pygmalion, we have constructed an illusion around medical culture, a sort of statue of perfection. We cling to the belief that doctors "save lives," when in fact, we simply delay death temporarily, only to visit the inevitable another day. Our patients place great trust in our healthcare providers and systems, because in moments of great fear and vulnerability, they need us. They want us to be perfect, to offer them hope and relieve suffering when they are most in need. Many somewhat blindly agree to risky operations and potentially dangerous treatments based on trust in our skills and judgment. But if we err, or worse, fail, many feel betrayed that they did not get the promise of a cure or improved health they expected.

Throughout this book, we have deconstructed this model, to reveal how capricious our efforts at diagnosis can be. Even our thought process, the heart of diagnostic work and the product of all our study, is subject to unconscious bias and failed logic. Our cognitive skills suffer from fatigue, emotions, circadian cycles, and limited recall. Our diagnostic tools are imperfect; even our best tests have less than perfect diagnostic performance. Our diagnostic processes often require numerous steps, all with some failure rate. And there is one fact that is simply beyond our control—the reality that biological systems have variable expressions of disease that makes our efforts all the more difficult. None of this is our fault, but what can be faulted is our failure to acknowledge these limitations. Our growing awareness of diagnostic errors has created a crack in our model and revealed what was present all along: diagnosis is not certain, and our understanding and skills not perfect.

We should not despair. All the good of modern healthcare is not the less for our labor. But it is less than what can be achieved. We hope that the content of these pages helps breathe life into a new model for medical thought, one that helps mature the discipline of medicine, grounded in a better understanding. Each of the topics we have focused on provides a foundation from which to build anew, to improve and mature medical care. Our model, like Pygmalion's cold, lifeless statue, can be made to be more true to flesh and blood, more realistic to the nature of biology.

It is not wrong to strive for perfection. But it is also not wise to consider perfection the standard by which we judge ourselves. The Japanese Zen concept of *wabi sabi* may provide some useful insight—namely that nothing is perfect, and nothing is ever finished. The drive to perfection is a process, but not an endpoint. If so, we should recognize that our drive to improve diagnosis should be an endless process, never quite done and always deserving our focus. We hope that our struggle to describe the work of diagnosis, and the fallibilities of human cognition and healthcare systems, can inform and inspire further work to drive progress forward in this important endeavor.

Glossary

abductive logic: a form of reasoning that begins with an observation from which a theory (or plausible explanation) is developed that best fits the available information. Abductive reasoning acknowledges uncertainty and makes use of information that may be imperfect or incomplete. Examples of abductive reasoning include medical diagnosis (seeking a diagnosis for a clinical syndrome) and jury verdicts (determining guilt in a criminal trial).

absent grandmother syndrome: a casual term that describes the tendency of parents of young children to seek medical attention for routine benign problems for which many experienced laypeople (such as grandmothers) might otherwise provide guidance. The need to seek professional help may be evidence of the lack of an experienced grandmother figure or other personal resources.

ad hominem fallacy: a logical fallacy, *see* Appendix II.

affective bias: a cognitive bias, *see* Appendix I.

aggregate bias: a cognitive bias, *see* Appendix I.

ambiguity: a logical fallacy, *see* Appendix II.

ambiguity effect: a cognitive bias, *see* Appendix I.

analytical reasoning: deliberate, thoughtful, considered, conscious thinking; contrasts with a shoot-from-the-hip approach driven by intuition.

anchoring bias: a cognitive bias, *see* Appendix I.

anecdotal: a logical fallacy, *see* Appendix II.

apohenia: the human tendency to see patterns in random objects. Examples include seeing objects in cloud formations or the outline of mythical figures in constellations; also known as pareidolia.

appeal to authority/emotion/nature: logical fallacies, *see* Appendix II.

appropriate use criteria: evidence-based criteria for the use of medical services (testing and treating) that take into account potential benefits and risks. Tests and treatments are considered appropriate if there is a margin of benefit that clearly exceeds their risk.

appropriateness guidelines: suggested guides to testing that take into account how effective a test is in answering a clinical question and when testing is clinically indicated, and considers whether the cost of testing can be justified. The American College of Radiology publishes appropriateness guidelines for diagnostic imaging to advise clinicians in the selection and use of radiology tests.

art of medicine: the aspects of medicine not well described simply as the application of medical knowledge; this includes interpersonal skills, empathy, compassion, communication, and advocacy (to name a few).

ascertainment bias: a cognitive bias, *see* Appendix I.

asymmetric paternalism: paternalism is the practice of making decisions for or limiting options to someone else (in our context, a patient) by a person of authority presumably acting in the patient's best interests and in accordance with their expressed desires. Asymmetric paternalism refers to the practice of providing information and options to someone based on the degree of their autonomy, ability to understand, or desire to engage in decisions about their health. Physicians may use asymmetric paternalism in determining when (or if) they should inform a patient that they are terminally ill, or limiting options for end-of-life care based on what they think is in the patient's best interests.

attentional bias: a cognitive bias, *see* Appendix I.

augenblick diagnosis: diagnosis made literally "in the blink of an eye" by pattern recognition and largely by the instantaneous recognition of prototypical or pathognomonic features.

authority gradient: the tendency for persons of less seniority, experience, or perceived expertise to fail to challenge or question the actions of another individual whom they view as having more authority. A common example is the failure of a co-pilot to challenge the actions of pilot; in medicine, a student may question their own judgment and defer to a more senior physician; a nurse may fail to challenge the actions of a physician. Authority gradients may contribute to harm in the course of medical care.

availability bias: a cognitive bias, *see* Appendix I.

bandwagon effect: *see* Appendices I and II.

base rate neglect: *see* Appendices I and II.

begging the question: a logical fallacy, *see* Appendix II.

belief bias: a cognitive bias, *see* Appendix I.

belief engine: the hard wired disposition to believe rather than not, to trust rather than distrust. Also refers to the tendency to find causal connections between objects, events, or phenomena in our environment.

bias: a tendency to respond based on perception or belief.

bias inoculation: introducing trainees to specific cognitive and affective biases in the context of clinical examples and providing advice for how to avoid or prevent errors in clinical reasoning.

black-or-white: a logical fallacy, *see* Appendix II.

blank slate: the concept that human minds are born with a blank slate, without any influences from our evolutionary past. This philosophy is known as empiricism, and those who hold it are empiricists.

blind spot bias: a cognitive bias, *see* Appendix I.

bounded rationality: a type of decision style that acknowledges the need to move forward with thinking and deciding despite uncertainty and incomplete knowledge. Rather than wait for perfect knowledge, bounded rationality assumes that many decisions need to be made within the constraints of available time and knowledge.

brutish automatism: an expression used by Lawrence Durrell in his excursion into Buddhism (*A Smile in the Mind's Eye: An Adventure into Zen Philosophy*, 2012) to denote the evolutionary origins of our decision making.

burden of proof: a logical fallacy, *see* Appendix II.

calibration: the process by which an individual makes sound judgments that are reasonably free from favoritism, bias, stereotyping, and other factors that can distort reasoning (as described in Chapter 3), or the ability to adjust (or the adjustment of, or revision of) one's thinking based on feedback from the results of prior actions or decisions (as illustrated in the *dual process model* in Figure 3.1).

clinical efficacy (effectiveness): the ability of a test to change management or impact outcome measures. As diagnostic tests are increasingly used to not only detect disease, but also monitor and change treatment, these measures convey how well a test performs.

clinical epidemiology and biostatistics: the coursework that is typically given medical students that includes research methodology and statistics, how to interpret study results, and skills in translating research findings into clinical practice.

clinical judgment: the sum total of one's knowledge, cognitive processes, and experience applied to clinical decisions. The concept of clinical judgment is used to explain how an expert knows what he does, even when he can't fully articulate how he arrives at his conclusions. Although clinical judgment by experts is respected, it is sometimes criticized by those who argue for decisions grounded primarily in evidence and guided by an evidence-based approach.

cognitive bias: systematic ways in which individuals misjudge what is true, often through the use of heuristics.

cognitive bias mitigation: attempts to reduce the overall impact of cognitive bias on decision making.

cognitive bias plus: the collective influence of cognitive biases and logical fallacies as well as influences from conflicts of interest and ethical violations that lead to distorted reasoning and erroneous decisions.

cognitive contamination: the idea that one's ideas might be contaminated by hearing another's opinion such that they might not complete their own assessment without bias or influence from anything but the evidence.

cognitive dissonance: discomfort from holding contradictory values or opinions, or conflict that arises when one's behavior is in conflict with one's values.

cognitive engineering: the idea of teaching, testing, and reinforcing ideas about cognitive bias in the context of real-life examples and repeating it regularly throughout the training period, analogous to vaccination with booster shots.

cognitive forcing strategies: a rule or guideline to help individuals avoid the negative consequences of common cognitive biases.

cognitive indolence: analogous to cognitive laziness, a desire to minimize cognitive effort.

cognitive manager: the ability of the brain to cognitively monitor itself and change its output arises from metacognition—the ability to consciously think about thinking. This is accomplished through Type 2 processes referred to in the dual process model as a *cognitive manager role.*

cognitive miser: a function illustrated in the *dual process model* that describes the tendency to default to Type 1 processes to preserve cognitive energy and resources.

cognitive tutoring systems: software that detects and provides feedback to individuals about their tendency for cognitive biases, then provides specific strategies to help debias or retrain them to avoid the influence of those biases.

cognitive vaccination: teaching, testing, then reinforcing again and again to engrain awareness and recognition of particular cognitive biases.

collective conscious: a set of common ideals and values within a society.

collective consciousness: describes the fact that individuals have awareness of things in general for which they may have no individual experience, a kind of general awareness. This concept describes (though doesn't fully explain) why people who live in places without snakes still have a phobia of snakes.

collective rationalization: the tendency of groups to re-enforce shared points of view and fail to consider new or contrary evidence; excessive conformity within a group structure that tends to discount new ideas.

commission bias: a cognitive bias, *see* Appendix I.

competency: the development of knowledge and skills necessary to acquire expertise. Before training, individuals generally are *unconsciously incompetent* (they don't know what they don't know). Novices gradually transition to a state of *conscious incompetence* (having entered the medical arena they are acutely aware of their lack of skill). As they develop skills they acquire confidence and become *consciously competent*; with time, practice, and experience they gradually become *unconsciously competent* as they go about their daily work. The final phase of expertise involves *reflective competence*, reflecting on what one knows and how one knows it, where knowledge gaps lie, and how to teach others what one knows.

complementary alternative medicine (CAM): efforts at prevention, diagnosis, and treatment outside the usual and highly regulated system of professional medical practice. Complementary medicine includes many options that may be used alongside traditional measures. Alternative methods may be used to replace standard medicine. Both methods are absent the usual rigorous scientific standards that substantiate orthodox medicine.

composition/division: a logical fallacy, *see* Appendix II.

comprehensive assessment of rational thinking (CART): a test to evaluate rational thinking.

confirmation bias: a cognitive bias, *see* Appendix I.

congruence bias: a cognitive bias, *see* Appendix I.

contemplative stage: a phase in the *transtheoretical model of change* in which an individual contemplates change.

contrast effects: a cognitive bias, *see* Appendix I.

corridor (or hallway) consultation: *see* Curbside consultation.

cost-effectiveness ratio (CE ratio): the cost of an intervention expressed by outcome of lives saved by that intervention (or adjusted for improved quality of life). *Cost-effectiveness analysis* is used most often to compare different interventions to determine which one(s) achieve(s) the most improvement in outcome (in lives and quality of health) for the cost. (See Figure 13.3.).

CReME (clinical reasoning in medical education): an initiative in medical schools in the United Kingdom to develop and share resources for teaching clinical reasoning.

critical thinking: reasoning that is marked by the ability to observe, ask questions, analyze, synthesize, evaluate, make valid conclusions, assess the strength and validity of arguments, and demonstrate skill in understanding different perspectives.

crystallized intelligence (Gc): the lifelong accumulation of information and skills, the things you know.

cum hoc fallacy: the tendency to confuse correlation with causation. A sports enthusiast may correlate his team's victory to the fact that he wore his favorite socks on the day of victory—and be driven to wear those socks for subsequent games hoping to improve his team's chances. *See* Appendix II.

curbside consultation (also referred to as corridor consultation): a casual and informal request for advice from a professional outside an established patient-physician encounter, dangerous because it is often done without the usual rigor of proper questioning and physical examination, and without the usual degree of thought. Curbside consultations may be convenient, but lack the standards that a formal evaluation provides.

cyberchondria, cyberchondriasis: an unfounded escalation of concern over common symptoms based on information read on the Internet that may result in undue anxiety and further risk of harm from self-diagnosis and self-treatment.

Dalhousie model of clinical reasoning: a model for educating students in cognitive processes and clinical decision making showing the types of knowledge and skills that need to be incorporated into standard medical education, including content from cognitive psychology, critical thinking, and metacognition. *See* Figure 14.2.

data blinding: a method to prevent individuals from being influenced by others before they make their own opinion or draw their own conclusions. This is meant to provide a type of *cognitive independence*.

decisions outcome inventory (DOI): an instrument used to determine decision-making competence.

declarative knowledge: factual knowledge that is known by an individual. Differs from procedural knowledge which is the skill to do. Declarative knowledge is knowing "what"; procedural knowledge is knowing "how to do."

deductive logic: reasoning from known general premises to generate specific valid conclusions. If the premises of a deductive argument are true, the conclusion is almost certainly true. In the mathematical example below, the premises are true and certain, and the conclusion guaranteed to be true.

Premise 1: If $x = 3$.
Premise 2: And $y = 5$.
Conclusion: Then $x + y = 8$.

deductive syllogism: a natural language equivalent of a deductive proof. For example:

Premise 1: All men are mortal.
Premise 2: Sam is a man.
Conclusion: Sam is mortal.

deliberation-without-attention approach: an approach in which decisions are made without reaching a conscious level. It's not that thinking doesn't occur at all; rather, thinking may occur at an unconscious level.

diagnosis error: a failure to arrive at a "correct" diagnosis in a timely manner such that available interventions can be offered. There are many nuances to the term depending upon the situation, including classifications such as avoidable or preventable. The National Academy of Sciences, Engineering, and Medicine defines diagnostic error as: "the failure to establish an accurate and timely explanation of the patients health problems(s) or communicate that explanation to the patient." Terms such as missed diagnosis (in which a diagnosis was not made, perhaps only determined at autopsy), misdiagnosis (an alternative diagnosis was made in error), and delayed diagnosis (implying it took longer than desired to reach the conclusion) are included in the general definition for diagnosis error.

diagnosis momentum: the tendency to accept a diagnostic label provided by another without further critique; the label may be accepted without critically reviewing the facts or demanding sufficient proof, possibly as a means to reserve resources, time, or energy. *See also* Appendix I.

diagnostic efficacy: the ability of a test to discriminate between a population of individuals with disease versus one without disease. Also referred to as *diagnostic efficiency,* characterized by receiver operator curves and measured in terms of sensitivity, specificity, predictive values (positive and negative), and likelihood ratios. Some reserve the use of the terms *efficacy, clinical efficacy,* and *effectiveness* to refer to the test performance in real-life circumstances with less well-defined populations and outside formal research settings. (*See* clinical efficacy.).

diagnostic threshold (or threshold to test, threshold to treat): Decisions about when to test and when to treat should weigh the prior probability of disease, and consider the characteristics of diagnostic tests (sensitivity, specificity, risks of testing), and the potential benefits, risks, and costs of treatment. A threshold can be determined at which the benefit of treating removes the need for further testing; a treatment threshold may be met in which a diagnosis is presumed sufficient to treat. The original concept of thresholds was introduced by Pauker and Kassirer in 1980; their method involved a fairly rigorous mathematical analysis. The concept (if not their exact methods) has been widely applied in medical practice since their original description.

diagnostic trajectory: the concept that diagnosis often occurs over time, not necessarily in a single moment. Patients and clinicians benefit from recognizing where they are in the pathway (or trajectory) so that they continue moving their diagnostic work forward. The phrase also implies that it is reasonable and even necessary to provide a timeline for diagnostic work, such that common or dangerous conditions are detected early, while understanding that less common, or atypical presentations should be considered in due course.

dialectical reasoning: dialectical thinking allows someone to examine or hold opposing thoughts. It is the process of arriving at truth through a process of comparing and contrasting both sides of an argument. Whereas the Western tradition relies on formal analytical procedures and propositional logic in argument, the Eastern tradition emphasizes a more conciliatory position, promotes empathy, and is more likely to reduce conflict. Examining the opposite of something, allows us to gain a better understanding of the world and ourselves. Thus, dialectical thinkers might include thinking of passivity and aggression, seeing love and hate, or impulsivity and withdrawal, as well as considering different answers to questions around morality. Dialectical reasoning may lead to reduction in the impact of certain biases.

differential diagnosis: a list generated in the course of diagnostic evaluations that considers all the likely possible causes of a clinical symptom or syndrome.

differential psychology: the study of individual decision making and how individuals differ from one another in their thought processes and reasoning.

difficult patient: a patient that seems to resist, or make difficult, the attempts of clinicians to evaluate and/or treat them. Examples include those with personality disorders, patients who have trouble with compliance with treatment, or those with values or personal choices that may conflict with their healthcare team.

diffusion of responsibility: a psychological feature of groups that describes how individuals seem to feel and act less responsibly within a group than when acting alone; each one assumes someone else will act. This phenomenon may cause teams to fail, and explains how/why actions that might otherwise be taken may be neglected in teams that fail to assign or make explicit guidelines for handling tasks.

drive-by diagnosis: a rapid diagnosis made in the moment absent the usual rigor and discipline of a formal evaluation.

dual process theory, dual process model: described in detail in Chapter 3 and illustrated in Figure 3.1, the dual process model describes how cognition occurs through 2 different types of cognitive processing: a rapid intuitive pathway (system 1), and a second more deliberate analytical pathway (system 2) that is slow and requires more conscious cognitive effort.

dual process theory training: knowledge of and training in the dual process model, including the nature of cognitive bias and heuristics.

dysrationalia: the failure to think rationally despite normal (or above normal) intelligence.

editorial bias: any bias that impacts editorial decisions other than the objective quality and value of the study, free of influence from sponsorships (like pharmaceutical companies), language bias, or reputation of the author, and other influences.

ego bias: the tendency to overestimate the prognosis of one's patients contrasted with the expected outcome of similar patients. *See also* Appendix I.

ego depletion: a form of exhaustion in which an individual may compromise their decisions to save energy. In cognition, it may explain resorting to heuristics and short-cuts that an individual might not otherwise use.

e-iatrogenesis: a term coined to describe iatrogenic harm that comes to patients primarily from flaws in information technology. Best described for problems with electronic ordering of medications, it can also describe the impact of error introduced into the medical record.

empiricism: the concept that our brains are blank slates at birth, and all knowledge comes from experience.

encapsulation, encapsulated knowledge: a phase in the development of medical expertise in which students transition from detailed descriptions of pathophysiology and lengthy descriptions to an abbreviated package of salient details with sufficient explanatory detail for a diagnostic label or syndrome. This encapsulation reveals the relevant working knowledge necessary for decision making at the bedside.

enlightenment: a state of higher awareness or insight.

environment of evolutionary adaptedness (EEA): a model of understanding how the brain and cognition developed based on behaviors that likely benefited survival.

e-patient (and the e-patient movement): a healthcare consumer who fully participates in his medical care and considers himself an equal partner with his doctor(s). The phrase *e-patient* was first coined by Tom Ferguson, the founder of the Society for Participatory Medicine. An e-patient is *e*quipped, *e*nabled, *e*mpowered, and *e*ngaged and encouraged to use electronic (and Internet) resources.

epistemic rationality, or evidential rationality: how well our beliefs map onto the real world with evidence; the use of facts and evidence to support thinking.

error management theory (EMT): the theory that proposes that thinking failures are the result of evolved, naturally selected patterns of behavior that served us well in our evolutionary past, and for which we are now hard wired.

error of ignorance: error due to lack of knowledge.

error of implementation: error in the application of knowledge.

evidence-based medicine: the conscientious, explicit, and judicious use of current best evidence in making decisions about the care of individual patients. The practice of evidence based medicine means integrating individual clinical expertise with the best available external clinical evidence from systematic research.

evolution of cognition: a model of cognition that argues that our modern brains are influenced by our evolutionary past as well as our contemporary environment. A model of evolution of cognition describes levels of cognitive functioning beginning with primitive functioning and evolving into more complex and conscious thinking, evolving from instinct to primitive processing, to unconscious processing, and finally to sophisticated processing.

evolutionary probability gambling: given the inherent uncertainty in many environments, it might be argued that some gambling behavior is an attempt at predicting a reward in an uncertain environment. Where events are random, but a particular pattern appears to emerge, may lead to erroneous predictions.

evolutionary psychology: a field of psychology that argues that some of our current decisions and behaviors are hard wired—the product of evolutionary pressures in our ancient environment hundreds of thousands of years ago.

executive control: a feature of the dual process model in which system 2 may override the tendency to use intuition, or system 1 processes; this function is utilized when an individual examines their thinking and slows down to default to a more rigorous approach perhaps recognizing a more difficult or atypical problem that requires more effort.

false cause: a logical fallacy, *see* Appendix II.

false negative: a negative test result despite the presence of the condition being tested.

false positive: a positive test result despite the absence of the condition being tested.

feedback sanction: *see* Appendix I.

fixed-action pattern: a particular pattern of behavior seen across a species that is set off by a stimulus and, once started, completed.

flesh and blood decision making: decision making that occurs *in vivo*, under real-life circumstances, that is often subject to suboptimal circumstances and influenced by ambient conditions.

fluid intelligence (Gf): the ability to use knowledge to solve new problems, deal with new situations, and find patterns.

forcing function: a rule that requires an individual to stop before making a final decision or acting. Forcing functions can be mechanical (a mechanical device that prevents an action), procedural (such as a forced time out before surgical procedures), or cognitive (a rule to stop and reconsider or expand alternatives). They can be general or specific. Forcing functions are considered particularly effective ways of cognitive bias mitigation.

framing effect: a cognitive bias, *see* Appendix I.

fundamental attribution error: the tendency to place blame on an individual rather than explore the circumstances or situations that may contribute to their behavior. *See also* Appendix I.

Gambler's fallacy: *see* Appendices I and II.

gender bias: a cognitive bias, *see* Appendix I.

general intelligence (G): the mental capacity that governs the ability to acquire and use information.

"geography is destiny": *see* triage cueing and Appendix I.

groupthink: the tendency for decisions made by a group to become overly focused on group dynamics and harmony and less on the accuracy of their decisions. Groupthink can lead to suboptimal decisions.

health utility index: a measure that accounts for how much value is lost from the remainder of one's life as a consequence of a condition. The healthcare index may be used to calculate the quality adjusted life years (QALY) for those who survive but have diminished quality of life. The index is typically measured on a scale from 1 (normal) to 0 (death or vegetative state). It can also be used to measure how much quality may potentially be gained from an intervention.

heuristics: information processing rules that allow an abbreviated form of decision making (a short-cut) to get a reasonable conclusion. Heuristics save time and effort.

Hickam's dictum: the dictum that argues counter to Occam's razor (find the simplest explanation), instead arguing that a patient can have as many conditions as he pleases.

hindsight bias: a cognitive bias, *see* Appendix I.

hypothesis-hopping: the act of jumping between system 1 and system 2, or between different hypotheses, in the course of making a diagnosis or a complex decision.

hypothetico-deductive reasoning/model: a formal model of reasoning that involves the deliberate, thoughtful consideration of all the possible hypothetical causes of a clinical syndrome, then tests and reframes the likely causes until the best solution is found. This model is slow and deliberate, but highly accurate.

identity protective cognition: the tendency for people to interpret evidence in such a way that it aligns with the views and values of a particular group they identify with. *See also* myside bias.

illness scripts: recognizable combinations of symptoms and physical signs in a defined clinical context that form discrete chunks of information that clinicians come to recognize to typify certain diagnoses. An illness script includes an understanding of pathophysiology, recognition of the distinguishing features of the condition, and an abbreviated concise problem representation that forms the basis for recognizing a disease or condition. The development of illness scripts is a phase in the acquisition of expertise.

inattentional blindness: a form of perceptual blindness that occurs when individuals focus on one task and fail to see an unrelated object that would otherwise appear obvious. This phenomenon was famously demonstrated in an Internet video when people who were focused on passing a basketball around a circle were asked to count the number of times people with white shirts touched the ball; many failed to notice a gorilla (or rather, a person in a gorilla suit) pass right by them. This phenomenon was also demonstrated by Etam et al. when they placed an object of a gorilla in a CT done to detect pulmonary nodules; radiologists focusing on the search for a nodule failed to see the artifact. People are strangely unaware of this perceptional blindness of objects around them when focused on another task, such as driving while texting.

incidentaloma: an unexpected incidental finding detected on studies done for purposes unrelated to the abnormality detected (the incidentaloma). The term incidentaloma is most commonly used to describe an adrenal adenoma, but has been liberalized to include any unsuspected abnormality detected that is of questionable significance. Clinicians sometimes feel the need to do more testing to better characterize and evaluate incidentalomas, thus driving up the cost of care and adding to the risk of (mostly) unnecessary tests and procedures.

inductive logic: a form of logic that begins with specifics and tries to match it to a more general category or draw a more general conclusion; an example is beginning with a symptom and reaching a conclusion about the cause. It has been described as the logic of personal experience. Unlike deductive reasoning, induction always has some degree of uncertainty.

information bias: a cognitive bias, *see* Appendix I.

information gap: a deficiency in information that may contribute to suboptimal decisions about clinical care. The gap is the difference between what information is needed and what is known in the moment when decisions must be made.

instinct: a fixed and predictable response driven by a stimulus that doesn't appear to involve any conscious thought. The behavior of animals, other than humans, is largely instinctive; for example, the phenomenon of migratory birds.

instrumental rationality: thinking behaviors that get us what we most want given available resources.

integrated (integrative) medicine: an approach that attempts to combine conventional medicine with complementary and alternative methods.

intelligence: the ability to acquire and apply information. Measured by intelligence quotient (IQ). General intelligence (G) is further delineated into crystalline intelligence (Gc) and fluid intelligence (Gf).

intelligent design: the idea that the universe, and living things, could not have originated by means other than the actions of an intelligent being.

interview illusion: the over-weighted belief that on a very small sample of someone's behavior in an interview, the interviewers feel confident that the impression formed can outweigh a significantly larger amount of evidence that might include letters of recommendation from people who usually have had much wider exposure to the candidate's behavior over a much longer period, the Graduate Record Exam score, the college grade average (which summarizes academic performance over a four-year period and may include grades from thirty or more academic courses), which collectively are shown to provide a more reliable estimate of the person. The illusion shares some overlap with fundamental attribution error.

intuition, intuitive model of thinking: a form of reasoning that is fast, automatic, and with little or no deliberate thought. It is largely driven by rapid recognition of a familiar and typical condition. Face recognition is one example in which one easily and immediately recognizes the form of a family member without having to stop and consider exactly their distinguishing features and who they are.

lay referral system: the network of family, friends, and acquaintances that may offer medical guidance and recommendations absent little if any formal training in medicine.

linear sequential unmasking (LSU): an approach to data acquisition and evaluation used in forensic science to minimize the impact of bias, beginning with evaluation of trace evidence as free as possible of context or knowledge of the crime, and evolving through successive stages. At each phase, the scientist conducts his study, draws conclusions, and states his level of certainty.

loaded question: a logical fallacy, *see* Appendix II.

logical fallacy: an error in reasoning that leads to an invalid argument or conclusion.

magical thinking: the belief of an association between actions and events that cannot be explained by reason or science.

medical enlightenment: the growing acceptance and awareness of cognitive bias and factors that interfere with rationality; these concepts are gradually finding their way into medical education and providing new insights about how we think, and how we diagnose. Some might call this a new era of medical enlightenment.

medical narrative: a description of a patient's symptoms in their own words in the context of how they experience disease and their illness.

medical students' disease: a type of hypochondriasis in which medical students become concerned, even obsessed, about symptoms they read about that they may have experienced; some come to fear they have any number of the conditions they learn about however rare or unusual.

meliorism, meliorists: the concept (and those who agree) that human reasoning is not always rational but can be improved (ameliorated) through education. Meliorists see major discrepancies between normative reasoning and descriptive models of how many people actually think.

memory biases: cognitive biases that augment or impair memory, or alter the content of the recall of a past event. *See also* Appendix I.

metacognition: the ability to detach from the moment, see the broader view, and think about one's thinking. It is sometimes referred to as cognitive awareness, reflection, self-regulation, or mindfulness, and is another way to describe the executive control feature of the dual process model.

metacognitive awareness inventory: an instrument designed to measure metacognitive skills.

middle ground: a logical fallacy, *see* Appendix II.

mindful attention awareness scale: a test used to assess mindfulness.

mindfulness: a state of focusing awareness on one's thoughts, feelings, and bodily sensations in a nonjudgmental manner in a given moment.

mindware: specific tools described by David Perkins that can be defined as rules, procedures, strategies and other forms of knowledge (knowledge of probability, logic, and scientific inference) that are stored in memory and can be retrieved in order to optimize problem solving and make more rational decisions. The absence of this knowledge is referred to as a *mindware gap* and distortions of the knowledge are known as *mindware contamination*, for example, cognitive biases and/or logical fallacies.

multiple alternatives bias: a cognitive bias, *see* Appendix I.

myside bias: the tendency for people to accept facts and form opinions based on their own prior opinions and attitudes. *See also* Appendix I.

naïve realism: proposes that we can only perceive the world around us through our senses, and therefore our conscious experience is not of the real world but instead our internal representation of it. If we believe that what we see (or perceive) of the world around us is objective, then others who see things differently must be irrational or biased.

nativism: the concept that our bodies and brains are the products of millions of years of evolution, and that some behaviors and thought patterns are acquired from our ancestors.

naturalistic bias: is an appeal to Nature such that anything which arises from Nature is good for you; that is, because it is from Nature it is good, and if it isn't, it must be unnatural and therefore bad.

need for cognition (NFC): a trait that recognizes the desire and enjoyment of effortful cognitive activity. People who score high on NFC seem to enjoy learning and deliberate cognitive activity. They may be less prone to error since they spend proportionately more time in Type 2 thinking and, therefore, less likely to be miserly with their cognitive effort.

normative reasoning: reasoning considered to display optimal judgment and decision making or the normal or correct way of doing something; rational thought is a normative notion in that normative reasoning is based on an ideal standard or model. Normative thinkers are generally better, more rational thinkers. Systematically failing to reason normatively is described as irrational.

nosophobia: an irrational fear of contracting a disease.

not invented here bias: the tendency to reject an idea or approach because it was not developed within one's professional discipline or domain. *See also* Appendix I.

no true Scotsman: a logical fallacy, *see* Appendix II.

nudging: a gentle push to steer healthcare providers toward better or best choices. Examples include engineering controls that require clinicians to state a clinical question as they order an x-ray, then complete a checklist of criteria to guide appropriate imaging.

Occam's razor: the dictum that the simplest explanation is likely the best. In medicine, this principle argues that a single diagnosis should be sought to account for most, if not all, the symptoms and problems a patient has.

omission bias: a cognitive bias, *see* Appendix I.

order effects: a cognitive bias, *see* Appendix I.

outcome bias: a cognitive bias, *see* Appendix I.

overconfidence: the tendency to think you know more than you do, or undue confidence in one's knowledge or skills; considered by some to be the most widespread cognitive bias in medicine. *See also* Appendix I.

overtriage: a triage decision that over-estimates the severity of a medical problem or the need for urgent attention.

panglossians: those who believe that the brain has developed optimal information processing, and thus see intuition as an effective means of functioning with the numerous decisions to be made daily. Panglossians may argue that we should trust our instincts and intuitions more.

pareidolia: seeing a familiar pattern where one does not exist.

paternal libertarianism: gently guiding individuals to a choice that is in their best interest while maintaining and respecting their freedom of choice.

pathognomonic: a symptom, physical finding, or test result that is so specific and so highly characteristic that it's presence, accurately detected, is virtually diagnostic of a condition.

pattern recognition: the recognition of a particular feature in the patient's presentation (Augenblick diagnosis), a cluster of signs and symptoms, or a syndrome or disease by typical manifestations that are highly specific and easily recognizable by an experienced eye (*see also* illness script).

personal incredulity: a logical fallacy, *see* Appendix II.

personalized medicine (also known as precision medicine): an evolving model for medical care based on a patient's genomic profile, environment, and lifestyle that determines their unique risk of disease and/or detects specific characteristics of their disease state(s). This model contrasts with a more traditional model of care that evaluates and treats patients based on evidence from large populations, assuming that they will, on average, be similar to the study population.

playing the odds (frequency gambling): a cognitive bias, *see* Appendix I.

pleistocene era: the evolutionary time period during which modern human species (*Homo erectus, Home sapiens*) evolved—about 2.5 million years ago to about 12,000 years ago.

post hoc fallacy: a logical fallacy, *see* Appendix II.

posterior probability error: a cognitive bias, *see* Appendix I.

precontemplative stage: a stage characterized by lack of awareness and disinterest in change. Individuals who are uninformed, or not motivated, are not likely to be receptive to efforts directed at change.

preference for intuition versus deliberation scale (PID): a tool used to determine an individual's decision-making style.

premature closure: the tendency to prematurely decide without considering all the evidence or exploring all possibilities. *See also* Appendix I.

primacy: the tendency to recall better what we hear first (primacy) in a series of statements; memory is U-shaped, with the ability to recall best for the first (primacy) and last (recency) items of a conversation.

primitive processing: a level of cognitive function (above instinct) that involves innate automatic responses but also involves some cognitive input to recognize co-variation of events, frequencies, and inferences.

principle of parsimony: the simplest explanation that sufficiently explains most if not all of the findings and requires the fewest assumptions is likely true. Similar to Occam's razor.

proactive coping inventory: an instrument used to measure individual skills in coping with stress.

procedural knowledge: the knowledge or skill in how to do something, or the ability to apply knowledge.

prospective hindsight: a technique of looking into the future and asking oneself what one's future-self might think of actions in the moment, and whether they might wish to reconsider an action or decision in anticipation of future regret.

psych-out errors: errors or misjudgment based on a bias toward patients with psychiatric illness, or mistakenly ascribing a psychiatric diagnosis to a patient with a medical disease. *See also* Appendix I.

publication bias: bias that impacts a decision to submit or accept a research study for publication based on factors other than scientific merit. One example is the tendency to reject studies that show no difference in outcome, or those that simply are not interesting enough. Authors themselves may be less likely to submit studies that support the null hypothesis. The bias impacts conclusions drawn from meta-analytic studies. *See also* Appendix I.

quality adjusted life year (QALY): a measure of change in quality and duration of life, usually defined as lessened by an adverse healthcare problem or event, or improved by some intervention, that differs from one's usual state of health or projected natural life.

rational-experiential inventory (REI): a personality test that measures an individual's disposition toward experiential (Type 1) versus rational (Type 2) decision making.

rationality: reasoning based on facts or evidence; the conformity of one's beliefs with one's reasons to believe. Rational decisions contrast with decisions that are negatively impacted by cognitive bias, logical failure, and other cognitive failings, in which case they may be suboptimal or irrational.

the great rationality debate: describes the ongoing polarization in the cognitive sciences literature between Meliorists and Panglossians. This has recently spilled over into medicine. Medical Meliorists believe that a significant proportion of diagnostic failures are due to failed cognition (cognitive biases, logical failures) and that remedial action might be taken by educational interventions as well as improving information management. Medical Panglossians in contrast, *see* no difference between descriptive and normative models of clinical reasoning, and believe that too much emphasis has been placed on cognitive biases and, in fact, intuitive decision making is equal to or may outperform rational thinking.

rationality quotient: a test (not yet developed) that would measure rationality (analogous to the intelligence quotient for measuring intelligence), the tendency to think and decide rationally. In 2016 the prototype of a comprehensive test of rational thought was published, the Comprehensive Assessment of Rational Thinking (CART).

rationality training: courses offered in medical and other domains for training in rationality. The work of cognitive psychologists and others over the last 40 years has contributed substantially to our present understanding of the concept of rationality.

re-biasing: a strategy to replace one bias with another, a sort of forcing function to avoid common biases. One may have a bias to neglect or incompletely assess intoxicated patients in the emergency department. One example of re-biasing is with a cognitive forcing strategy, a rule that inebriated patients must be fully disrobed and examined to prevent missing other injuries or conditions.

recency: the ability for improved recall of items that are provided last (recency) in a conversation, contrasted with items that occur in the middle. *See also* Appendix I.

reinforcement learning theory: the dominant theory during the twentieth century about behavior change—that learning occurs on the basis of reward or punishment. Behavior which is reinforced is likely to be repeated (strengthened) whereas that which is not will be extinguished (weakened). In its simplest form, Pavlov demonstrated that pairing rewards or punishment with a particular stimulus like the sound of a bell would change behavior (classical conditioning). Skinner took this to a new level believing that in order to understand behavior we needed to look at the causes of action and its consequences. Operants, for Skinner, were intentional actions that have an effect on the immediate environment, and his approach was termed "operant conditioning." If we want a dog to learn a new trick we cannot explain the trick and expect it to understand, we reward it for successive approximations to what the trick looks like and punish it for deviations away from what we are looking for.

representativeness restraint: a cognitive bias, *see* Appendix I.

reflective competence: a final phase in competence when an expert can reflect on their unconscious competence, question how they arrived there, and how they can teach others.

reflective coping: an analytical style of thinking that focuses on optimal decision making, measured by the Proactive Coping Inventory.

Reflecting coping includes brainstorming, skill in analyzing problems and generating hypothetical plans.

reflective practice, reflection: a process in which an individual stops and thinks about his actions, analyzes them, and then draws new knowledge from that experience that can be used in subsequent practice.

Ringelman effect (also known as social loafing): the observation that when individuals are asked to pull a rope, the amount of energy expended by each one decreases with the successive addition of new team members.

rule of parsimony: the idea that the simplest explanation is likely the best; in medicine, the rule of parsimony suggests that the simplest most elegant diagnosis is the one which best accounts for all findings. The rule of parsimony attempts to find a single unifying diagnosis that accounts for all the patient's symptoms.

rule out worst-case scenario: a rule that one should, at the very least, consider the worst (most serious) condition to be tested and excluded before assuming a benign explanation.

search satisficing: an amalgam of the words satisfy and suffice first described by Herbert Simon; satisficing is the tendency to call off the search once sufficient effort is felt to have been invested. This search stopping point usually occurs with the first significant finding, and may lead to a second (or third) abnormality being missed. Also referred to, by some, as *search satisfying*, as in the satisfaction of search. *See also* Appendix I.

second victim: the concept that when a patient (the first victim) suffers an adverse event or has a poor outcome because of some flaw in care, the healthcare provider may become a second (and often unrecognized) victim, subject to remorse, guilt, lowered self-esteem, depression, and even grief.

selective recall: the tendency to recall events in such a way that they conform to one's preexisting beliefs. *See also* Appendix I.

self-diagnosis: the tendency of individuals to decide likely explanations for their condition without seeking professional help.

serial position: referring to the effect on memory of objects present in a series, with the ability to recall favoring items listed first (primacy) and last (recency); sometimes referred to as *order effects*. *See also* Appendix I.

shallow (narrow) thinking: making a judgment from generalities without concern for specifics; failing to think more about the specifics.

slippery slope: a logical fallacy, *see* Appendix II.

slowing down strategies: specific rules intended to force an individual or team into analytical mode just long enough to reflect and be sure to complete a full cognitive check of the situation. A time out in the operating room is one example of a strategy to help avoid wrong site surgery.

snake phobia: a universal fear of snakes even in those who have never seen one; a phobia attributed to the impact of evolution on cognition.

social biases: a prejudice or bias toward an individual based on age, gender, race, weight, and/or lifestyle choices that may be conscious or unconscious and expressed in speech, writing, or behavior. *See also* Appendix I.

social loafing: the tendency for individuals to contribute less to a goal as individuals are added to the task.

social stigma: the tendency to judge someone critically (disapprove or discount them) based on some characteristic that distinguishes them from others, including age, gender, weight, appearance, poverty, education, and so on.

Solomon questionnaire: a questionnaire developed to help individuals describe the type of decisions they make and test what strategies they apply in those decisions, as well as the emotions they experience in making those decisions.

somatization: the expression of psychiatric illness by physical symptoms that lack an identifiable medical explanation.

sophisticated processing: a level of cognitive functioning that involves meaning and affect, and forms the basis of differences in individual decision making. This form of thinking is attributed mostly to humans, although there are examples of other species that show aspects of sophisticated processing.

special pleading: a logical fallacy, *see* Appendix II.

standard social science model: the approach to education in the medical sciences developed in the twelfth century on which our current system rests; it is based on an empiricist view that neglects evolutionary psychology and its implications for understanding patients' and providers' emotions, thoughts, and behaviors.

standardized diagnostic interviews: a standardized process that can force an individual to objectively collect and record facts to avoid error from short-cuts in information gathering.

standing rules: general rules that attempt to guard against common short-cuts or biases that may contribute to diagnostic errors.

stopping rules: rules that help avoid premature stopping and insist upon certain rigorous steps. For example, a physical exam of a fracture should include the joint above and below the identified injury (to avoid missing a second or third related injury).

strong inference (strategy): a strategy that requires that alternative ideas be considered to avoid bias. Checklists are a type of strong inference strategy that causes one to ask, "What else might this be?".

structured data acquisition: an explicit structured approach toward the clinical interview in Psychiatry for DSM disorders (SCID: Structured Clinical Interview for DSM disorders). It is believed to improve diagnostic performance by nullifying biases.

sunk cost bias: the unwillingness to give up or abandon an idea because of an investment of interest, time, or ego. Diagnosticians may be overly invested in a diagnosis and unwilling to let go of their theory even when new facts lead otherwise. *See also* Appendix I.

Sutton's slip: *see* Appendix I.

system 1: a mode of decision making that is fast, reflexive, and mostly unconscious; often referred to as intuition.

system 2: an analytical mode of thought that is deliberate and thoughtful, but slow.

telehealth: an umbrella term used to describe a variety of options for the delivery of healthcare using telecommunications. It can include exchange of emails, use of patient portals, and methods of exchanging information when a patient and the healthcare services they need are remote from one another.

telemedicine: a term used to describe a variety of services provided remotely through the sharing of images. Unlike "telehealth," which refers to telecommunication with patients, telemedicine refers mostly to the specialized services provided between providers. Teleradiology (remote interpretation of images) and telepathology (remote review of pathology specimens) are just two examples that describe how specialists remote from the site of care can review and interpret medical tests.

telephone triage: triage done on the basis of a phone call to help direct a patient to a resource in a time appropriate for their condition and situation, a process that is limited to information the patient provides. This is generally thought to be inferior to an in-person evaluation in which a professional can see, touch, and interact directly with the individual.

the fallacy fallacy: a logical fallacy, *see* Appendix II.

the Texas sharpshooter: a logical fallacy, *see* Appendix II.

transtheoretical model of change: a model that explains the phases that individuals experience as they evolve or change, including precontemplative, contemplative, and final maintenance stages.

triage cueing: the tendency for the initial triage in healthcare systems to influence the diagnostic thinking of clinicians who subsequently encounter the patient. For example, a complaint of back pain triaged to a low acuity area may mislead clinicians from considering serious alternative (non-musculoskeletal) explanations. Also referred to as *"geography is destiny." See* Appendix I.

Tu quoque: a logical fallacy, *see* Appendix II.

type 1 processing: a type of cognition described in the dual process model for thought processes that are fast and automatic and virtually without conscious thought. Type 1 processes are also described as intuition. In modern living, people make most of their daily decisions using type 1 processes, through serial associations. Thus, one gets up in the morning, eats breakfast, showers, and gets to work much of the time with little engagement of type 2 processing, and therefore minimal cognitive effort.

type 2 processing: also referred to as analytic reasoning; a type of cognition that is slow, deliberate, and conscious. Type 2 processing is used to solve problems or make complex decisions.

unconscious processing: a level of cognitive functioning that is automatic and unconscious, but involves tacitly learning about the environment and may involve a memory for important stimuli.

undertriage: a triage assessment that under-estimates the severity of a medical problem and the need for urgent attention.

universals: features of human culture, society, language, and behavior that underlie all human activities and which have been found among all ethnographic groups studied. Human universals number in the hundreds and have been described in some detail by the anthropologist Donald Brown.

unfreezing: the first phase of cognitive debiasing when an individual may be informed or become aware of a cognitive bias and first become open to change.

unpacking principle: a cognitive bias, *see* Appendix I.

verbal priming: Priming occurs when exposure to one stimulus or perceptual pattern influences the response to another stimulus, especially if it is in the same sensory modality. Thus, verbal priming works best with verbal cues. Priming effects may underlie certain biases in medicine. For example, if a patient is referred to as a drug user, the term may more easily evoke negative stereotyping of the patient, ascertainment bias, and a lessened standard of care.

vertical line failure: a cognitive bias, *see* Appendix I.

visceral bias: a cognitive bias, *see* Appendix I.

yin-yang out: a cognitive bias, *see* Appendix I.

young earth: the idea that the Earth and all living things were created less than 10,000 years ago.

white coat effect: a spurious elevation of blood pressure attributed to the stress of being in a doctor's office or healthcare setting.

will to believe: the desire to believe that treatments (orthodox or not), or actions one takes, cause a cure or improvement. This may in part account for the placebo effect in which treatments of no proven benefit are followed by symptomatic improvement.

zebra retreat: a cognitive bias, *see* Appendix I.

Appendix I: Cognitive and Affective Biases

affective bias: some degree of affect enters into all decision making; thus, most if not all cognitive biases are said to contain some affect, but affective bias is recognized when there is an inordinate intrusion of affect (either positive or negative) into the decision-making process that results in a compromise of rationality (*see also* visceral bias).

aggregate bias or fallacy: when physicians believe that aggregate data, such as those used to develop clinical practice guidelines, do not apply to individual patients (especially their own), they are invoking the aggregate fallacy. The belief that their patients are atypical or somehow exceptional may lead to errors of commission; for example, ordering x-rays or other tests when guidelines indicate none are required.

ambiguity effect: ambiguity is associated with uncertainty. The ambiguity effect is due to decision makers avoiding options when the probability is unknown. In considering options on a differential diagnosis, for example, this would be illustrated by the tendency to select options for which the probability of a particular outcome is known, over an option for which the probability is unknown. The probability might be unknown because of lack of knowledge, or because the means to obtain the probability (a specific test or imaging) is unavailable.

anchoring: anchoring is the tendency to fixate on specific features of a presentation too early in the diagnostic process, and to base the likelihood of a particular event on information available at the outset (i.e., the first impression gained on first exposure, the initial approximate judgment). This may often be an effective strategy. However, this initial impression exerts an overly powerful effect in some people, and they fail to adjust it sufficiently in the light of later information. Anchoring can be particularly devastating when combined with confirmation bias (*see* confirmation bias). Anchoring may lead to a premature closure of thinking.

ascertainment: ascertainment bias occurs when the physician's thinking is pre-shaped by expectations or by what the physician specifically hopes to find; that is, we see what we expect to see. Thus, a physician is more likely to find evidence of congestive heart failure in a patient who relates that he or she has recently been non-compliant with his or her diuretic medication, or more likely to be dismissive of a patient's complaint if he or she has already been labeled as a "frequent flyer" or "drug-seeking." Gratuitous or judgmental comments at hand-off rounds and other times can do much to seal a patient's fate. Ascertainment bias characteristically influences goal-directed, "top-down" processing. Stereotyping and gender biases are examples of ascertainment bias.

attentional bias: the tendency to believe there is a relationship between two variables when instances are found of both being present. More attention is paid to this condition than when either variable is absent from the other (*see also* selective recall).

availability: availability is the tendency for things to be judged more frequent if they come readily to mind. Thus, things that are common will be more readily recalled. The heuristic is driven by the assumption that the evidence that is most available is the most relevant. Thus, if an emergency physician saw a patient with headache that proved to be a subarachnoid hemorrhage (SAH), there will be a greater tendency to bring SAH to mind when the next headache comes along. Availability is one of the main classes of heuristic and underlies recency effect (*see* recency effect). Availability may influence a physician's estimates of base rate of an illness. Non-availability ("out of sight, out of mind"), occurs when insufficient attention is paid to that which is not immediately present ("zebras"). Novices tend to be driven by availability, as they are more likely to bring common prototypes to mind, whereas experienced clinicians are more able to raise the possibility of the atypical variant or zebra.

bandwagon effect: the tendency for people to believe and do certain things because many others are doing so. Groupthink is an example, and it can have a disastrous impact on team decision making and patient care.

base rate neglect: the tendency to ignore the true prevalence of a disease, either inflating or reducing its base rate, and distorting Bayesian reasoning. However, in some cases, clinicians may (consciously or otherwise) deliberately inflate the likelihood of disease, such as in the strategy of "rule out the worst-case scenario" to avoid missing a rare but significant diagnosis.

belief bias: the tendency to accept or reject data depending on one's personal belief system, especially when the focus is on the conclusion and not the premises or data. Those trained in logic and argumentation appear less vulnerable to the bias.

blind spot bias: people often can easily recognize bias in the decisions of others but are not particularly aware of it in their own decisions. It appears to be due mostly to the faith they place in their own introspections. This bias appears to be universal across all cultures.

commission bias: results from the obligation toward beneficence, in that harm to the patient can only be prevented by active intervention. It is the tendency toward action rather than inaction. It is more likely in overconfident physicians. Commission bias is less common than omission bias and may be augmented by team pressures or by the patient.

confirmation bias: the tendency to seek confirming evidence to support a diagnosis rather than look for disconfirming evidence to refute it, despite the latter (falsification of the hypothesis) often being the more scientifically sound strategy. In difficult cases, confirming evidence feels good, whereas disconfirming evidence undermines the hypothesis and means that the thinking process may need to be restarted, that is, looks like more work, requiring more mental effort. Confirmation bias may seriously compound errors that arise from anchoring, where a prematurely formed hypothesis is inappropriately bolstered.

congruence bias: similar to confirmation bias but refers more to an over-reliance on direct testing of a given hypothesis and a neglect of indirect testing. Again, it reflects an inability to consider alternative hypotheses.

contrast effects: occurs when the value of information is enhanced or diminished through juxtaposition to other information of greater or lesser value. Thus, if an emergency physician was involved in a multiple-trauma case and subsequently saw a patient with an isolated extremity injury, there might be a tendency to diminish the significance of the latter.

diagnosis momentum: once diagnostic labels are attached to patients, they tend to become stickier and stickier. Through intermediaries (patients, paramedics, nurses, physicians), what might have started as a possibility gathers increasing momentum (without gathering increasing evidence) until it becomes definite, and little or no attention is paid to other possibilities. This bias is distinguished from premature closure, which occurs within an individual, as opposed to a diagnosis, which is passed from person to person. Also known as "diagnostic creep."

ego bias: in medicine, ego bias is systematically overestimating the prognoses of one's own patients compared with those of a population of similar patients. More senior physicians tend to be less optimistic and more reliable about patients' prognoses.

feedback sanction: the ultimate goal of individual decision making is well-calibrated decision making. This absolutely depends on timely and accurate feedback. Absent feedback provides no incentive to change, and inaccurate or delayed feedback is often of little value. Ineffectual feedback is an important sanction on the calibration of decision making.

framing effect: how diagnosticians see things may be strongly influenced by the way in which the problem is framed. The particular way in which a patient's symptoms are expressed and the influence of context may significantly influence how the problem is perceived. Also, a physician's perception of risk to the patient may be strongly influenced by whether the outcome is expressed in terms of the possibility that the patient might die or might live. In terms of diagnosis, physicians should be aware of how patients, nurses, and other physicians frame potential outcomes and contingencies of the clinical problem to them.

fundamental attribution error: the tendency to be judgmental and blame patients for their illnesses (dispositional causes), rather than examine the circumstances (situational factors) that might have been responsible, can distort the diagnostic process. In particular, psychiatric patients, minorities, and other marginalized groups are vulnerable to this bias. Cultural differences exist in terms of the respective weights attributed to dispositional and situational causes.

gambler's fallacy: attributed to gamblers, this fallacy is the belief that if a coin is tossed 10 times and comes up heads each time, the 11th toss has a greater chance of being tails (even though a coin has no memory). An example would be a physician who sees a series of patients with chest pain in clinic or the emergency department, diagnoses all of them with an acute coronary syndrome, and assumes the sequence will not continue. Thus, the pretest probability that a patient will have a particular diagnosis might be influenced by preceding but independent events. The gambler's fallacy is contrasted with posterior probability error (*see* probability error), where, for different reasons, the belief is that the sequence will not reverse but continue.

gender bias: the tendency to believe that gender is a determining factor in the probability of diagnosis of a particular disease when no such pathophysiological basis exists. Generally, it results in an overdiagnosis of the favored gender and underdiagnosis of the neglected gender.

geography is destiny (*see* triage cueing)

hindsight bias: hindsight is defined as the ability to understand an event or situation only after it has happened. Thus, much valuable learning can occur through retrospective analysis. However, hindsight bias occurs when knowing the outcome influences the perception of past events and prevents a realistic appraisal of what actually occurred. In the context of diagnostic error, it may compromise learning through either an underestimation (illusion of failure) or overestimation (illusion of control) of the decision maker's abilities.

information bias: the tendency to believe that the more evidence one can accumulate to support a decision the better. It is important to anticipate the value of information and whether or not it will be useful in making a decision, rather than collect information because we can, or for its own sake, or out of curiosity.

memory biases: there are many memory biases. They are cognitive and affective biases that influence whether a memory will be recalled or not, how long it might take, and whether or not the content of the memory has been adulterated.

multiple alternatives bias: if a physician had decided on a choice between two working hypotheses, but additional information emerges that raises additional and reasonable possibilities on the differential, the bias would predict that the tendency to avoid conflict and added uncertainty inclines the physician back to choosing among the original hypotheses. Thus, a multiplicity of options on a differential diagnosis may lead to significant conflict and uncertainty. The process may be simplified by reverting to a smaller subset with which the physician is familiar, but this may result in inadequate consideration of other possibilities. One such strategy is the three-diagnosis differential: "it is probably A, but it might be B, or I don't know C." Although this approach has some heuristic value, if the disease falls in the C category and is not pursued adequately, it may result in an important diagnosis being missed.

myside bias: in everyday thinking, people are generally more receptive to facts based on their own prior opinions and attitudes and more likely to form opinions in a similar manner. They may generate evidence and evaluate evidence in a way that is biased toward their own opinions. A stronger myside bias appears when issues are related to current beliefs.

not invented here bias: it reflects a profound attitude-based bias toward knowledge (ideas, concepts, technologies) from a source that is considered external or outside of one's usual affiliation. There is, thus, a tendency to reject an idea or approach because it was not developed within one's professional discipline or domain.

omission bias: the tendency toward inaction, rooted in the principle of non-maleficence. In hindsight, events that have occurred through the natural progression of a disease are more acceptable than those that may be attributed directly to the action of the physician. The bias may be sustained by the reinforcement often associated with not doing anything, but it may prove disastrous. Omission biases typically outnumber commission biases.

order effects: information transfer is a U-function: we tend to remember the beginning part (primacy effect) or the end (recency effect) of a message or communication. These are referred to as serial position effects. Primacy effect may be augmented by anchoring. In transitions of care, in which information transferred from patients, nurses, or other physicians is being evaluated, care should be taken to give due consideration to all information, regardless of the order in which it was presented.

outcome bias: the tendency to opt for diagnostic decisions that will lead to good outcomes, rather than those associated with bad outcomes, thereby avoiding chagrin associated with the latter. It is a form of value bias in that physicians may express a stronger likelihood in their decision making for what they hope will happen rather than for what they really believe might happen. This may result in serious diagnoses being minimized.

overconfidence bias: the universal tendency to believe we know more than we do. Overconfidence reflects a tendency to act on incomplete information, intuition, or hunches. Too much faith is placed in opinion instead of carefully gathered evidence. The bias may be augmented by both anchoring and availability, and catastrophic outcomes may result when there is a prevailing commission bias.

playing the odds (frequency gambling): in equivocal or ambiguous presentations, it is the tendency to opt for a benign diagnosis on the basis that it is significantly more likely than a serious one. It may be compounded by the fact that the signs and symptoms of many common and benign diseases are mimicked by more serious and rare ones. The strategy may be unwitting or deliberate and is diametrically opposed to the rule of the worst-case scenario strategy (*see* base rate neglect).

posterior probability error: this occurs when a physician's estimate for the likelihood of disease is unduly influenced by what has gone before for a particular patient. It is the opposite of the gambler's fallacy in that the physician is gambling on the sequence continuing; for example, if a patient presents in the office five times with a headache that is correctly diagnosed as migraine on each visit, it is the increased likelihood that the patient will be diagnosed with migraine on the sixth visit. Common things for most patients continue to be common, and the potential for a non-benign headache being diagnosed is lowered through posterior probability.

premature closure: a powerful bias accounting for a high proportion of missed diagnoses It is the tendency to apply premature closure to the decision-making process, accepting a diagnosis before it has been fully verified. The consequences of the bias are reflected in the maxim: "When the diagnosis is made, the thinking stops." It may reflect other biases, such as overconfidence and possibly some laziness of thought coupled with a desire to achieve completion, especially under adverse conditions (cognitive overload, fatigue, sleep deprivation, and/or dysphoria). It is distinguished from diagnosis momentum in which there is also failure to verify, but which involves others.

psych-out error: psychiatric patients appear to be particularly vulnerable to biases generally, as well as other errors in their management, some of which may exacerbate their condition. They appear especially vulnerable to fundamental attribution error. In particular, comorbid medical conditions may be overlooked or minimized. A variant of psych-out error occurs when serious medical conditions (e.g., hypoxia, delirium, metabolic abnormalities, CNS infections, head injury) are misdiagnosed as psychiatric conditions.

publication bias: bias that impacts a decision to submit or accept a research study for publication based on factors other than scientific merit is referred to as publication bias or editorial bias. One example is the tendency to reject studies that show no difference in outcome, or those that simply are not interesting enough. Authors themselves may be less likely to submit studies that support the null hypothesis. Pharmaceutical companies are known to deliberately avoid publication of studies that are not commercially favorable. These publication biases distort the literature and therefore impact conclusions drawn from meta-analytic studies.

recency: the ability for improved recall of items that are provided at the end of a communication, contrasted with items that occur in the middle. *See* order effects.

representativeness restraint: representativeness is one of the most powerful of the major heuristics. It drives the diagnostician toward looking for prototypical manifestations of disease: "If it looks like a duck, walks like a duck, quacks like a duck, then it is a duck." Yet restraining decision -making along these pattern-recognition lines may lead to atypical variants being missed. Further, the representativeness heuristic tends to be insensitive to pretest probabilities and, therefore, may neglect prevalence.

search satisficing: the term "search satisficing" originates from two words, "satisfy" and "suffice." Search satisficing is the state of being satisfied that a sufficient search has occurred, such that further search may be called off. Essentially it is a stopping rule, that is, when something significant is found, stop searching. This often works well in everyday life, but can be disastrous in medicine, with additional diagnoses, second foreign bodies, other fractures, co-ingestants in poisoning, and other important findings being missed. Further, if a search yields nothing, diagnosticians should satisfy themselves that they have been looking in the right place.

selective recall: individuals may show a tendency to recall only certain aspects of something they have experienced with the result that their recall for the total experience is incomplete. Often, this may fit in with current attitudes and beliefs in a similar manner to *belief bias* and *myside bias*. It may also be due to selective attention, where certain things may be selectively recalled because the person was only paying attention to them; that is, no actual memory exists for those things that escaped attention originally.

serial position (*see* order effects): referring to the effect on memory of objects presented in a series, with the ability to recall favoring items listed first (primacy) and last (recency).

social biases: a prejudice or bias toward an individual based on age, gender, race, weight, socioeconomic status, lifestyle choices, and other characteristics. They may be conscious or unconscious and expressed in speech, writing, or behavior. While some individuals deny having biases, their biases can be revealed through other means.

sunk costs: the more clinicians invest in a particular diagnosis, the less likely they may be to release it and consider alternatives. This is an entrapment form of bias more associated with investment and financial considerations. However, for the diagnostician, the investment is time and mental energy, and, for some, ego may be a precious investment. Confirmation bias may be a manifestation of such an unwillingness to let go of a failing diagnosis.

Sutton's slip: Sutton's law is a clinical law based on the diagnostic strategy of "going for where the money is." It takes its name from the Brooklyn bank robber, Willie Sutton. When asked by the judge at his trial why he robbed banks, Sutton is alleged to have said, "Because that's where the money is" (in actuality, he didn't say this; it was said by a reporter writing up the trial). Going for the obvious makes sense, but it is often associated with persistent behavior attempting to diagnose the obvious, failing to look for other possibilities, and calling off the search once something is found (*see* search satisficing). When treatment is tightly coupled to the diagnosis, and the "obvious" diagnosis has been accepted, the outcome may be catastrophic; for example, the initial presentation and electrocardiogram findings in aortic dissection may mimic those of acute myocardial infarct and, in the interests of saving time, thrombolysis may be initiated. Sutton's law is also characterized by Occam's razor, the principle of parsimony in philosophy and psychology, and by the popular acronym KISS (keep it simple, stupid). Applications of Sutton's law, Occam's razor, and KISS may often be successful and may avoid costly, time-delaying diagnostic tests. However, whenever they are used, there should be an awareness of the associated pitfalls. Sutton's slip is the error associated with Sutton's law.

triage cueing (geography is destiny): the triage process occurs throughout the healthcare system, from the self-triage of patients, to the triage nurse in the emergency department, to the selection of an appropriate specialist by the referring physician. In the emergency department, triage is a formal process that results in patients being sent in particular directions, which cues their subsequent management. Many biases are initiated at triage, leading to the maxim "Geography is destiny." Once a patient is referred to a specific discipline, the tendency of practitioners in that discipline to look at the patient only from their own perspective is a further bias referred to as *deformation professionelle.*

unpacking principle: a clinician's rationality improves when the decision is based on relevant information. The failure to elicit all relevant information (unpacking) in establishing a differential diagnosis may result in significant possibilities being missed. The more specific a description of an illness that is received, the more likely the event is judged to exist. If patients limit their history giving, or physicians otherwise limit their history taking, unspecified possibilities may be discounted and diagnostic failure more likely.

vertical line failure: routine, repetitive tasks often lead to thinking in silos—predictable, orthodox styles that emphasize economy, efficacy, and utility. Though often rewarded, the approach carries the inherent penalty of inflexibility. In contrast, lateral thinking styles create opportunities for diagnosing the unexpected, rare, or esoteric. An effective lateral thinking strategy is simply to pose the question: "What else might this be?" It is not the ability to conjure up rare or exotic diagnoses that is important, but, instead, the capability to step outside the apparent constraints of the problem domain boundaries. This is especially important in those situations where the data or findings do not quite fit together. Lateral thinking in the appropriate situation often characterizes those with clinical acumen and those who can avoid vertical line failure.

visceral bias: the influence of affective sources of error on decision making has been widely underestimated. Visceral arousal leads to poor decisions. Countertransference, both negative and positive feelings toward patients, may result in diagnoses being missed. Some attribution phenomena such as fundamental attribution error (*see* fundamental attribution error) may have their origins in countertransference.

yin-yang out: when patients have been subjected to exhaustive and unavailing diagnostic investigations, they are said to have been worked up the yin-yang. The yin-yang out is the tendency to believe that nothing further can be done to throw light on the dark place where, and if, any definitive diagnosis resides for the patient; that is, the physician is let out of further diagnostic effort. This may prove ultimately to be true, but to adopt the strategy at the outset is fraught with the chance of a variety of errors.

zebra retreat: zebra retreat occurs when a rare diagnosis (zebra) figures prominently on the differential diagnosis, but the physician retreats from it for various reasons, resulting in the diagnosis being delayed or missed. There are a number of barriers to pursuing rare diagnoses; for example:

1. the clinician may anticipate inertia in the system such that there might be resistance to, or lack of support for, pursuing the diagnosis, or that there will be difficulty in obtaining special and costly tests to confirm the diagnosis;

2. the clinician may be self-conscious about seriously entertaining a remote and unusual diagnosis, and gaining a reputation for being esoteric;

3. the clinician might fear that he or she will be seen as unrealistic and wasteful of resources;

4. the clinician may have under or overestimated the base rate for the diagnosis;

5. the anticipated time and effort to pursue the diagnosis might dilute the clinician's conviction;

6. the clinician is underconfident;

7. team members may exert coercive pressure to avoid wasting the team's time;

8. inconvenience of the time of day or weekend and difficulty getting access to specialists;

9. unfamiliarity with the diagnosis might make the clinician less likely to go down an unfamiliar road;

10. fatigue or other distractions may tip the clinician toward retreat.

Any one or a combination of these reasons results in a failure to verify the initial hypothesis.

Adapted from Croskerry, P., *Acad Med.* 78(8), 777–78, 2003, with permission; and Croskerry, P. et al., *Patient Safety in Emergency Medicine*, Wolters Kluwer, 2009, with permission; and Croskerry, P: Achieving quality in clinical decision making: Cognitive strategies and detection of bias. *Acad Emerg Med.* 2002. 9. 1187–1200. Copyright John Wiley & Sons, Philadelphia, with permission.

Appendix II: Logical Fallacies

ad hominem fallacy: trying to undermine an argument by attacking the person rather than their logic.

ambiguity: using double meanings or ambiguities of language to mislead or misrepresent the truth.

anecdotal: using personal experience or an isolated example instead of a valid argument, especially to dismiss statistics.

appeal to authority: saying that because an authority thinks something, it must therefore be true.

appeal to emotion: manipulating an emotional response in place of a valid or compelling argument.

appeal to nature: making the argument that because something's "natural" it is therefore valid, justified, inevitable, good, or ideal.

bandwagon: appealing to popularity or the fact that many people do something as an attempted form of validation. Also known as *appeal to popularity, argument by consensus, argumentum ad populum,* and *authority of the many.*

base rate neglect: the tendency to misjudge the likelihood of something by failing to take into account all the relevant information; in medicine, failing to take into account the prevalence of a condition in the population.

begging the question: a circular argument in which the conclusion is included in the premise.

black-or-white: where two alternative states are presented as the only possibilities, when in fact more possibilities exist. Also known as *either/ or* fallacy and *false dilemma.*

burden of proof: saying that the burden of proof lies not with the person making the claim, but with someone else to disprove.

composition/division: assuming that what's true about one part of something has to be applied to all, or other, parts of it.

cum hoc, ergo propter hoc: the tendency to think that if two events happen at the same time or in close proximity, then one has caused the other. A variant of the tendency to think that association implies causation.

false cause: presuming that a real or perceived relationship between things means that one is the cause of the other.

gambler's fallacy: believing "runs" occur to statistically independent phenomena such as roulette wheel spins or coin tosses.

genetic: judging something good or bad on the basis of from where or from whom it comes.

loaded question: asking a question that has an assumption built into it so that it can't be answered without appearing guilty.

middle ground: saying that a compromise, or middle point, between two extremes must be the truth.

no true Scotsman: making what could be called an appeal to purity as a way to dismiss relevant criticism or flaws of an argument.

personal incredulity: saying that because one finds something difficult to understand, it's not true.

post hoc fallacy: a tendency to ascribe cause and effect based on their temporal relationship. The fact that B followed A might lead one to conclude erroneously that A caused B.

slippery slope: asserting that if we allow A to happen, then B, C ... Z will ultimately happen too; therefore, A should not happen.

special pleading: moving the goalposts or making up exceptions when a claim is shown to be false.

strawman: misrepresenting someone's argument to make it easier to attack.

the fallacy fallacy: presuming a claim to be necessarily wrong because a fallacy has been committed.

the Texas sharpshooter: cherry-picking data clusters to suit an argument, or finding a pattern to fit a prescription.

tu quoque: avoiding having to engage with criticism by turning it back on the accuser—answering criticism with criticism.

Adapted in part from the poster: "Thou shalt not commit logical fallacies." From: yourlogicalfallacyis.com. With permission.

Index